Cellular and PCS/PCN Telephones and Systems

An Overview of Technologies, Economics, and Services

By Lawrence Harte and Steve Prokup

Published by:

APDG Publishing

4736 Shady Greens Drive
Fuquay-Varina, NC 27526 USA
(919) 557-2260, (800) 227-9681
Fax (919) 557-2261
emal: bookworm@nando.net
web: www.cybercom.net/~apdg

All rights reserved. No part of this book may be reproduced or transmitted in any form or by any means, electronic or mechanical, including photocopying recording or by any information storage and retrieval system without written permission from the authors, except for the inclusion of brief quotations in a review.

Copyright © 1996 by APDG Publishing
First Printing 1996

Library of congress: 96-083575

International Standard Book Number: 0-9650658-1-2

ACKNOWLEDGEMENTS

We thank all the gifted people who gave technical and emotional support for the creation of this book. In many cases, published sources were not available on this subject area. Experts from manufacturers, service providers, trade associations, and other telecom related companies gave their personal precious time to help us and for this we sincerely thank and respect them.

Of the numerous manufacturers representatives that have dedicated their time to creating the standards that makes digital cellular a reality, we thank: David DeVaney of Astronet; Lucian Dang from Casio Manufacturing Corporation; Richard D. Lane and Christine Scaplen of Reflection Technology Inc.; Susan Shelton from Cincinnati Microwave; Shaul Berger of DSP Group; Carlton Peyton, Eric Stasik, Henrik Hogberg, Lars Wilhelmsson, Ray Henry, Ron Bohaychuk, Sandeep Chennakeshu, Ted Ericsson, Tony Gorse, and Tony Sammarco from Ericsson; Stan Kay of Hughes Network Systems; Mat Kirimura with Japan Radio Company; Imamura-San from Mitsubishi; Steve Jones of NEC; Mike Wise with Oki Telecom; Tom Crawford of Qualcomm; Prem Sood from Sharp; and Marybeth Flanders with Telular.

Thanks to service provider representatives: Chris Lawrence, Jim Lipsit, and Mike Bamburak from AT&T Wireless Services; John Boduch of BellSouth Cellular Corp.; P.F. Ng with Cantel; David Danaee' from Claircom; Jim Durcan of Comcast; Richard Dreher with DCR; Chris Leach of GTE Mobilnet; Jerry Adams from Iridium, Inc.; Hilbert Chan of Mobility Canada; Rob Hellstrom with Optus Communications; Malcolm White of ReadyCom; Ken Corcoran from Southwestern Bell; and Bill McClellan of Sprint Cellular.

Thanks also to the following research and consulting firm experts: Dave Mountain of APDG; Richard Levine of Beta Laboratories; Pat Sweeney from Bishop Company; David McCron, Michael Vadon, Russell Logan of BRC Consultancy Ltd.; M.B. Pautet and Michel Mouly from Cell & Sys; Dave Crowe of Cellular Networking Perspectives; Scott Goldman of Communications Now; Michael Sommer at Eastern Communications Telecom Consulting; Elliott Hamilton with EMCI; Herschel Shosteck of Herschel Shosteck Associates; David Balston at Intercai Mondiale; Deborah Jones (telecom consultant); Marty Nelson from Wireless Link; Sasan Ardalan at XCAD; and Konstantin Zsigo of Zsigo Wireless Data Consultants

Specific mention must go to our trade press, industry associations, and education experts: Eric Schimmel of the Telecommunications Industry Association; Sang-Lin Han from Chung Nam National University; Rhonda Wickham of Intertec/Cellular Business; and Maxine Carter-Lome from Probe/Wireless for the Corporate User.

Thanks to the vendors, distributors, and technology development company experts: Lee Horsman from Allen Telecom; Paul Wilkinson of Audiovox; Gregg Aurelius Associates with Southern Marketing; Jill Roumeliotis from ArrayComm, Inc; Gary Schober of Berkley Varitronics; Al Baughman with LCC Inc.; and George Peponides from PCSI.

We say thanks to our financial experts and industry analyst reviewers who include Jeff Hines and Walt Piecyk of Paine Webber, Tony Robertson from Robertson Stevens, Jeffrey Schlesinger of UBS Securities, and Linda Gossack with the US Department of Commerce.

The editors and illustrators who made this book read and look good include Jacquelyn Gottlieb (designer and editor), Annette Balint (editor), Bill Stevens (designer), Dave Smith of Smith and Strozier (editor), Mike Unger (illustrator), and Chris Rauh (semiconductor expert editor).

About the Authors

Lawrence Harte is the president of APDG, a provider of expert information to the telecommunications market. Mr. Harte has over 17 years of experience in the electronics industry including company leadership, product management, development, marketing, design, and testing of telecommunications (cellular), radar, and microwave systems. He has been issued patents relating to cellular technology and authored over 75 articles on related subjects. Mr. Harte earned his Bachelors degree from University of the State of New York and an MBA at Wake Forest University. During the IS-54 TDMA cellular standard development, Mr. Harte served as an editor for the Telecommunications Industries Association (TIA) TR45.3, the digital cellular standards committee.

Steve Prokup is vice president of engineering for ReadyCom, a provider of cellular voice paging service. Mr. Prokup has spent the last several years developing and managing software for cellular systems in the North American market. He has worked with several different cellular standards including analog and digital cellular systems. He has been involved with the development of an IS-136 TDMA cellular phone, including support of the industry specification validation efforts. Mr. Prokup earned his Bachelors degree in Computer Science from Ohio State University. He previously worked on software development for Department of Defense applications. Mr. Prokup is an inventor of several technologies relating to cellular telephony and has authored numerous papers describing digital cellular technology.

The creation of this book would not have been possible without the support and understanding of our families:

I dedicate this book to Jacquelyn Gottlieb—the love of my life, Virginia Harte— a great mother and coach, Lawrence William and Danielle Elizabeth—my children, and all the rest of my loving family. - Lawrence

I dedicate this book to Nancy, Christopher, Sara, and Alexander for their assistance, patience, and understanding. Without their help this would not have been possible. I love you all. - Steve

PREFACE

Despite the success of cellular systems, much of the necessary information used to obtain an understanding of the technology and market is hard to find. Due to the significant recent changes in the industry, searching for this information takes a significant amount of time. This book provides this hard to find information which is organized to allow quick searching.

In 1996, there were over 75 million cellular customers worldwide and predictions show that over 300 million customers will be using cellular systems by the year 2,000. The new cellular, PCS/PCN phones and systems that are being introduced to the market have advanced features, services, and cost advantages over the older cellular technologies. This book is a guide which provides a big picture of these new technologies, features, costs, and services for handsets and systems.

Cellular and PCS/PCN Telephones and Systems; An Overview of Technologies, Economics, and Services offers a balance between marketing and technical issues. It covers what's new in cellular technology, explains how it works using over 300 illustrations, and describes the marketing aspects of the new technologies and services. Many of the industry buzzwords are defined and explained. Over 100 industry experts have reviewed the technical content of this book.

To help meet the growing demand for cost effective cellular service and advanced features, several digital cellular technologies are available in over 60% of the world markets. Theoretically, analog cellular systems might have provided for cost-effective expansion indefinitely, but for a variety of reasons, they have not. Furthermore, each digital cellular technology's unique advantages and limitations offer important economic and technical choices for managers, engineers, and others involved with cellular systems. Cellular and PCS/PCN Telephones and Systems; An Overview of Technologies, Economics, and Services provides the background for a good understanding of the technologies, issues, and options available.

The chapters in this book are organized to help technical and non-technical readers alike to find the information they need. Chapters are divided to cover specific technologies, economics, and services. The chapters may be read either consecutively or individually.

Chapter 1. Summarizes the original Advanced Mobile Phone Service (AMPS) analog cellular system and its evolved EIA-553 standard. The technologies throughout the book evolved from AMPS. This chapter is an excellent introduction for newcomers to cellular technology.

Chapter 2. Describes digital technology and compares analog and digital cellular systems. The chapter includes an overview of Time Division Multiple Access (TDMA), Code Division Multiple Access (CDMA), Narrowband AMPS (NAMPS), and Global System for Mobile Communication (GSM) cellular systems.

Chapter 3. Explains the first US digital standard, TDMA IS-54, including its radio frequency (RF) channel structure, signaling, and call processing, including a brief history of why various functions were developed. This chapter explains the differences between AMPS and IS-54.

Chapter 4. Covers the TDMA standard IS-136 with its new Digital Control Channel (DCC). This chapter explains the differences between IS-54 and IS-136 standards.

Chapter 5. Explains CDMA technology, its channel structure, signaling, and call processing. This chapter explains the major differences between CDMA and AMPS, including soft handoff, tight power control, and the elimination of frequency planning.

Chapter 6. Describes IS-89 NAMPS, including its RF channel structure, signaling, call processing, and the major differences between AMPS and NAMPS.

Chapter 7 Explains the Global System for Mobile Communication (GSM) digital standard, including its RF channel structure, signaling, call processing, and uses in Personal Communication Services (PCS) and Personal Communications Network (PCN) systems.

Chapter 8. Provides the fundamental building blocks of digital cellular telephones (subscriber equipment), including the radio, digital signal processing, audio sections, battery considerations, and differences between analog and digital subscriber unit designs.

Chapter 9. Includes digital cellular network requirements, equipment, implementation methods, a high-level overview of the Public Switched Telephone Network (PSTN), cellular network interconnections (such as IS-41), cellular system equipment, and system planning.

Chapter 10. Reviews cellular system economics, including costs of digital subscriber units and system equipment, and an analysis of cellular network capital and operational costs. Revenue producing services, distribution channels, churn, and activation subsidy are explained.

Chapter 11. Describes future cellular technologies including Integrated Dispatch Enhanced Network (iDEN), Cellular Digital Packet Data (CDPD), Extended Time Division Multiple Access (E-TDMA), cellular voice paging. Spatial Division Multiple Access (SDMA), Wireless Office Telephone Systems (WOTS), and satellite cellular systems.

Chapter 12. A non-technical description of advanced digital services common to all digital technologies, summarizing potential new services and how they may be implemented.

Appendixes are provided which include acronyms, definitions, sources of specifications, and a listing of world cellular systems.

This book is based on current standards, but the TIA, Electronics Industry Association (EIA) and European Telecommunications Standard Institute (ETSI) are continually improving and changing specifications. For more specific information on each technology, be sure to consult the latest revision of each applicable standard.

TABLE OF CONTENTS

Introduction, 1

1. Analog Cellular (AMPS), 9
    1.1. System Overview, 9
        1.1.1. SubscriberUnit, 10
        1.1.2. Base Station, 12
        1.1.3. Mobile Telephone Switching Office, 13
    1.2. System Attributes, 14
        1.2.1. Frequency Reuse, 14
        1.2.2. Capacity Expansion, 15
        1.2.3. System Control, 17
        1.2.4. Subscriber Unit Identity, 18
    1.3. System Operation, 19
        1.3.1. Access, 20
        1.3.2. Paging, 21
        1.3.3. Hand-off, 21
    1.4. System Parameters, 22
        1.4.1. Frequency Allocation, 22
        1.4.2. Duplex Channels, 23
        1.4.3. Modulation Type, 24
        1.4.4. RF Channel Structure, 25
        1.4.5. RF Power Classification, 26
        1.4.6. Radio Interference, 26
    1.5. Signaling, 28
        1.5.1. Supervisory Audio Tone, 30
        1.5.2. Signaling Tone, 31
        1.5.3. Dual Tone Multi Frequency (DTMF), 31
    1.6. Call Processing, 33
        1.6.1. Initialization, 34
        1.6.2. Idle Mode, 35
        1.6.3. System Access, 36
        1.6.4. Conversation, 37

2. Digital Cellular, 41
    2.1 CTIA Requirements, 41
    2.2 CEPT Requirements, 43
    2.3 Increased Capacity, 43
    2.4 New Features, 45
    2.5 Equipment Availability, 45
    2.6 Digital Transmission Quality, 45
    2.7 Digital Technology, 46
        2.7.1 Voice Digitization, 46

2.7.2 Speech Coding, 47
2.7.3 Channel Coding, 48
2.7.4 Modulation, 50
2.7.5 RF Amplification, 51
2.8 Spectral Efficiency, 51
2.9 System Efficiency, 52
2.10 Authentication, 53
    2.10.1 North American Authentication, 53
    2.10.2 GSM Authentication, 55
2.11 Digital Cellular Technologies, 56
    2.11.1 Narrowband AMPS (NAMPS), 56
    2.11.2 IS-54 Time Division Multiple Access (TDMA), 58
    2.11.3 IS-136 TDMA Digital Control Channel (DCC), 59
    2.11.4 IS-95 Code Division Multiple Access (CDMA), 60
    2.11.5 Global System for Mobile Communications (GSM), 62
2.12 Dual Mode Cellular, 63

3. TDMA (IS-54), 65

3.1 System Overview, 66
    3.1.1 Speech Data Compression, 67
    3.1.2 Radio Channel Structure, 68
3.2 System Attributes, 70
    3.2.1 Frequency Reuse, 70
    3.2.2 Capacity Expansion, 70
    3.2.3 Secondary Dedicated Control Channels, 71
    3.2.4 Optional Reversed Paging and Access Channels, 71
    3.2.5 Dynamic Time Alignment, 73
    3.2.6 Mobile Assisted Hand-off, 74
    3.2.7 Multipath, 75
    3.2.8 Doppler Effect, 76
    3.2.9 Slot Structure, 76
        3.2.9.1 Forward Data Slot, 76
        3.2.9.2 Reverse Data Slot, 77
        3.2.9.3 FACCH Data Slot, 78
        3.2.9.4 Shortened Burst, 78
3.3 Signaling, 79
    3.3.1 Control Channel Signaling, 79
        3.3.1.1. Digital Capability Indicator (PCI), 79
    3.3.2 Analog Voice Channel Signaling, 79
    3.3.3 Traffic Channel Signaling, 80
        3.3.3.1 Slow Associated Control Channel (SACCH), 80
        3.3.3.2 Fast Associated Control Channel (FACCH), 81
        3.3.3.3 DVCC, 82
        3.3.3.4 DTMF Signaling, 83
3.4 System Parameters, 84

        3.4.1 Frequency Allocation, 84
        3.4.2 Duplex Channels, 85
        3.4.3 Modulation Type, 85
        3.4.4 RF Power Classification, 87
    3.5 System Operation, 88
        3.5.1 Access, 89
        3.5.2 Paging, 89
        3.5.3 Hand-off, 90
    3.6 Call Processing, 91
        3.6.1 Initialization, 91
        3.6.2 Idle, 92
        3.6.3 System Access Task, 93
        3.6.4 Conversation Mode, 95

4. TDMA Digital Control Channel (IS-136), 97
    4.1 System Overview, 99
        4.1.1 New Digital Control Channel, 100
        4.1.2 Logical Control Channels, 102
    4.2 System Attributes, 103
        4.2.1 Multi-Function Radio Channels, 103
        4.2.2 Digital Control Channels, 103
        4.2.3 Public, Private, and Residential Systems, 104
        4.2.4 Paging Classes, 104
        4.2.5 Short Message Services, 105
        4.2.6 Digital Only Phones, 105
        4.2.7 Data Transmission, 106
        4.2.8 SPACH ARQ (Automatic Retransmission Request), 106
        4.2.9 Monitoring of Radio Link Quality (MRLQ), 106
        4.2.10 Control Channel Reselection, 107
        4.2.11 Mobile Assisted Channel Allocation (MACA), 107
        4.2.12 Slot Structure, 108
    4.3 System Parameters, 110
        4.3.1 Duplex Channels, 110
        4.3.2 Modulation Type, 111
        4.3.3 Special Encoding Methods, 111
        4.3.4 Channel Structure, 112
    4.4 System Operation, 115
        4.4.1 Access, 115
        4.4.2 Paging, 116
        4.4.3 Hand-off, 117
    4.5 Call Processing, 118
        4.5.1 Initialization, 118
        4.5.2 Idle, 120
        4.5.3 System Access Task, 121
        4.5.4 Conversation Mode, 121

5. CDMA (IS-95), 125

5.1 System Overview, 126
    5.1.1 New Wide RF Channel, 127
5.2 System Attributes, 128
    5.2.1 Frequency Reuse, 129
    5.2.2 Capacity Expansion, 130
    5.2.3 System Control, 130
    5.2.4 Soft Hand-off, 130
    5.2.5 Variable Rate Speech Coding, 131
    5.2.6 Primary and Secondary Channels, 132
    5.2.7 Discontinuous Reception (Sleep Mode), 132
    5.2.8 Soft Capacity, 133
    5.2.9 Precise RF Power Control, 133
    5.2.10 Digital Control Channels, 135
    5.2.11 Frequency Diversity, 137
    5.2.12 Time Diversity, 137
5.3 Signaling, 138
    5.3.1 AMPS Control Channel Signaling, 138
    5.3.2 CDMA Control Channel Signaling, 138
    5.3.3 AMPS Voice Channel Signaling, 139
    5.3.4 CDMA Digital Traffic Channel Signaling, 140
5.4 System Parameters, 141
    5.4.1 Frequency Allocation, 141
    5.4.2 Duplex Channels, 142
    5.4.3 Modulation Type, 143
    5.4.4 RF Power Classification, 144
5.5 System Operation, 145
    5.5.1 Access, 145
    5.5.2 Paging, 146
    5.5.3 Hand-off, 147
5.6 Call Processing, 148
    5.6.1 Initialization, 148
    5.6.2 Idle, 149
    5.6.3 System Access Task, 151
    5.6.4 Conversation Mode, 152

6. NAMPS (IS-88), 157

6.1 System Overview, 158
6.2 System Attributes, 160
    6.2.1 Frequency Reuse, 160
    6.2.2 Sub-Band Digital Audio Signaling, 160
    6.2.3 Mobile Reported Interference, 161
    6.2.4 Capacity Expansion, 162
    6.2.5 RF Sensitivity, 162
    6.2.6 Co-Channel Interference Rejection, 163

6.2.7 Subscriber Unit Identity, 164
6.3 Signaling, 165
    6.3.1 Control Channel Signaling, 165
    6.3.2 AMPS Voice Channel Signaling, 165
    6.3.3 NAMPS Voice Channel Signaling, 166
        6.3.3.1 Digital Supervisory Audio Tone (DSAT), 166
        6.3.3.2 Digital Signaling Tone (DST), 166
        6.3.3.3 Sub-Band Digital Audio Messaging, 167
6.4 System Parameters, 168
    6.4.1 Frequency Allocation, 168
    6.4.2 Duplex Channels, 169
    6.4.3 Modulation Type, 169
    6.4.4 Adjacent Channel Interference Guardbands, 170
    6.4.5 Alternate Channel Interference, 171
    6.4.6 RF Power Classification, 171
6.5 System Operation, 172
    6.5.1 Access, 173
    6.5.2 Paging, 174
    6.5.3 Handoff, 174
6.6 Call Processing, 175
    6.6.1 Initialization, 176
    6.6.2 Idle Mode, 176
    6.6.3 System Access, 179
    6.6.4 Conversation, 180

7. Global System for Mobile Communications (GSM), 183
7.1 System Overview, 185
    7.1.1 Speech Coding, 186
    7.1.2 Radio Channel Structure, 188
7.2 System Attributes, 190
    7.2.1 Frequency Reuse, 191
    7.2.2 Time Division Multiple Access, 191
    7.2.3 Dynamic Time Alignment, 191
    7.2.4 Slow Frequency Hopping, 193
    7.2.5 Multi-Function Traffic Channels, 194
    7.2.6 Discontinuous Reception (Sleep Mode), 196
    7.2.7 Power Control, 197
    7.2.8 Mobile Assisted Hand-over, 198
    7.2.9 Multi-path and Equalization, 199
    7.2.10 Slot Structure, 200
    7.2.11 Short Message Service (SMS), 202
    7.2.12 Subscriber Identity Module (SIM), 203
7.3 Signaling, 203
    7.3.1 Control Channel Signaling, 203
    7.3.2 Traffic Channel Signaling, 206

7.4 System Parameters, 208
    7.4.1 Frequency Allocation, 208
    7.4.2 Duplex Channels, 209
    7.4.3 Modulation Type, 210
    7.4.4 RF Power Classification, 210
7.5 System Operation, 211
    7.5.1 Access, 211
    7.5.2 Paging, 212
    7.5.3 Hand-over, 213
7.6 Call Processing, 214
    7.6.1 Initialization, 215
    7.6.2 Idle, 215
    7.6.3 Initial Assignment/Access, 217
    7.6.4 Dedicated Mode, 218

8. Digital Cellular Telephones, 221

8.1 User Interface, 223
    8.1.1 Audio Interface, 223
    8.1.2 Display, 223
    8.1.3 Keypad, 223
    8.1.4 Accessory Interface, 224
8.2 Radio Frequency Section, 225
    8.2.1 Transmitter, 225
    8.2.2 Receiver, 225
    8.2.3 Antenna Section, 226
8.3 Signal Processing, 226
    8.3.1 Speech Coding, 227
    8.3.2 Channel Coding, 227
    8.3.3 Audio Processing, 228
    8.3.4 Logic Section, 228
    8.3.5 Subscriber Identity, 229
8.4 Battery Technology, 229
8.5 Accessories, 231
    8.5.1 Hands Free Speakerphone, 232
    8.5.2 Data Transfer Adapters, 232
    8.5.3 Voice Activation, 233
    8.5.4 Battery Chargers, 233
    8.5.5 Software Download Transfer Equipment, 233
    8.5.6 Antennas, 234

8.6 IS-54 TDMA Dual Mode Subscriber Units, 234
8.7 IS-136 TDMA Dual Mode Subscriber Unit, 236
8.8 IS-95 CDMA Dual Mode Subscriber Unit, 236
8.9 IS-88 NAMPS Dual Mode Subscriber Unit, 238
8.10 GSM Subscriber Unit, 240

9. Cellular System Networks, 243

9.1 Cell Sites, 244
    9.1.1 Radio Antenna Towers, 245
    9.1.2 Transmitter, 246
    9.1.3 Receiver, 246
    9.1.4 Controller, 247
    9.1.5 RF Combiner, 247
    9.1.6 Receiver Multi-Coupler, 247
    9.1.7 Communication Links, 248
    9.1.8 Antenna Assembly, 249
    9.1.9 Scanning or Locating Receiver, 250
    9.1.10 Power Supplies and Backup Energy Sources, 250
    9.1.11 Maintenance and Diagnostics, 250

9.2 Mobile Switching Center, 252
    9.2.1 Controllers, 253
    9.2.2 Switching Assembly, 253
    9.2.3 Communication Links, 254
    9.2.4 Operator Terminals, 255
    9.2.5 Backup Energy Sources, 255
    9.2.6 Home Location Register, 256
    9.2.7 Visitor Location Register, 256
    9.2.8 Billing Center, 256
    9.2.9 Authentication Center, 256

9.3 Public Switched Telephone Network, 256

9.4 Cellular Network System Interconnection, 257
    9.4.1 Inter-System Hand-off, 259
    9.4.2 Roamer Validation, 260
    9.4.3 Authentication, 261
    9.4.4 Automatic Call Delivery, 262
    9.4.5 Subscriber Profile, 263

9.5 Cellular System Planning, 264
    9.5.1 Radio Propagation, 265
    9.5.2 Strategic Planning, 267
    9.5.3 Frequency Planning, 268
    9.5.4 System Testing and Verification, 268
    9.5.5 System Expansion, 269

9.6 Network Options, 270
    9.6.1 Integrated and Overlay Systems, 270
    9.6.2 Sub Rate Multiplexing, 270
    9.6.3 Broadband Linear Amplified Base Stations, 272
    9.6.4 Distributed Switching, 272
    9.6.5 Cell Site Repeaters, 273
    9.6.6 Network Communication Links, 274

10. Cellular Economics, 275
  10.1 Subscriber Unit Equipment Costs, 276
    10.1.1 Development Costs, 276
    10.1.2 Cost of Production, 278
    10.1.3 Patent Royalty Cost, 279
    10.1.4 Marketing Cost, 279
    10.1.5 Post Sales Support, 280
    10.1.6 Manufacturers Profit, 281
  10.2 System Equipment Costs, 281
    10.2.1 Development Costs, 281
    10.2.2 Cost of Production, 282
    10.2.3 Patent Royalty Cost, 284
    10.2.4 Marketing Cost, 284
    10.2.5 Post Sales Support, 285
    10.2.6 Manufacturers Profit, 285
  10.3 Network Capital Costs, 285
    10.3.1 Cell Site, 286
    10.3.2 Mobile Switching Center, 289
  10.4 Operational Costs, 289
    10.4.1 Leasing and Maintaining Communications Lines, 289
    10.4.2 Local and Long Distance Tariffs, 290
    10.4.3 Billing Services, 291
    10.4.4 Operations, Administration, and Maintenance, 292
    10.4.5 Land and Site Leasing, 292
    10.4.6 Cellular Fraud, 293
  10.5 Marketing Considerations, 294
    10.5.1 Service Revenue Potential, 294
    10.5.2 System Cost to the Service Provider, 296
    10.5.3 Voice Service Cost to the Consumer, 296
    10.5.4 Data Service Cost to Consumer, 296
    10.5.5 Subscriber Unit (Mobile Phone) Cost to Consumer, 297
    10.5.6 Consumer Confidence, 298
    10.5.7 New Features, 298
    10.5.8 Retrofitting, 298
    10.5.9 Churn, 299
    10.5.10 Availability of Equipment, 299
    10.5.11 Distribution and Retail Channels, 300

11. Future Cellular and PCS Technology, 305
  11.1 Personal Communication Services, 306
  11.2 Integrated Dispatch Enhanced Network (iDEN), 311
  11.3 Extended TDMA (E-TDMA), 311
    11.3.1 ETDMA Operation, 312
  11.4 Voice Paging, 313
  11.5 Cellular Digital Packet Data (CDPD), 315

11.5.1 System Elements, 316
11.5.2 System Operation, 317
11.6 Spatial Division Multiple Access, 320
11.7 Wireless Office Telephone Systems, 322
11.7.2 Telego, 322
11.7.2 Cellular Auxilliary Personal Communication Service (CAPCS), 323
11.8 Satellite Cellular, 324

12. Digital Services,
12.1 Multi-Media, 327
12.1.1 Bearer Services, 328
12.1.2 Tele Services, 328
12.2 System Features, 328
12.2.1 Caller Identification, 328
12.2.2 Message Waiting, 329
12.2.3 Short Message Service (SMS), 330
12.2.4 Point to Point SMS, 331
12.2.5 Point to Multi-Point SMS, 332
12.2.6 Broadcast SMS, 333
12.2.7 Private Systems, 333
12.2.8 Facsimile Over Cellular, 334
12.2.9 Voice Privacy, 335
12.2.10 Increased Battery Life, 336
12.2.10.1 Extended Standby Time, 336
12.2.10.2 Enhanced Talk Time, 337
12.2.11 Authentication, 338
12.2.12 Data Transmission, 339
12.2.12.1 Circuit Switched Data, 340
12.2.12.2 Packet Switched Data, 341
12.2.13 Video, 341
12.2.14 Position Location, 342
12.3 New Applications, 344
12.3.1 Advertising, 344
12.3.2 Weather and Traffic Reports, 344
12.3.3 Direction Routing/Maps, 345
12.3.4 Telemetry/Monitoring, 345
12.3.5 Fax Delivery, 345
12.3.6 Video Transfer, 346
12.3.7 File Transfer, 347
12.3.8 Location Monitoring, 347
12.3.9 Point of Sale Credit Authorization, 348
12.3.10 Still Image Capture, 349
12.3.11 Interactive Information Access, 349
12.3.12 Two Way Paging, 349

12.3.13 News Services, 350
12.3.14 Sports Betting, 350
12.3.15 Remote Vending, 351

Appendix I - Definitions, 353

Appendix II - Acronyms, 365

Appendix III - Standards, 369

Appendix IV - World Cellular System Listing, 371

Introduction

Although mobile radio has been used for about 70 years and the cellular concept was conceived in the 1940's, public mobile cellular radio was not introduced in the U.S. until 1983. Today's cellular system requires technologies such as broadcast radio, simplex (two way, one speaker at a time) and duplex (two way, simultaneous) transmission, Public Switched Telephone Network (PSTN) connection, trunking connection ability, and high spectral efficiency. In 1996, every one out of ten people in the United States, (over 30 million subscribers) had a cellular phone [1].

Figure I.1, Microportable Cellular Telephone
Source: Motorola

Early in this century, the high power requirements and sheer bulk of vacuum tube technology limited mobile radio to shipboard use. Automotive mobile radio systems of the 1920's used 6-volt batteries with limited storage capacity, and under the stress of mechanical vibration, tubes performed poorly and unreliably [2].
In 1928, Detroit police department employed one of the first feasible automotive mobile radio systems [3]. The system broadcast from a central location and could only be received by the mobile radios. The policeman would have to respond to the call by using a wired call box.

Economic depression in the early '30's delayed the development of two-way mobile applications until 1933 when the Bayonne, New Jersey police department began using a simplex Amplitude Modulation (AM) push-to-talk system. A Frequency Modulation (FM) mobile system was not introduced until 1940, when the Hartford, Connecticut State Police employed a two-frequency simplex system (one-way at a time) [4]. In 1946, a mobile radio system was connected to the land line telephone network for the first time in St. Louis. This "urban" system supported just three channels [5].

In the mid 1950's, RCA developed the first full duplex (simultaneous two-way) mobile transmission system for the Philadelphia police. The system was controversial because its specifications were beyond the technology of the time, and a limited form of mobile FM duplex transmission was used [6].

Early radio systems required dedicated radio channels for a small group of users, a practice that limited the efficiency of the radio spectrum. In 1964, the Improved Mobile Telephone Service (IMTS) MJ system was the first real step towards mobile telephony as we know it today. It supported automatic channel selection, trunking, and full duplex transmission [7]. Trunking is an automatic process that allows a large group of users to share several communications channels.

Early wireless telephone systems were hampered by the lack of available frequency bands, this limited the number of radios that one system could support. Without today's crystal-controlled stability, early mobile applications required bandwidths up to 20 times today's standards to allow for frequency drift [8]. The FCC's policy was to serve the greatest number of users, a policy which led them to assign much of the available commercial spectrum to broadcast systems [9]. In 1976, although trunking and more stable transmitters did increase spectrum efficiency, the small spectrum allocation still limited the number of people who could be served, and many thousands of people were placed on waiting lists to get a mobile phone [10].

Mobile radios have been in demand for many years, but because of their high cost, few people bought them until the early 1980's. Market studies had shown that many more people would buy them at lower prices [11], and the studies have been validated now that the costs of cellular technology have declined.

Cellular Mobile Radio System

It is generally known that the cellular concept originated at Bell Laboratories in 1947 [12] and the first automatic cellular commercial system started operation October 1983 in Chicago. However, in 1949, another early cellular system began as a taxicab dispatch system in Detroit. The system used many small coverage areas (cell sites) to serve hundreds of square miles. Frequencies were re-used in alternate cells, and drivers manually switched between channels at predetermined locations[13].

According to the 1979 Bell Systems Technical Journal, the basic goals for a cellular system are:

(1) Large subscriber capacity
(2) Efficient use of spectrum
(3) Nationwide compatibility
(4) Widespread availability
(5) Adaptability to traffic density
(6) Service to vehicles and portables
(7) Support of special services
(8) Telephone quality of service
(9) Affordability

The cellular concept employs a central switching office to interconnect small radio coverage areas into a larger system. To maintain a call when the mobile station moves to another coverage area, the system switches the mobile radio's channel to the new cell site. The cellular concept also allows a frequency to be used by more than one subscriber at a time because subscribers using the same channel will not interfere if they are far apart. Cellular systems take advantage of this by breaking the coverage area into many small cells. Each cell supports several radio channels. Adjacent cells use different frequencies to avoid inter-

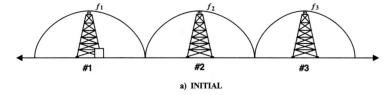
a) INITIAL

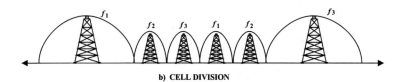
b) CELL DIVISION

Figure I.2, Cellular System Expansion by Cell Splitting

ference, but widely separated cells can reuse the same frequencies. This allows the system to repeatedly reuse radio channels and increase the number of subscribers they can serve. The cell site radio coverage area is determined by the base station's transmitter power: lowering power decreases the coverage area.

When a cell site has reached its maximum number of radio channels, the cellular system can expand by adding more radio towers with smaller coverage areas. Figure I.2 (a) shows how 3 frequency channels provide radio (RF) coverage to specific areas. These coverage areas have been divided in figure I.2 (b) so that 6 channels cover the same area. More channels in the same area allow more users to access the system at the same time.

In theory, adding towers to the system results in immense expandability. In practice, however, several factors cause significant problems and limit expansion. Chief among these is the high cost of construction and of obtaining cell site land rights [14].

To maintain a call while moving throughout several cell site areas, the call is transferred between adjacent cell sites while in progress. This call transfer, called switching or handoff, was not implemented until March 1977, when the FCC authorized a test system in Chicago. Until that time, sufficiently fast high-capacity switching systems were not available.

Analog Cellular

When the current U.S. cellular system was introduced in 1983, it was termed Advanced Mobile Phone Service (AMPS), now defined by the Electronics Industries Association (EIA) specification EIA-553, Base Station to Mobile Station Compatibility Standard. To work in the U.S. system, mobile and base station units must be manufactured to this specification.

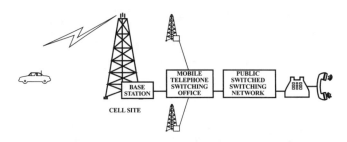

Figure I.3, AMPS Cellular System

Cost

Figure I.3 shows the AMPS analog cellular system which allows only one mobile phone to communicate on each radio channel. When all of the available radio channels have been installed in a cell site, the system requires additional cell sites to increase the total number of available channels. Start-up costs for each cell site are about $ 500,000 to $ 1,000,000 [15]. The added cell sites require additional telephone lines between the cell sites and MTSO. These lines are often leased from the local telephone company. As a result of adding cell sites and telephone lines, as analog systems expand, the average system cost per subscriber increases with the number of subscribers.

Digital Cellular

If the cellular system works well today, why do we need a new system? The answers are simple. No guarantee exists that a cellular system operator can build a tower wherever one is needed. In addition, minimum cell sizes, and fixed and operational costs of cell sites make expansion difficult.

In 1988, the Cellular Telecommunications Industry Association (CTIA) commissioned the Advanced Radio Technology Subcommittee (ARTS) to review alternative technologies to allow for cost-effective cellular expansion in the United States [16]. In 1989, cellular manufacturers and service providers started developing a new technology called Digital Cellular to allow cellular expansion at reasonable cost. This resulted in the creation of the User Performance Requirements (UPR) which defined the necessary technology for the next generation of cellular technology.

To help ease the transition from analog to digital cellular service, the CTIA UPR specified that initial phones reaching the market should have Dual Mode capability. Dual Mode cellular transceivers are designed to allow for a gradual transition to digital. They consist of both Analog (FM) and digital transmission sections. When digital service is available, dual mode transceivers automatically attempt to access digital channels. If no digital channel is available, they try to access an analog channel.

The first digital system accepted by the CTIA allows users to share a radio channel by time division. Each user is assigned specific time slots for transmission and reception. This system is called Time Division Multiple Access (TDMA).

A second digital system accepted by the CTIA allows users to share a radio channel by code division. Each user is assigned a specific PN code for transmission and reception. This system is called Code Division Multiple Access (CDMA).

Another CTIA-accepted access system increases capacity by dividing AMPS system channels into smaller radio channels 1/3 the original AMPS channel bandwidth, and assigns a user to each channel. This system is called Narrowband AMPS (NAMPS).

In the 1980s, the early success of European cellular systems also demonstrated the need for a new type of cellular system.. A new system was required that could meet the long term system service and cost objectives for European cellular service providers. In 1982, the Conference of European Posts and Telecommunications (CEPT) held a meeting to begin the standardization process for a single next generation European cellular system. This meeting established the Groupe Special Mobile (GSM) standards body During 1985, the Consultative Committee of International Telegraph and Telephones (CCITT) created a list of technical recommendations that would be used to create the specifications. From these requirements, an action plan was created to coordinate the creation of the Global Standard for Mobile Communications (GSM) specifications.

World Compatibility

Since the official introduction of the U.S. AMPS cellular standard in 1983, more than 52 countries have adopted it. Widespread use of this standard allows the same subscriber unit to operate in many different countries. For example, North and South America are U.S. cellular compatible. In the 1980s, Europe was more diversified than the U.S., with many types of incompatible cellular systems. However, with the introduction of GSM systems, Roaming between almost all European countries and over 10 others is now possible. Appendix IV contains a summary of cellular systems adopted worldwide.

Many systems in other countries have adopted U.S. digital standards, but digital service is not likely to expand rapidly everywhere. Some countries will not need to convert to digital for many years because it is still easy for them to expand analog capacity and because digital equipment costs are relatively high.

Upgrading to Digital Cellular

Cellular subscribers might want to convert to digital cellular because it provides better voice quality, reduced blockage, and fewer dropped calls, but the bottom line is that digital cellular is more cost effective for the cellular service provider (carrier). At reasonable cost, existing cell sites can be converted to offer digital service which allows more users to simultaneously access on the system. The less it costs to expand access capacity, the lower the cost for everyone.

Commercial digital cellular service began in Europe in 1991 and in the United States in 1992. It can take several years to convert to digital service. Depending on the product, conversion for the existing subscriber may be as simple as replacing the radio transceiver although it is likely the entire mobile phone must be replaced. Because of added complexity, the initial digital handheld phones were larger and more expensive than analog equivalents.

The higher cost of digital phones creates a marketing challenge. Digital features such as longer talk time for portables, improved voice quality, voice privacy, and other benefits, have been used to sell the new technology to the consumer. Some cellular service providers have passed some of their system equipment cost savings onto their customers by lowering airtime charges.

References

[1]. Heads up, Daily News Service, 27 February, 1995.
[2]. Fred Link, former president and owner of Link Radio, personal interview, 17 February 1991.
[3]. George Calhoun, "Digital Cellular", p.26 Artech House, 1988.
[4]. Edward Singer, "Land Mobile Radio Systems", p.10, Prentice Hall, New Jersey, 1989.
[5]. Bell System Technical Journal, January 1979, Vol. 58, No. 1, American Telephone and Telegraph Company, Murray Hill, New Jersey, p.3.
[6]. Fred Link, former president and owner of Link Radio, personal interview, 17 February 1991.
[7]. Bell System Technical Journal, January 1979, Vol. 58, No. 1, American Telephone and Telegraph Company, Murray Hill, New Jersey, p.3.
[8]. Fred Link, former president and owner of Link Radio, personal interview, 17 February 1991.
[9]. George Calhoun, Digital Cellular, Artech House Publsihing, 1989.
[10]. William Lee, "Cellular Communications", McGraw Hill 1989, p.2.
[11]. Bell System Technical Journal, January 1979, Vol. 58, No. 1, American Telephone and Telegraph Company, Murray Hill, New Jersey.
[12]. ibid.
[13]. Fred Link, former president and owner of Link Radio, personal interview, 17 February 1991.
[14]. Dawn Stover, "Cellular Goes Digital", Popular Science, January 1990.
[15]. ibid.
[16]. CTIA Winter Exposition, Jessee Russell, "Technology Update", Reno Nevada, February 6, 1990.

Chapter 1
Analog Cellular (AMPS)

1. Analog Cellular (AMPS)

This chapter provides a high-level overview of the Analog FM (EIA-553) cellular system, including semi-technical descriptions, functional block diagrams, and rudimentary cellular operation for those not familiar with it. Experienced cellular engineers will find the description very basic.

1.1. System Overview

In 1971, AT&T proposed a cellular radio telephony system [1] that could meet the FCC's requirement to serve a large number of subscribers with a limited spectrum allocation. The AT&T proposal was the backbone of the US commercial cellular system that started service in Chicago, October 1983. Since that time, cellular radio telephony has evolved into the Analog FM (EIA-553) cellular system we have today.

Today's cellular system consists of three basic elements: A Subscriber Unit (mobile radio), Base Station (BS), and Mobile Telephone Switching Office (MTSO). Figure 1.1 shows a basic cellular system in which the service area is divided into smaller cells. A Subscriber Unit communicates by radio signals to the Base Station within its cell. The Base Station converts these radio signals for transfer to the MTSO via land line or alternate communications links. The MTSO routes the call to another Subscriber Unit in the system or the appropriate land line facility. These three elements are integrated to form a ubiquitous coverage radio system that can connect to the Public Switched Telephone Network (PSTN).

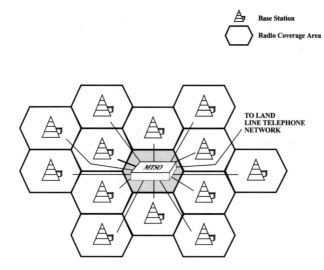

Figure 1.1, Cellular System Network

1.1.1. Subscriber Unit

A Subscriber Unit contains a transceiver, control head, and antenna assembly (see figure 1.2). The transceiver converts audio to RF and RF into audio. The control head's keypad communicates the subscriber's commands to the transceiver and its display informs the subscriber about the phone's status. The antenna assembly focuses and converts RF energy for transmission and reception.

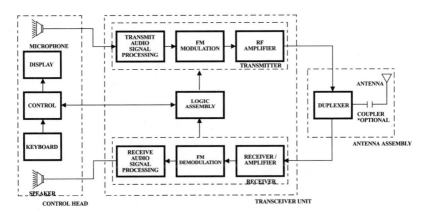

Figure 1.2, Subscriber Unit Block Diagram

Subscriber Units must conform to EIA-553 (Electronic Industries Association, EIA-553, "Mobile Station - Land Station Compatibility Specification", 1990) and IS-19B (Electronic Industries Association, EIA Interim Standard IS-19-B, Recommended Minimum Standards for 800 MHz Cellular Subscriber Units, May 1988). These standards contain parameters for the RF channel structure, signaling, and call processing functions designed to ensure compatibility with the cellular system network.

Mobile cellular radios have many industry names that vary by type of radio. Hand-held cellular radios may be called "portables," those mounted in cars "mobiles," and those mounted in bags, "bag phones." We will refer to any type of mobile cellular radio as a Subscriber Unit (SU).

Subscriber Units are classified by power output. A class I Subscriber Unit is capable of 6 dBW (3 Watts), a class II delivers 2 dBW (1.6 Watts), and class III (portables) provide 2 dBW (.6 watts) of power. Class III units conserve the portables' limited battery storage and reduce emissions by allowing the Base Station to send commands to adjust the unit's power output.

During the past 10 years, the evolution of portable cellular telephones has annually reduced the phones' weight, size, and cost. Figure 1.3 shows the progression of portable telephones over the last few years.

Figure 1.3, Portable Subscriber Unit Evolution
Source: Ericsson

1.1.2. Base Station

A Base Station is the radio portion of the cell site. It consists of transceivers, control sections, an RF combiner, communications links, a scanning receiver, backup power supplies, and an antenna assembly (see figure 1.4). The transceiver sections are like the Subscriber Unit transceiver in that they convert audio to RF and RF to audio. The control sections coordinate the Base Station's overall operation. The RF combiner allows the separation of radio channels between multiple transceivers and the antenna assembly. Communication links route audio and control information between the Base Station and MTSO. The scanning receiver measures signal strength on the cellular radio channels to decide when to hand Subscriber Units off to other Base Stations. The backup power supply maintains operation when primary power is interrupted. Most sections of the Base Station are duplicated to maintain functioning if equipment fails.

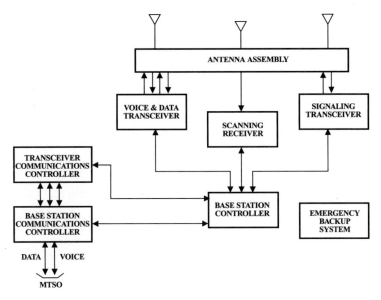

Figure 1.4, Cellular Base Station Block Diagram

1.1.3. Mobile Telephone Switching Office

The MTSO consists of controllers, a switching assembly, communication links, an operator terminal, a subscriber database, and backup energy sources (see figure 1.5). The controllers guide the MTSO (and the whole cellular system) by creating and interpreting commands to and from the Base Stations. A switching assembly routes voice connections from the Base Stations to the PSTN land lines. Communication links may be copper wire, microwave, or fiber optic. An operator terminal allows a human operator to supervise and maintain the system. A subscriber database contains billing records and records of features that customers have specified for their cellular service. Backup energy sources power the MTSO during primary power interruptions. Like the Base Station, the MTSO has duplicate circuits and backup energy sources to maintain the system during failures.

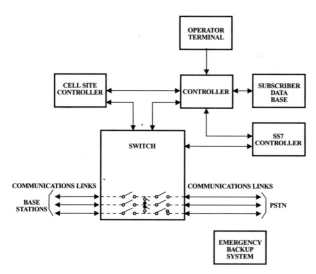

Figure 1.5, Cellular Mobile Telephone Switching Office

1.2. System Attributes

The cellular system meets the original Bell System goals [2] through frequency reuse, cost effective capacity expansion, and coordinated system control. To conserve the limited amount of radio spectrum, cellular systems reuse the same channels many times within the coverage area. The technique, called frequency reuse, makes it possible to expand system capacity by increasing the number of channels that are effectively available for subscribers. As the subscriber moves through the system, the Mobile Telephone Switching Office (MTSO) centrally transfers calls from one cell to another and maintains call continuity.

1.2.1. Frequency Reuse

In early mobile radio systems, one high-power transmitter with a modest allocation of frequency spectrum served a large geographic area. However, the small frequency allocation limited the number of radio channels, and because each cellular radio requires a certain bandwidth, capacity remained so low that the few available channels could not meet demand. In 1976, New York City's 12 radio channels supported 545 subscribers while 3,700 people remained on the waiting list [3].

To add radio channels where spectrum allocations are limited, cellular providers must reuse frequencies. One strategy for reusing frequencies exploits the fact that signal strength decreases exponentially with distance. Therefore, two subscribers who are far enough apart can use the same radio channel without interference (see figure 1.6).

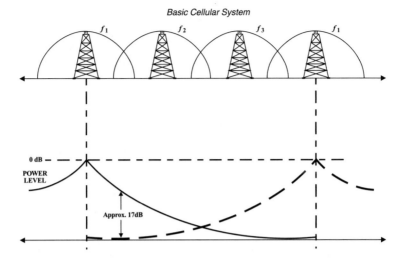

Figure 1.6, Frequency Reuse

To separate subscribers using the same radio channels, cellular system planners separate cell sites that use the same radio channels by specified distances. The distances between sites are initially planned by general RF propagation rules. However, it is difficult to account for all of the propagation factors encountered once the system begins to operate, so towers are rarely positioned precisely. Power levels must usually be adjusted later on.

The acceptable distance between cells that use the same channels is determined by the distance to reuse (D/R) ratio. The D/R ratio is the ratio of the distance between cells using the same radio frequency (D) to the radius of the larger cell (R). In today's system, a typical D/R ratio is 4.6: a channel being used in a cell with a 1 mile radius would not interfere with the same channel being reused at a cell 4.6 miles away.

1.2.2. Capacity Expansion

As cellular systems mature, they must serve more subscribers, either by adding more radio channels in a cell, or by adding new cells. To add radio channels, cellular systems use several techniques in addition to strategically locating cell sites that use the same frequencies. Directional antennas and underlay/overlay transmit patterns improve signal quality by focusing radio signals into one area to reduce interference to other areas. The reduced interference allows for more frequency reuse. Directional antennas can sector a cell so that a single radio channel covers only a portion of the cell area (e.g. 1/3 or 120 degrees). Such sectoring reduces interference with the other cells in the area. Figure 1.7 shows three 120 degree sectors.

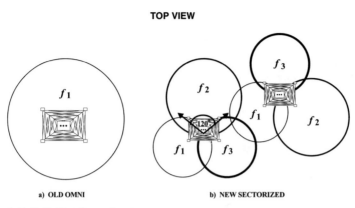

Figure 1.7, Cell Site Sectorization

Another technique, called cell splitting, helps to expand capacity gradually. Cells are split by adjusting the power level or using directional antennas to cover a reduced area (see figure 1.8). Reducing a coverage area by changing the RF boundaries of a cell site has the same effect as placing cells farther apart, and allows new cell sites to be added. However, the boundaries of a cell site vary with the terrain and land conditions, especially with variations in foliage. Coverage areas increase in autumn and winter as leaves fall from trees.

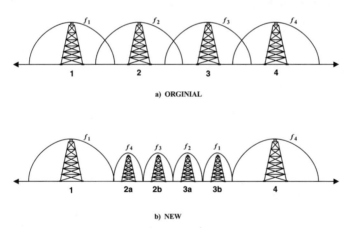

Figure 1.8, Cell Splitting

Currently, an analog system radio channel serves only one subscriber at a time, so system capacity is directly proportional to the number of radio channels available. Most subscribers use the system for only a few minutes a day, so on a give day, many subscribers will use a single channel (typically, 20 - 32) [4]. Generally, a cell with 50 channels can support 1000 - 1600 subscribers.

As a new cellular system expands, it eventually begins to exceed the limited number of callers it can effectively serve with a limited number of radio channels, and callers experience too many system busy signals, known as blocking. To expand, the system must effectively add available radio channels within the coverage area. Systems accomplish this by adding more cells with smaller coverage areas, thereby reducing interference and allowing radio channels to be reused.

Figure 1.8 illustrates capacity expansion through the reuse of radio channels. The original cells in Figure 1.8 cover an entire cellular market. As the system expands, cells 2 and 3 split into smaller cells 2a, 2b, 3a, 3b. The splitting allows for frequency reuse that increases the system's capacity. In Figure 1.7, the original system uses radio channel (f1) in cell number 1. The new system with split cells reuses radio channel (f1) in two cells: 1 and 3b. Further cell splitting can provide more capacity as required.

System planning must also account for present and future coverage requirements. After the FCC awards a license, cellular carriers have five years to provide coverage to more than 90% of the licensed territory [5]. To allow for 90% coverage in five years, and to ensure efficiency and competitiveness, carriers plan and design the system in advance.

1.2.3. System Control

As early as 1949, the Detroit taxi system employed frequency reuse to provide communications over a large service area. However, as taxis moved from one service area to another, they had to reestablish voice communications by manually retuning their radios to a new channel [6]. Today, cellular systems automatically switch channels to maintain smooth transitions as Subscriber Units move between cells. The MTSO's switching equipment transfers calls from cell to cell, and connects calls to other Subscriber Units or the land line telephone network. The MTSO also creates and interprets signals to control the Subscriber Units via Base Stations so that Subscriber Units automatically switch from channel to channel as they move from one coverage area to another. This switching is known as hand-off.

Figure 1.9 shows the cellular hand-off process. Initially, Base Station #1 communicates with the Subscriber Unit (t1). Where the Subscriber Unit's signal strength decreases below a set limit, it becomes necessary to transfer the call to a neighboring cell, Base Station #2. To make the transfer, Base Station #1 sends a hand-off command to the Subscriber Unit (t2), which tunes to a new radio channel (428) and transmits a tone (SAT) indicating it is operating on the channel (t3). The system senses that the Subscriber Unit is ready to communicate on channel 428, and the MSC switches the call to Base Station #2 (t4). The conversation then continues uninterrupted (t5). This process usually takes less than 1/4 of a second.

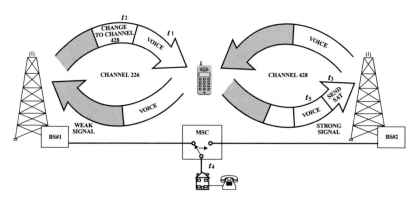

Figure 1.9, Cellular System Hand-off

1.2.4. Subscriber Unit Identity

Each Subscriber Unit must be unique in the eyes of the cellular system, and be served by options specified by the subscriber. The AMPS system uniquely identifies Subscriber Units with a number assignment module (NAM) and an electronic serial number (ESN).

The NAM contains unique Subscriber Unit information such as its telephone number, home system identifier, access classification, and other subscriber features. The Subscriber Units 10 digit phone number is called a mobile identification number (MIN). To receive calls, the Subscriber Unit listens for its MIN on the paging channel, and to transmit its identity, the Subscriber Unit sends its MIN. To help determine whether the Subscriber Unit is operating in its home system, the NAM also contains a home system identifier (HSID) to compare with the one broadcast on the control channel. If the comparison reveals a mismatch, the Subscriber Unit is operating in a system other than home, causing the control head to display ROAM, and alerting the subscriber to changed billing rates. Visited systems often charge a premium for service. The NAM also stores an access overload class (ACCOLC) code (0-15) to selectively inhibit groups of Subscriber Units from transmitting when the system is busy. Higher level access classes are reserved for emergencies. Early Subscriber Units programmed NAM information into a standard Programmable Read Only Memory (PROM) chip. However, manufacturers soon made NAM information programmable via the handset keypad due to high costs for PROM chips and special programming devices.

Each AMPS Subscriber Unit is created with a unique eleven-digit electronic serial number (ESN). The first three digits represent the manufacturer, and the last eight are a serial number. The combined MIN and ESN uniquely identify a valid subscriber. However, duplicate MINs and ESNs are technically possible, so

Chapter 1

advanced authentication programs were developed to provide a reliable unique identification system (discussed in chapter 2).

1.3. System Operation

Cellular systems use two types of channels, control channels and voice channels. Control channels allow the Subscriber Unit to retrieve system control information and compete for access. Voice channels primarily transfer voice informa-

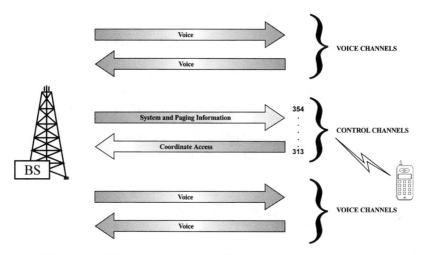

Figure 1.10, Control Channels and Voice Channels

tion, but also send and receive some digital control messages. In figure 1.10, note that channels 313 to 354 are dedicated as control channels to coordinate access to the voice channels.

When it is first powered on, a Subscriber Unit initializes by scanning the prede-

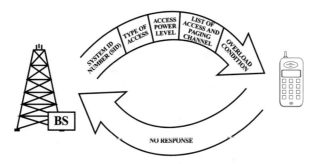

Figure 1.11, Cellular System Broadcast Information

termined set of control channels and tuning to the strongest one. Figure 1.11 shows that during this initialization mode, the Subscriber Unit retrieves system identification and setup information.

19

After initialization, the Subscriber Unit enters idle mode and waits either to be paged for an incoming call or for the user to place a call (access). When a call is to be received or placed, the Subscriber Unit enters system access mode and tries to access a control channel. If the attempt is successful, the control channel sends an Initial Voice Channel Designation (IVCD) message to indicate an open voice channel. The Subscriber Unit tunes to the designated voice channel and enters conversation mode. The voice channel uses Frequency Modulation (FM) similar to commercial broadcast FM radio. The voice channel also sends some control messages, very briefly replacing voice information with short bursts (blank and burst) of digital information.

1.3.1. Access

The attempt to obtain service from a cellular system is called "access," and Subscriber Units compete to obtain it. The access attempt begins when the Subscriber Unit receives a command indicating an incoming call or when the user places a call. Before and during the access attempt message, both the Subscriber Unit and the Base Station monitor the control channel's busy/idle sta-

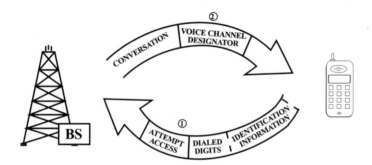

Figure 1.12, Cellular System Call Origination Radio Channel Access

tus. If the control channel is available, it transmits within a prescribed time limit. If it does not transmit within the time limit, the system assumes that another Subscriber Unit has the Base Station's attention and the attempt stops.
As Subscriber Units randomly attempt access to the cellular network, the control channel serves as the access point to the system.. To prevent multiple Subscriber Units from initiating access simultaneously, a seizure collision avoidance procedure has been developed. Four elements make up the process: the busy status of the channel, system response time interval, random time delays, and maximum number of automatic access attempts.

Busy status of the channel: The forward control channel is interleaved with dedicated bits that indicate whether the Base Station is busy. The Subscriber Unit monitors these bits before attempting access to the system, and waits until they indicate that the Base Station is not busy. **System response time interval:** When the Base Station is busy during an access attempt, the control channel indicates busy within a prescribed time period. If the control channel indicates busy before or after the time period, the Subscriber Unit assumes the Base Station is responding to a competing Subscriber Unit and inhibits transmission of the access attempt. **Random time delays:** Following an unsuccessful access attempt, the Subscriber Unit waits a random time before trying again. The random wait prevents two or more competing Subscriber Units from repeatedly attempting access again and again at identical intervals. **Maximum number of automatic access attempts:** The system limits the number of automatic attempts to prevent overloading when many Subscriber Units constantly attempt access.

If the Subscriber Unit gains access, the system sends an Initial Voice Channel Designation (IVCD) message to assign a voice channel. Figure 1.12 displays the access process when a call is placed from the Subscriber Unit. The access message is sent with the dialed digits. If the system is available, it sends a voice channel designator message to assign a voice channel. If access fails, the Subscriber Unit waits for a random period of time before trying again.

1.3.2. Paging

Subscriber Units are notified of incoming calls through the paging process. A page is a control channel message (phone number) containing the mobile identification number (MIN) to indicate that an incoming call is to be received. Figure

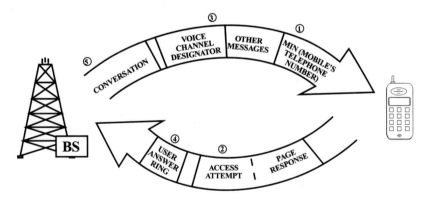

Figure 1.13, Cellular System Paging

1.13 shows that when the user presses SEND in response to the page, the Subscriber Unit attempts access to the system. The type of access request message that the Subscriber Unit sends identifies the access attempt as a response to a page.

1.3.3. Hand-off

As a Subscriber Unit moves away from the cell that is serving it, the cellular system must eventually transfer service to a nearer cell. Figure 1.14 illustrates the process, called hand-off. To determine when hand-off is necessary, the serving Base Station continuously monitors the Subscriber Unit's signal strength. When signal strength falls below a minimum, the serving Base Station requests adjacent Base Stations to measure that specific Subscriber Unit's signal strength (step 1). The adjacent Base Stations tune to the Subscriber Unit's current operating radio channel and measure the Subscriber Unit's signal strength. When a nearer adjacent Base Station measures sufficient signal strength (step 2), the serving

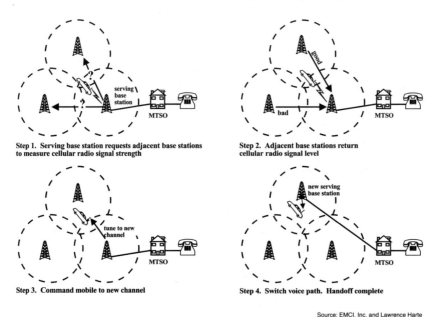

Figure 1.14, Hand-off Messaging

Base Station commands the cellular radio to switch to the nearer Base Station (step 3). After the cellular radio starts communicating with the new Base Station, the communication link carrying the land line voice path is switched to the new serving Base Station to complete the hand-off (step 4).

1.4. System Parameters

The cellular system was developed with the technology constraints of the 1970s. Frequency bands suitable to cost-effective equipment design were limited, the selected modulation type had to conform to a hostile mobile radio environment, and the control structure had to support large numbers of available voice channels.

1.4.1. Frequency Allocation

All cellular carriers in the US must obtain a cellular license from the Federal Communications Commission (FCC). The FCC's licensing system is designed to support many users with limited frequency spectrum. In the 1970s, available spectrum was constrained to above 800 MHz [7]. Equipment design limitations and poor radio propagation characteristics at frequencies above one Ghz led the FCC to allocate 825-890 MHz for cellular radio.

In 1974, an additional 40 MHz of spectrum was allocated for cellular service [8], and in 1986, 10 more MHz of spectrum was added to facilitate expansion [9].

To provide for competition, the FCC allocated 50 MHz of radio spectrum for each cellular service area. Cellular service areas are shared between two cellular companies, called A and B carriers. Today, the US is divided into 734 cellular service areas, each with an A and a B carrier. The A carrier does not have a controlling interest in the local telephone company; the B carrier (often a Bell operating company) can have a controlling interest in a local telephone company. The A and B carriers are each licensed to use 25 MHz of radio spectrum. Each carrier divides their 25 MHz into no more than 416 radio channels.

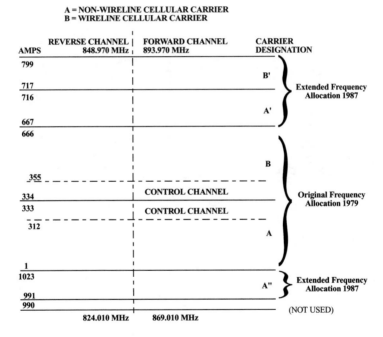

Figure 1.15, Cellular Radio Channel Frequency Allocation

Twenty-one of these channels are for control and paging, and 395 are for voice. Figure 1.15 illustrates cellular channel allocations:

1.4.2. Duplex Channels

To support simultaneous transmission and reception (no need for push to talk), the Base Stations transmit on one set of radio channels (869 - 894 MHz), called forward channels, and receive on another set (824 - 849 MHz), called reverse channels. Forward and reverse channels in each cell are separated by 45 MHz. Figure 1.16 (a) illustrates forward and reverse channels. A Base Station transmits to the Subscriber Unit on the forward channel at, for example, 875 MHz.

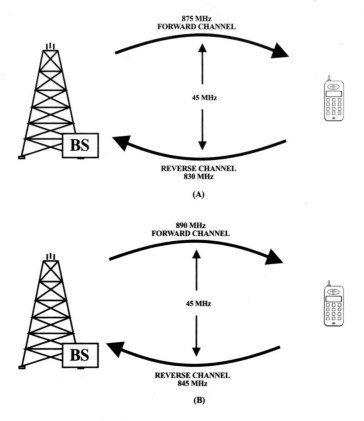

Figure 1.16, Duplex Radio Channel Spacing

The Subscriber Unit then transmits to the Base Station at 830 MHz on the reverse channel. If, alternatively (b), the Base Station transmits at 890 MHz, the Subscriber Unit transmits at 845 MHz.

1.4.3. Modulation Type

RF Modulation converts information into a radio signal. The modulation input signal containing information is the <u>baseband signal</u>. The RF carrier that transports the information is the <u>broadband signal</u>.

Frequency Modulation (FM) is the modulation method used for voice, and Frequency Shift Keying (FSK) is the method used for digital signaling. To optimize modulation for the mobile environment operating at 800 MHz radio, specially audio processing enhances high frequencies and minimizes transients caused by 800 MHz signal fades [10].

1.4.4. RF Channel Structure

In early mobile radio systems, a Subscriber Unit scanned the few available channels until it found an unused one, but today, no Subscriber Unit could scan the 832 channels in a reasonable time [11]. Therefore, control channels were established to quickly direct a Subscriber Unit to an available channel. Today, of the 416 channels available per system, 21 are control channels and cannot be used as voice channels. The remaining 395 channels are either voice or control channels.

A Subscriber Unit communicates with the cellular system by sending signaling messages. Signaling message formats vary between control channels and voice channels. Control channel signaling is all digital and voice channel signaling is a mixture of digital messages and audio tones.

Control Channels

The dedicated control channels carry the following four types of messages to allow the cellular radio to listen for pages and compete for access:

o <u>overhead messages</u> continuously communicate the system identification (SID) number, power levels for initial transmissions, and other important system registration information
o <u>pages</u> tell a particular cellular radio that a call is to be received
o <u>access information</u> is exchanged between the Subscriber Unit and the system to request service
o <u>channel assignment commands</u> establish the radio channels for voice communications.

Voice Channels

After a Subscriber Unit is assigned a voice channel, FM signals transmit the voice, and FSK signals transmit control messages. Control messages that are sent on the voice channel include:

o <u>hand-off messages</u> instruct the Subscriber Unit to tune to a new channel
o <u>alert</u> tells the Subscriber Unit to ring when a call is to be received
o <u>maintenance</u> commands monitor the status of the Subscriber Unit
o <u>flash</u> requests a special service from the system (such as 3 way calling)

1.4.5. RF Power Classification

Subscriber Units are classified by maximum RF power output. AMPS cellular phones have three classes of maximum output power. Class 1 maximum power output is 6 dBW (4 Watts), Class 2 is 2 dBW (1.6 Watts), and Class 3 is -2 dBW

RF Power	Class 1	Class 2	Class 3
Maximum Power	6 dBW (4 Watts)	2 dBW (1.6 Watts)	-2 dBW (.6 Watts)
Minimum Power	-22 dBW (6 mW)	-22 dBW (6 mW)	-22 dBW (6 mW)

Table 1.1
AMPS RF Power Classification

(.6 Watts). However, actual Subscriber Units' RF power outputs vary because Base Station commands adjust output in increments of 4 dB down to the minimum for all AMPS Subscriber Units of -22 dBW (6 milliwatts). Table 1.1 shows the AMPS Subscriber Unit RF Power classifications.

1.4.6. Radio Interference

Radio interference limits the number of radio channels that can be used in a single cell. Three types of interference are important for cellular systems. These are co-channel, adjacent channel, and alternate channel interference.

Co-channel Interference

Co-channel interference occurs when two cellular radios operate on the same radio channel and are too near each other. Co-channel interference at a particular location can be measured by comparing the desired signal's radio energy with that of the interfering signal. Today's analog systems are designed to limit interfering signal strength to below 2 percent of desired signal strength. At this level, the desired signal is nearly undistorted.

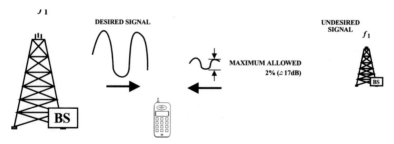

Figure 1.17, Co-Channel Interference

To minimize interference, cellular carriers frequently monitor the received signal strength by regularly driving test equipment throughout the system to determine whether the combined interference from cells using the same channel exceeds the

2 percent level of the desired channel. This information is used to determine whether radio channels and/or power levels at each cell need to be changed. Figure 1.17 illustrates co-channel interference.

Radio technology that could exceed two percent co-channel interference without distortion would allow greater frequency reuse and increase system capacity. Such higher tolerance is another advantage of next-generation technologies.

Adjacent Channel Interference

Adjacent channel interference occurs when a radio channel interferes with a channel next to it (e.g. channel 412 interferes with 413). Each radio channel is 30 kHz wide, but a low level of radio energy is transmitted at outside the 30 kHz

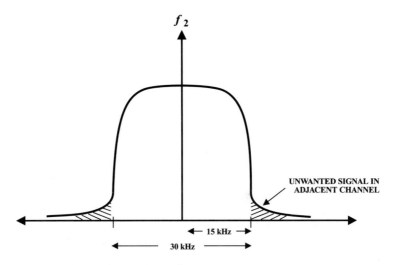

Figure 1.18, Adjacent Channel Interference

band. A Subscriber Unit at full power can produce enough low-level radio energy outside the channel bandwidth to interfere with adjacent channels. A 90 kHz channel separation would be enough to protect against adjacent channel interference, but for frequency planning reasons (discussed in chapter 9), 630 kHz (21 channels) separates each AMPS radio channel from the others in its Base Station or sector. Figure 1.18 illustrates adjacent channel interference.

Alternate Channel Interference

In a phenomenon called alternate channel interference, radio frequencies separated by two channels can interfere with each other. Figure 1.19 illustrates the bandpass spectral mask, a chart of allowable alternate and adjacent channel interference levels.

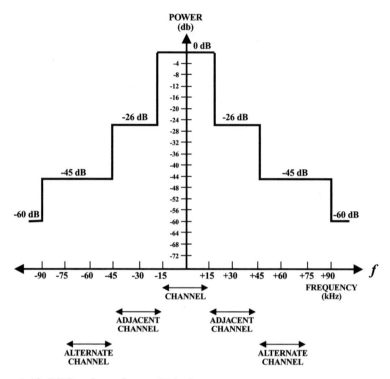

Figure 1.19, RF Bandpass Spectral Mask

1.5. Signaling

The control channel sends information by Frequency Shift Keying (FSK) at a rate of 10 kbps. To allow self-synchronization, the information is Manchester encoded, forcing a frequency shift (bit transition) for each bit input [12]. Orders are sent as messages of one or more words.

To help coordinate multiple Subscriber Units accessing the system, busy/idle indicator bits are interlaced with the other bits. Before a Subscriber Unit attempts access, it checks the busy/idle bits to see if the control channel is serving another Subscriber Unit. This system, called Carrier Sense Multiple Access (CSMA), helps avoid collisions during access attempts. Figure 1.20 illustrates coordination of access using busy/idle bits.

Figure 1.21 illustrates transference of a control channel message. First, a dotting sequence of alternating 1's and 0's is sent to indicate that a message is about to begin. The alternating bits produce a strong, easily-detected 5 kHz frequency component [13]. A synchronization word follows the dotting sequence to define the exact starting point of the coming message. Message words follow the synchronization word. Radio channels can fade and introduce errors, so the message words are repeated 5 times to ensure reliability. Of the five repeats, a majority

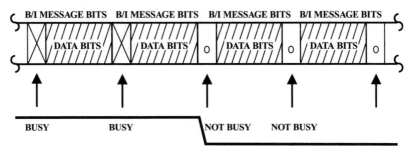

Figure 1.20, Busy/Idle Bit Signaling

vote of 3 out of 5 words can be used to eliminate corrupted messages.

Message and signaling formats on the control channels vary between forward and reverse channels. The forward channel is synchronous and the reverse channel is asynchronous.

Forward Control Channel

On the forward control channel, ten words follow the dotting/sync word sequence. Words are alternated A,B,A,B, etc. A words are designated for mobile units with even phone numbers. B words are designated for Subscriber Units with odd phone numbers. A forward channel word is 40 bits. Each word

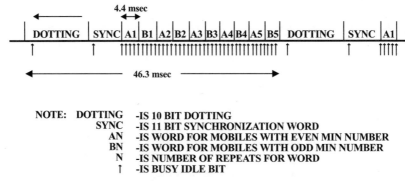

Figure 1.21, Forward Control Channel Signaling

includes BCH error correction/detection, containing 28 bits with parity of 12 bits.

Reverse Control Channel

On the reverse channel, 5 words follow the dotting sequence. A reverse channel word is 48 bits. Each reverse channel word includes BCH error correction/detec-

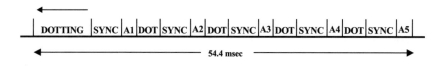

```
NOTE:  DOTTING  -IS 101 BIT DOTTING
       SYNC     -IS 11 BIT WORK SYNCRONIZATION WORD
       AN       -IS 48 BIT WORD
       N        -IS NUMBER OF REPEATS FOR WORD
       DOT      -IS SHORT 37 BIT DOTTING
```

Figure 1.22, Reverse Control Channel Signaling

tion, containing 36 bits with parity of 12 bits. Messages on the reverse channel are sent in random order, and coordinated using the Busy/Idle bits from the forward control channel.

Voice Channel Signaling

The analog voice channel passes user information (usually voice information) between the Subscriber Unit and the Base Station. Signaling information must also be sent to provide physical layer control. Signaling on the voice channel can be divided into in-band and out-of-band signaling. In-band signaling occurs when audio signals between 300-3000 Hz either replace or occur simultaneously with voice information. Out-of-band signals are above or below the 300-3000 Hz range, and may be transferred without altering voice information.

Signals sent on the voice channel include Supervisory Audio Tone (SAT), Signaling Tone (ST), Dual Tone Multi-Frequency (DTMF), and blank and burst FSK digital messages.

1.5.1. Supervisory Audio Tone

The SAT tone provides a reliable transmission path between the Subscriber Unit and Base Station. The SAT tone is transmitted along with the voice to indicate a closed loop. The tone functions much like the current/voltage used in land line telephone systems to indicate that a phone is off the hook[14]. The SAT tone may be one of the three frequencies: 5970, 6000, or 6030 Hz. A loss of SAT implies that channel conditions are impaired. If the SAT tone is interrupted for longer than about 5 seconds, the call is terminated.

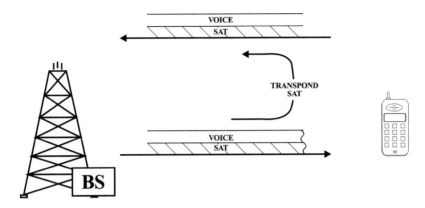

Figure 1.23, Transponding SAT

SAT can also mute the effects of co-channel interference. Interfering signals have a different SAT frequency than the one designated by the system for the call in progress. The incorrect SAT code alerts the Subscriber Unit to mute the audio from the interfering signal.

Re-transmission of SAT is also used to locate the Subscriber Unit's position. An approximate propagation time can be calculated by comparing the phase relationship between the transmitted and received SAT tones. This propagation time is correlated to the distance from the Base Station. However, multipath propagation (radio signal reflections) makes this location feature inaccurate and only marginally useful [15]. Only the re-transmission of SAT as a pilot tone is critical to operation.

1.5.2. Signaling Tone

The Signaling Tone (ST) is a 10 kHz tone burst used to indicate a status change. It confirms messages sent from the Base Station, and is similar to a land line phone going on or off hook [16].

1.5.3. Dual Tone Multi Frequency (DTMF)

Touch-tone (registered trademark of AT&T) signals (DTMF) may be sent over the voice channel. DTMF signals are used to retrieve answering machine messages, direct automated PBX systems to extensions, and a variety of other control functions. Bellcore specifies frequency, amplitude, and minimum tone duration for recognition of DTMF tones [17]. The voice channel can transmit DTMF tones, but varying channel conditions can alter the expected results. In poor radio conditions and a fading environment, the radio path may be briefly interrupted, sometimes sending a multiple of digits when a key was depressed only once.

Blank and Burst Messages

When signaling data is about to be sent on the voice channel, audio FM signals are inhibited and replaced with digital messages. This voice interruption is normally too short (34-54 msec) to be heard. The bit rate for messages is 10 kbps, and messages are transmitted by Frequency Shift Keying (FSK). Like control channel messages, these messages are repeated and a majority vote is taken to see which messages will be used.

To inform the receiver that a digital signaling message is coming, a 101 bit dot-

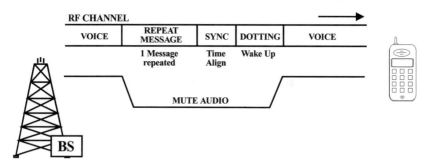

Figure 1.24, Voice Channel Message

ting sequence produces a 5 kHz tone preceding the message. After the dotting sequence, a synchronization word follows to identify the exact start of the message. Figure 1.24 illustrates a voice channel message transmission.

Blank and burst signaling differs on the forward and reverse voice channels. On the forward voice channel, messages are repeated 11 times to ensure control information is reliable even in poor radio conditions. On the reverse voice channel, words are repeated only 5 times. Words on the forward voice channel contain 40 bits, and 48 bits on the reverse voice channel. Both types of words have 12 bits of BCH error detect/correct parity. Figure 1.25 displays how signaling on the voice channel varies between the forward and reverse channels.

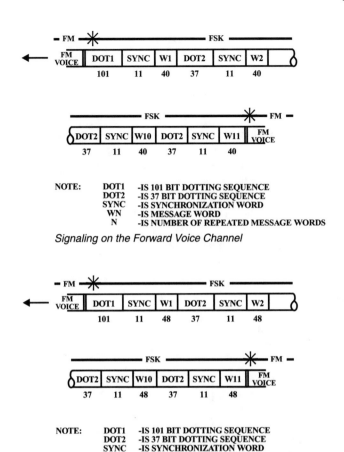

Figure 1.25, Signaling on the Voice Channels

1.6. Call Processing

Calls are processed by executing command messages sent to and from the cellular system. To operate within a system, a Subscriber Unit must follow a specific sequence for processing signaling messages. Functional operations are divided into four tasks; initialization, idle, access, and conversation. A task is the series of operations that the Subscriber Unit accomplishes to fill a functional requirement. The Subscriber Unit first completes the initialization task to obtain system parameters. It then remains in the idle task awaiting new system information, pages, or a user command. When a call is initiated or is ready to be received, the Subscriber Unit begins the system access task to compete for a voice channel. After the system has assigned a voice channel, the Subscriber Unit enters the conversation task.

In addition to these sequential tasks, other functions occur in parallel. To avoid interfering with other subscribers, timers are set and continuously monitored. If a timer fails, the Subscriber Unit will not transmit.

1.6.1. Initialization

Control channels continuously send system overhead information to provide Subscriber Units with parameters for to establishing communication and to inform subscribers of the cellular system status. Figure 1.26 illustrates the initialization process. When the Subscriber Unit is first turned on, it scans a group of dedicated control channels and locks onto the strongest signal. It then transfers system parameters, such as SID (System Identification), number of paging channels, and ROAM status. ROAM status is determined from information in the overhead messages that is input to the memory of the Subscriber Unit.

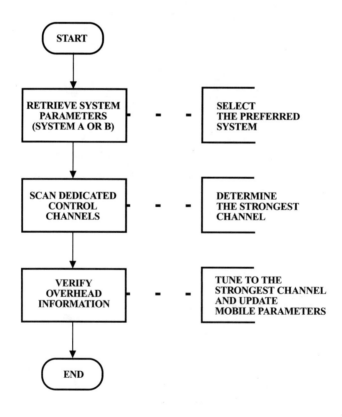

Figure 1.26, Call Processing Initialization Mode

1.6.2. Idle Mode

During the idle mode, the Subscriber Unit continually monitors the control channel overhead information messages for changes in system information, to obtain pages, to determine if a call is to be initiated, or if a short message has been received. Figure 1.27 displays the idle mode call processing. Overhead messages contain system information which is necessary for access to the system which is temporarily stored in the Subscriber Unit. System information can change on the control channel so the Subscriber Unit must continually update its system parameter informaiton.

The paging channel is continually monitored to determine if an incoming call is to be received. Page messages contain the Subcriber Units MIN. If the Subscriber Unit has been paged, it will setup its response message (a flag) to indicate it will access as a response to a page message.

The Base Station may command to Subscriber Unit to responds to commands (system orders) independent of the user's knowledge. This includes system registration commands. When the Subscriber Unit receives an order, it will setup its response message (a flag) to indicate the it will access as a response to a system order.

The Subscriber Unit also continually monitors the status of its control head (keypad) has changed to determine if the user has initiated a call. If the user has initiated a call, the Subscriber Unit will set its response message to indicate it will acess the system as a call origination.

While monitoring a contol channel during the idle mode, the Subscriber Unit may move away from the serving cell site until the signal level of the cotrol channel is below an acceptable level. When this happens, the Subscriber Unit will scan for other control channels and tune to the strongest signal. During the scanning process (sometimes up to 3 seconds), the Subscriber Unit will not be able to receive pages or messages.

As the cellular system technology evolves, new messages may be created (such as short message commands) that older Subscriber Units cannot understand. In the event the Subscriber Unit receives a message it cannot understand, it will discard the message. This allows the cellular systems to evolve and offer new features and services without causing eroneous operation of older Subscriber Units.

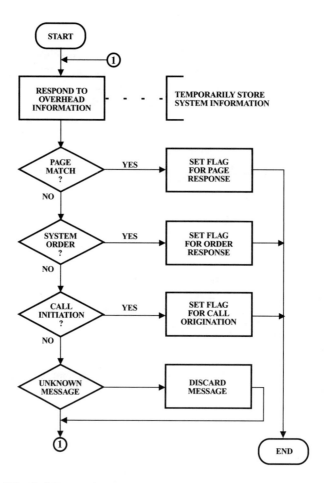

Figure 1.27, Idle Call Processing Mode

1.6.3. System Access

The Subscriber Unit attempts access to the system when the operator initiates a call, or whenever the unit receives a page (incoming call), orders, or a registration request. The access attempt contains a message that indicates which of these types of access is required. Registration requests, which are sent by the Base Station, require the Subscriber Unit to indicate that it is operating in the system. Registering active Subscriber Units in each cell area limits paging requirements because paging messages are sent only on control channels of cells in the vicinity of the last registration.

Control channels may be divided into separate paging and access channels where one control channel delivers pages and a different control channel coordinates access. Separating paging and access channels allows access channels to handle more service requests independent of paging requirements. However, assigning paging and access to different channels does not increase system capacity. The forward channel must carry all access signaling and the reverse channel carries all paging, leaving system capacity unaffected. As a result, paging and access are usually combined in one control channel. Figure 1.28 illustrates the steps in processing a call when a Subscriber Unit accesses the system.

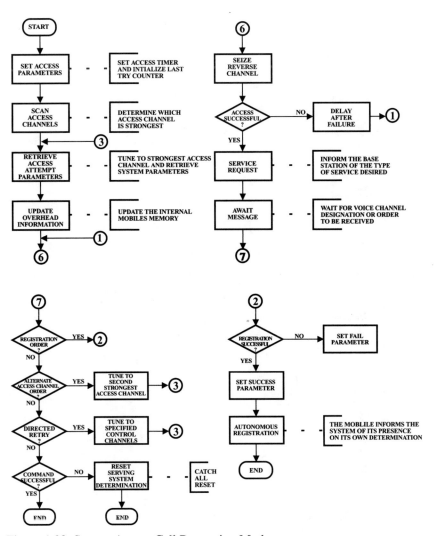

Figure 1.28, System Access Cell Processing Mode

1.6.4. Conversation

When access is granted, an Initial Voice Channel Designation (IVCD) message from the Base Station sends the Subscriber Unit a voice channel number to which it must tune to initiate conversation. During conversation, the Base Station continues to control the Subscriber Unit's power level, hand-off, alerting, etc. Control is accomplished with blank and burst signaling, briefly replacing voice information with signaling commands. Figure 1.29 illustrates conversation call processing.

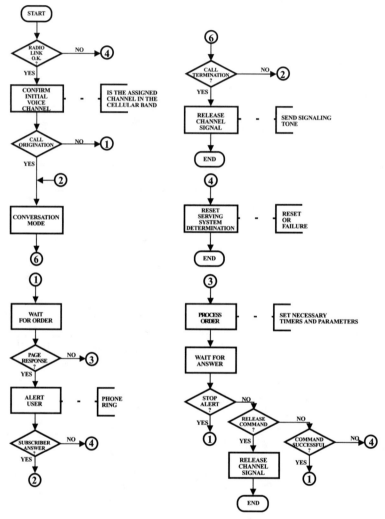

Figure 1.29, Conversation Call Processing

To ensure a reliable radio link is maintained throughout a cell, a radio link fade timer detects loss of radio continuity by monitoring the time that the SAT is interrupted. After 5 seconds without SAT, the radio link is determined to be broken. The timer system turns the Subscriber Unit off when the receive signal falls to a level below which the Base Station can no longer control it.

Discontinuous Transmission (DTX) is another feature that may be used in the conversation mode, particularly in portable Subscriber Units. To conserve power, the Subscriber Unit transmitter may be turned off during silent intervals in the conversation.

References

[1] Dr. George Calhoun, "Digital Cellular Radio", p.50-51, Artech House, MA. 1988.
[2] The Bell System Technical Journal, January 1979, Vol. 58, No. 1, American Telephone and Telegraph Company, Murray Hill, New Jersey.
[3] William Lee, "Mobile Cellular Telecommunications Systems", p.2, McGraw Hill, 1989.
[4] Personal interview, Pactel cellular systems manager, industry expert.
[5] FCC Regulations, Part 22, Subpart K, "Domestic Public Cellular Radio Telecommunications Service," 22.903, (June 1981).
[6] James Craig, "Dispatching a Large Taxi System", Communication Engineering, p.20-21, October 1953.
[7] Dr. George Calhoun, "Digital Cellular Radio", Artech House, MA. 1988.
[8] William Lee, "Mobile Cellular Telecommunications Systems", p.5, McGraw Hill, 1989.
[9] William Lee, "Mobile Cellular Telecommunications Systems", p.265 McGraw Hill, 1989.
[10] The Bell System Technical Journal, p.110-114, January 1979, Vol. 58, No. 1, American Telephone and Telegraph Company, Murray Hill, New Jersey.
[11] The Bell System Technical Journal, p.50, January 1979, Vol. 58, No. 1, American Telephone and Telegraph Company, Murray Hill, New Jersey.
[12] The Bell System Technical Journal, January 1979, Vol. 58, No. 1, American Telephone and Telegraph Company, Murray Hill, New Jersey.
[13] ibid.
[14] ibid.
[15] Ron Bohaychuk, personal Interview, Ericsson Radio Systems, 7 October 1990.
[16] The Bell System Technical Journal, p.47, January 1979, Vol. 58, No. 1, American Telephone and Telegraph Company, Murray Hill, New Jersey.
[17] Bellcore, "LSSGR; Signaling, Section 6", TR-TSY-000506, Rev 1, December, 1988.

Chapter 2
Digital Cellular

2. Digital Cellular

As more and more subscribers use a cellular system, the system must expand to accommodate them. If systems can expand by adding new cells, why change cellular technology? The answer is twofold. First, new technologies may provide less costly ways to expand than the existing Advanced Mobile Phone Service (AMPS). Second, AMPS lacks the capability to support many new services. This chapter presents a brief history of the need for advanced cellular services, and explains how the technologies were planned and developed. Following this overview, each new technology is introduced.

2.1 CTIA Requirements

In 1988, the Cellular Technology Industry Association (CTIA) Advanced Radio Technology Subcommittee (ARTS) was established to identify technology requirements that would lead to timely introduction of cost-effective technologies and new features. They created a User Performance Requirements (UPR) document which provided the goals for the new technology [1]. The UPR document assembled inputs from a Booz-Allen marketing study, assessments of cellular carrier needs, and consultations with manufacturers. Cellular service providers indicated that they wanted a new technology product life cycle of eight to ten years. Manufacturers worked with the CTIA to define a series of specific milestones to be achieved to introduce products by 1991. The result was the product development timeline contained in the UPR. The UPR document did not focus on any one technology, but specified the following customer service requirements for all technologies:

* a tenfold increase in system capacity compared to AMPS
* dual mode during transition
* provide for new features (e.g. short message services)
* ensure that equipment would be available by 1991
* set standards for high quality service

The Telecommunications Industry Association (TIA) was asked to create a specification based on the UPR requirements. As a result, Interim Standard 54 (IS-54) which uses a Time Division Multiple Access (TDMA) digital technology was released in early 1991 [2]. TDMA equipment was demonstrated and tested in mid-1991 in Dallas and Sweden.

However, the TDMA IS-54 standard did not meet all UPR goals, and soon after IS-54 was created, several companies proposed new technologies to meet industry needs. As a result, during the past several years, several new industry standards have been accepted. These include IS-136 for a new generation of TDMA [3], IS-95 for Code Division Multiple Access (CDMA) [4] and IS-89 for Narrowband AMPS (NAMPS) [5] Each of these has inherent advantages over AMPS technology, and each could result in savings.

Regardless of the technology used, the FCC requires all cellular systems to maintain AMPS service [6]. Because AMPS service will remain universally available, the CTIA desired that next-generation cellular radios operate with both the new technology and the existing AMPS system, automatically defaulting to AMPS mode where new technology is unavailable. This dual capability, called "dual-mode," has become the standard, although the FCC does not require it. Dedicated single mode phones may be smaller and cheaper, but they lack ROAMING capability in systems without digital service.

Figure 2.1 illustrates how a dual mode cellular system provides both AMPS and at least one new digital cellular service. Dual mode cellular systems consist of dual mode Subscriber Units, Base Stations, and a Mobile Switching Center (MSC).

Dual mode Subscriber Units have both analog and digital capability. When digital channels are available, the cellular system prefers to assign the Subscriber Unit to a digital channel. If no digital channel is available, such as when the Subscriber Unit is in a system that does not have digital capability, an analog channel will be assigned.

Dual mode cell site Base Stations must also have both analog and digital channels. In addition to the new digital radio channels, dual mode Base Stations require modifications to their scanning receiver, communications interface, and RF amplifiers.

The Mobile Switching Center (MSC) performs the same function as the Mobile Telephone Switching Office (MTSO). Modifications to the MTSO to upgrade it to digital service capability include software, echo cancel hardware, and other

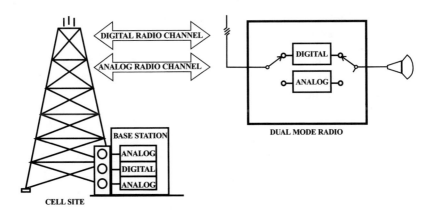

Figure 2.1, Dual Mode Cellular System

communications interface changes.

2.2 CEPT Requirements

In the 1980s, the early success of European cellular systems demonstrated the need for a new type of cellular system. Unlike the United States which enjoyed a standard AMPS cellular system, countries in Europe had created a variety of unique systems such as TACS, NMT-450, NMT-900, RC 2000, and CNet. A new system was required that could meet the long term system service and cost objectives for European cellular service providers. In 1982, the Conference of European Posts and Telecommunications (CEPT) held a meeting to begin the standardization process for a single next generation European cellular system. This meeting established the Groupe Special Mobile (GSM) standards body.

During 1985, the Consultative Committee of International Telegraph and Telephones (CCITT) created a list of technical recommendations that would be used to create the specifications. From these requirements, an action plan was created to coordinate the creation of the GSM specifications.

2.3 Increased Capacity

As a cellular carrier using the AMPS system continually adds new customers, more radio channels are needed. If the cell sites are not filled to capacity (typically 57) with radio channels, the carrier simply adds more. If cell sites are filled, carriers add more cell sites. Either way, to expand using the old technologies, cellular carriers must add radio channels.

The new cellular technologies allow capacity increases in a different way: by allowing more subscribers to share the same radio channel spectrum. The intensified use of radio spectrum is accomplished either by allowing more subscribers to share the same radio channel, or by packing more radio channels into a single cell site. To simultaneously serve multiple subscribers on the same radio channel, new technologies either assign time slots or unique codes to separate the calls. To add more radio channels per cell site, new technologies have created narrower channels. Ultimately, all of these techniques reduce the number of radio channels needed, and allow more subscribers to use a cellular service area. In this way, the new technologies reduce the average system equipment cost per customer.

To upgrade a cellular system for digital service, digital radio channels are added to the equipment racks at the Base Station. If all available channels are occupied by analog channels, some of the analog channels are replaced with digital channels. Digital cellular system manufacturers typically offer digital channels that simply replace an analog radio channel.

Cellular carriers can evaluate the potential system capacity factors of the new cellular technologies by reviewing two types of efficiency: radio channel efficiency and infrastructure efficiency. Radio channel efficiency is measured by the number of conversations (voice paths) that can be assigned per frequency bandwidth. Infrastructure efficiency is measured by the cellular system equipment and operating costs. Chapter 10 discusses the economic impact of these factors.

Figure 2.2 illustrates how a digital cellular system allows more users to share a cell site. The number of subscribers who can share a cell site is much greater than the number of available radio channels. Digital systems allow radio channels to be shared because not everyone talks at exactly the same time (except during traffic jams), and digital systems use the time much more efficiently. A typical digital system may add 20-32 subscribers to the system for every available voice channel [7]. An AMPS system cell site may have as many as 54 radio channels available, enough to serve about 1080 subscribers. The new technologies multi-

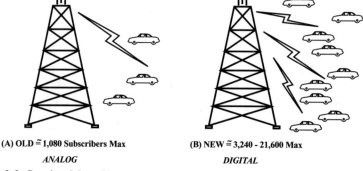

(A) OLD ≅ 1,080 Subscribers Max (B) NEW ≅ 3,240 - 21,600 Max
ANALOG *DIGITAL*

Figure 2.2, Serving More Users

ply this number by 3-20 times. If all of an AMPS system cell site's available channels were converted to digital, it could serve 3,240-21,600 customers.

2.4 New Features

All the proposed new technologies can provide similar subscriber features (e.g. calling number identification), although all possible features may not be included in current or proposed industry standards for each technology. All of the new technologies also allow for some advanced features not possible with AMPS technology, such as simultaneous voice and data transmissions. New features will bring new revenue potential, and the UPR contains the following wish list of new features:

* calling number identification (similar to paging)
* short message transmission (similar to dispatch)
* voice-activated control
* priority access (for emergency services)
* extension phone service
* lighter and more portable units
* enhanced voice privacy
* vehicle location
* ESN security (fraud protection)
* imaging (video) service.

2.5 Equipment Availability

The desired implementation timetable was perhaps the most critical requirement of digital cellular service. The UPR specified that the technology was to be available in the US by 1991. NAMPS service was available in 1991. US TDMA (IS-54) service started in 1992. GSM service became available in 1992. Code Division Multiple Access (CDMA) service was first offered in Hong Kong in 1995 [8].

When a new technology is specified, manufacturers face significant development time and costs both for system equipment and for Subscriber Unit (portable phones) equipment. Once equipment is available, installation can be complex, and the existence of multiple standards may limit availability of service over wide areas. These problems have resulted in a lack of new products, high costs, and market uncertainty. The result is that the new technologies had not captured a significant share of the market until 1995.

2.6 Digital Transmission Quality

As a radio signal passes through the air, distortion and noise enter the signal. A digital signal can be processed to enhance its resistance to distortion in three ways: signal regeneration, error detection, and error correction. Signal regeneration removes the added distortion and noise by creating a new signal without noise from a noisy one (see figure 2.3). Error detection determines if the channel impairments have exceeded distortion tolerances. Error correction uses extra bits provided with the original signal to recreate correct bits from incorrect ones.

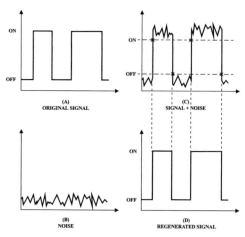

Figure 2.3, Digital Signal Regeneration

Figure 2.3 shows how noise 2.3 (b) is added to a digital signal 2.3 (a). By using ON/OFF threshold detection and conversion 2.3 (c), the original signal can be regenerated 2.3 (d).

2.7 Digital Technology

Most of the new cellular technologies use digital voice technology to achieve the goals of the UPR. Digital technology increases system efficiency by voice digitization, speech compression (coding), channel coding, phase modulation, and RF power control.

2.7.1 Voice Digitization

Figure 2.4 illustrates the conversion from an analog signal to a digital one. Speech into the microphone creates an analog signal. An audio bandpass filter removes high and low frequencies that interfere with digitization. The filtered signal is then sampled 8,000 times per second. For each sample, an 8 bit digital value is created. The resulting 64,000 bits per second represent the voice.

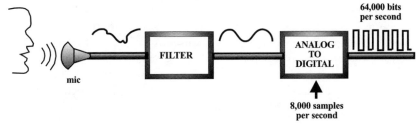

Figure 2.4, Voice Digitization

2.7.2 Speech Coding

About 64 kilobits per second (kbps) of data are required to reasonably digitize an analog waveform. Because transmitting a digital signal via radio requires about 1 Hz of radio bandwidth for each bps, an uncompressed digital voice signal would require more than 64 kHz of radio bandwidth. Without compression, this bandwidth would make digital less efficient than analog cellular, which uses only 25-30 kHz. Therefore, very high speech compression is necessary to increase cellular system capacity. Speech compression removes redundancy in the digital signal and attempts to ignore patterns that are not characteristic of the human voice. The result is a digital signal which represents the voice content, not a waveform.

The speech coder analyzes the 64 kbps speech information and characterizes it by pitch, volume, and other parameters. Figure 2.5 illustrates the speech compression process. As the speech coder characterizes the input signal, it looks up codes in a code book table which comes closest to the input signal. The compression process may be fixed or variable. For the TDMA system, the compression is 8:1. For CDMA, the compression varies from 8:1 to 64:1 depending on speech activity. GSM systems compress the voice by 5:1.

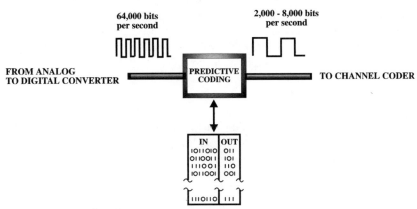

Figure 2.5, Speech Coding

High bit rate speech coders (small amount of compression) typically convert the waveform into a representative digital signal. Low bit rate speech coders (high amount of compression) analyze the waveform for key characteristics. In essence, low bit rate speech coders model the source of the waveform. This process makes low bit rate speech coders more susceptible to distortion from background noise and bit errors, poorer voice quality from a poor coding process model, and echoes from the speech coder processing time.

When there is a significant amount of background noise, distortion in the coding process occurs. Because the speech coder attempts to characterize the waveform as a human voice, the background noise is not in its code book. The speech coder will find the code that comes closest sound that matches the combined background noise and the human voice which creates distortion.

Because low bit rate speech coders model the waveform, a small number of bit errors can have a significant effect on the output waveform. For example, if several bits represent a unique sound, just a few bit errors may cause a completely different sound.

As a general rule, with the same amount of speech coding analysis, the fewer bits used to characterize the waveform, the poorer the speech quality. If the complexity (signal processing) of the speech coder can be increased, it is possible to get improved voice quality with fewer bits. The speech coders used in the digital cellular phones typically require 8 million instructions per second (MIPS) to process the voice signal. It has been estimated that it will take 4 times the amount of processing to reduce the amount of bits by an additional factor of two.

Voice digitization and speech coding take time. Typically, speech frames are digitized every 20 msec and input to the speech coder. The compression process, time alignment with the radio channel, and decompression at the receiving end all delay the voice signal. The combined delay can add up to 50-100 msec. Although such a delay is not usually noticeable in two-way conversation, it can cause an annoying echo when a speakerphone is used, or the side tone of the signal is high (so the user can hear themselves). However, an echo canceller can be used to process the signal and remove the echo.

2.7.3 Channel Coding

Once the digital speech information is compressed, control information bits must be added along with extra bits to protect from errors that will be introduced during radio transmission. Error protection consists of block coding and convolutional (continuous) coding. Control messages (such as power control) must be combined with speech information. Control messages are either time multiplexed (simultaneous) or they replace (blank and burst) the speech information.

Block coding adds bits to the end of a frame (usually after several hundred bits) of information. These bits allow the receiver to determine if all the information has been received correctly. Convolutional coding adds bits shortly after the information is sent (usually within 5 bits). This coding allows for quick checking and correction of information. Figure 2.6 shows how error detection and correction bits are added to the compressed speech. In all, error detection and correction bits add approximately 50% to the total number of bits used per subscriber. Error detection/correction reduces the number of bits available to users and decreases the system capacity.

Convolutional coders are described by the relationship between the number of bits entering and leaving the coder. For example, a 1/2 rate convolutional coder generates two bits for every one that enters. The larger the relationship, the more redundancy and better error protection. A 1/4 rate convolutional coder has much more error protection capability than a 1/2 rate coder.

Block CRC parity generation divides a given stimulus by a defined polynomial formula. The quotient and remainder are appended to the data stimulus to allow comparison when received. Figure 2.8 shows a CRC (Cyclic Redundancy Check) generator. A shift register and exclusive OR gates allow the division of the polynomial.

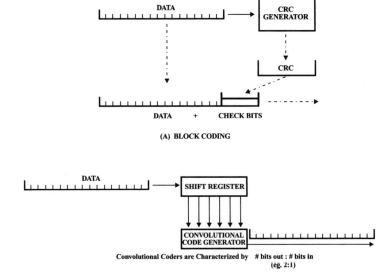

Figure 2.6 Block Error Detection and Convolutional Coding

As digital specifications were originally being developed, great emphasis was placed on a layered approach to their design. A key to this layered design was the concept of logical channels. A logical channel is simply a specification structure that separates different types of data from each other. This contrasts with the physical channel separation that is used on other systems such as EIA-553. The advantage of a logical channel separation is that it uses fewer RF frequencies to support the users. Theoretically, a single channel could be used. Different logical channels can be used to serve phones with different needs.

An example of logical channels is the control signals that are mixed with voice signals to direct the phone's communications with the Base Station. There are two types of control signaling, fast and slow. Slow signaling typically sends continuous channel quality measurements such as signal strength and the number of bits received in error in the last few frames. Fast signaling primarily sends channel assignment messages which must be acted on quickly. Fast messages replace the speech information for brief periods. Slow messages are divided and added (multiplexed) to the speech frames. Figure 2.7 shows the process of fast and slow message signaling.

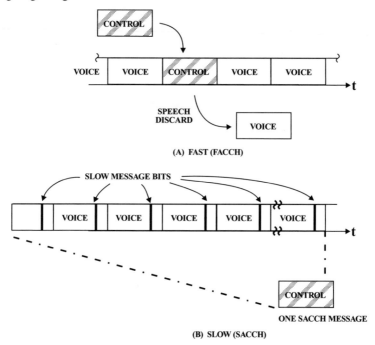

Figure 2.7 Fast and Slow Message Signaling

2.7.4 Modulation

AMPS and NAMPS use Frequency Modulation (FM) modulation. Frequency modulation is a process of shifting the radio frequency in proportion to the amplitude (voltage) of the input signal.

Digital technologies use phase modulation, a process that converts digital bits into phase shifts in the radio signal. Phase modulation is a result of shifting the carrier frequency higher and lower to introduce phase changes at specific points in time. Figure 2.8 displays a carrier frequency ($\omega\chi$) phase modulated by signal v(t). At time 1, the carrier is at the reference frequency. At time 2, the carrier is momentarily shifted above its reference for a short period to introduce a (-) phase shift. At time 3, the carrier is at the reference frequency. Time 4 requires the carrier to go below the reference frequency to introduce a (+) phase shift. At time 5, the carrier is again at reference frequency.

2.7.5 RF Amplification

A major difference between digital and analog technologies is that digital requires a linear RF amplifier. A linear amplifier distorts the signal less than the class C RF amplifiers used in AMPS cellular telephones. Unfortunately, the battery-to-RF energy conversion efficiency for linear amplifiers is 30-40% compared with 40-55% for class C RF amplifiers in AMPS phones [9]. Linear amplifiers require more input to produce the same RF energy output during transmission. Digital technologies overcome this limitation either by transmitting for shorter periods, or by precisely controlling power to transmit at lower average output power.

Figure 2.9 shows that the RF power amplifier can vary its output power. For AMPS, NAMPS, TDMA, and GSM, messages from the cell site adjust the Subscriber Unit's output power. For CDMA, Subscriber Unit output power is adjusted by a combination of the receive signal strength and fine adjustment messages from the cell site.

2.8 Spectral Efficiency

Spectral efficiency, or the number of users that can share a radio channel, varies depending on how the system is planned and used, but the new cellular technologies have been quoted to increase the spectral efficiency by 3-20 times AMPS [10]. The increase in spectral efficiency exists despite the fact that digital transmission of a complete audio signal is less efficient than analog transmission.

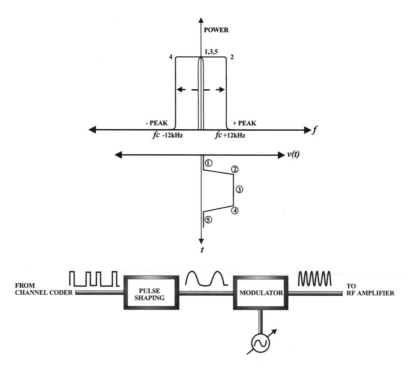

Figure 2.8 Radio Channel Modulation

Digital technologies realize their gains in spectral efficiency primarily through speech coding and voice activity detection.

For digital technologies, an adequate measure of spectral efficiency is the number of bits per second per user. Digital voice cellular technologies limit each user to less than 8,000 bits per second. As discussed in the speech coding section, when fewer bits per second per user are available, voice quality decreases. The basic tradeoff is increased capacity for reduced voice quality.

2.9 System Efficiency

Spectral efficiency is not a measure of the whole system's efficiency. System efficiency is dependent upon the amount and type of equipment required to serve a number of subscribers. System efficiency can be measured in the number of Base Stations, switching centers, computers to hold the subscriber database, and the leasing of communication lines between the Base Stations and the Mobile Switching Center.

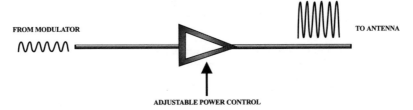

Figure 2.9, RF Power Level Control

Base Station radio transceivers become more efficient when they allow more users to share a single radio channel or when more radio channels can be installed. Mobile Switching Centers become more efficient by relying on mobile assisted handoff to reduce the burden of coordinating handoffs. Subscriber databases may actually grow larger with new service feature profiles. Communication lines become more efficient through voice compression that allows several users to share a single communications channel.

Each of the cellular technologies offers potential increased efficiencies for system equipment. The efficiency increase will be determined by both the technology (e.g. level of voice compression) and method of implementation.

2.10 Authentication

Authentication is the process of validating a mobile subscriber to determine if it is fraudulent, and if so, to deny access to the cellular system. Authentication functions by processing, transferring, and comparing previously stored information between the Subscriber Unit and the system. The North American (NAMPS, TDMA, and CDMA) authentication processes differ slightly from the European (GSM) authentication process.

A secret algorithm is at the heart of the authentication process. The algorithm defines a mathematical manipulation of data so that if two processors have the same initial values, they produce the same answer. The answer from the authentication algorithm is used to determine if a subscriber seeking access to the system is a valid registered unit.

2.10.1 North American Authentication

The authentication process in North American Systems (NAMPS, IS-54/IS-136 TDMA, and CDMA) is called the CAVE authentication algorithm. The CAVE algorithm operates on a group of data called the shared secret data (SSD). The SSD is contained in both the Subscriber Unit and cellular system. If either the Subscriber Unit or cellular system have an incorrect piece of the shared secret data, authentication fails. The SSD is 128 bits of data divided into two halves called SSD-A and SSD-B. SSD-A is used by the authentication process, and SSD-B by message encryption and voice privacy processes.

The key to authentication process is not the secrecy of the CAVE algorithm, but the initial values used when running the algorithm. Each subscriber receives a secret number called the A-KEY (or authentication key). The A-KEY is like the PIN number used with a bank card. The cellular subscriber enters the A-KEY on the keypad by typing A-K-E-Y (letters on the keypad), then pressing the function key twice. The Subscriber Unit does not store the A-KEY, but instead creates and stores a secret key called shared secret data (SSD). After the A-KEY is entered, it is known only to the subscriber and the network Home Location Register (HLR).

The cellular system begins the authentication process by sending AUTH over the control channel in the continuous System Parameter Overhead Message (SPOM). When the mobile unit receives the AUTH message, it executes the CAVE algorithm using the shared secret data and other values such as the mobile's ESN and dialed digits.

Subscriber Units add to the CAVE algorithm by automatically sending their call counts to the system, which keeps and increments the COUNT value during transactions between the mobile and Base Station. The COUNT value and the result of the CAVE algorithm are both sent to the Base Station to assist authentication.

After receiving the results of the mobile's authentication process, the Base Station compares the answer to its own calculations. If the values match, the call processing continues. Once a voice channel is assigned, the Base Station may update the mobile's SSD with a new value to be used in future transactions.

Authentication successfully prevents "cloning" of mobile phones because the A-KEY is never sent over the air. Even if the A-KEY were captured and duplicated, it would be very difficult for an invalid mobile to receive all the SSD updates and COUNT value increments that the "valid" Subscriber Unit was receiving.

Figure 2.10 shows the authentication process. A random number which changes periodically (RAND) is sent on the control channel as part of the secret key processing. The random number, shared secret data, and other information in the Subscriber Unit create an authentication response (AUTHR). The AUTHR is sent along with the first few digits of the RAND (RANDC), which is used to validate the Subscriber Unit. Sending the first few digits of the RAND is necessary because the RAND changes periodically and the AUTHR would be different.
In addition to being used for authentication, the CAVE algorithm is also used for message encryption and voice privacy. Message encryption scrambles non-voice messages sent between the Subscriber Unit and the Base Station. The Base Station controls which messages are encrypted.

The CAVE algorithm also creates the encryption mask used for voice privacy. The Subscriber Unit uses the mask to scramble the voice data it sends to the Base Station. With SSD information identical to that of the Subscriber Unit, the CAVE algorithm at the Base Station creates an identical mask that unscrambles the voice data. The Base Station then sends the unscrambled voice on to the land line network.

2.10.2 GSM Authentication

The GSM system is the authentication process used in Europe and other parts of the world. It uses the A3 authentication algorithm. The GSM A3 authentication algorithm is contained in a removable subscriber identity module (SIM) card. Unlike the CAVE authentication algorithm, which is standard for all Subscriber Units, the GSM A3 authentication process can change [11]. The SIM card contains a microprocessor which can store different authentication programs, so that different system operators can use different authentication algorithms.

The GSM algorithm processes data (RAND) with shared secret data (called Ki) to create a signed result (SRES). The Ki is stored in both the Subscriber Unit and

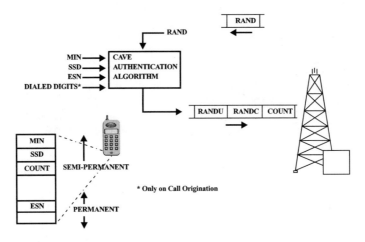

Figure 2.10, CAVE Authentication

cellular system. After receiving the results of the mobile's authentication process, the cellular system compares the answer to its own calculations. If the values match, the call processing continues. If either the Subscriber Unit or cellular system have an incorrect piece of the shared secret data the authentication process fails. The Ki key has a maximum length of 128 bits of data. Ki is also used to create the key used for voice privacy.

Figure 2.11 shows the authentication process used in the GSM system. A random number (RAND) is sent on the broadcast control channel as part of the secret key processing. This random number changes periodically. The random number, the Ki secret data, and other information in the Subscriber Unit are processed by the A3 authentication algorithm to create an Signed Response (SRES).

Unlike the North American CAVE authentication and voice privacy algorithm, the GSM system uses a different algorithm for message encryption and voice privacy. The A5 algorithm creates a message encryption mask for voice privacy. The encryption mask uses a Kc key, which is created at the beginning of each call with an A8 encryption algorithm. Throughout the call, the A5 algorithm uses the Kc key to scramble voice data sent to and from the Subscriber Unit. Since the cellular system has access to the same set of secret information, it generates the same encryption mask as the Subscriber Unit and uses it to unscramble the voice data before sending it to the land line network.

2.11 Digital Cellular Technologies

Five types of advanced technologies are available or specified today: Narrowband AMPS (NAMPS), IS-54 Time Division Multiple Access (TDMA), IS-136 TDMA, IS-95 Code Division Multiple Access (CDMA), and Global System for Mobile Communications (GSM).

2.11.1 Narrowband AMPS (NAMPS)

NAMPS is a frequency division multiple access (FDMA) system commercially introduced by Motorola in late 1991. It is currently deployed in countries all over the world. Like the existing AMPS technology, NAMPS uses analog FM radio modulation for voice transmissions. The distinguishing features of NAMPS are narrow radio channels and digital subaudible signaling. NAMPS uses a "narrow" channel bandwidth of 10 kHz, one third of the 30 kHz channel bandwidth used

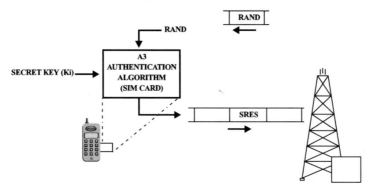

Figure 2.11, GSM Authentication Process

in the existing AMPS technology. In addition, NAMPS adds a low bit rate digital channel (200 bps) to the subaudible frequency range. The digital data and voice are sent simultaneously. Having a simultaneous digital channel provides for new features (such as short messaging) and communicates some AMPS control commands (such as SAT) as digital information.

NAMPS Subscriber Units cost less because they are relatively simple and similar to AMPS phones. NAMPS reduces system infrastructure costs by allowing more radio channels to be installed in each cell site, reducing the total number of cells in the system. Handoff system is simplified because Subscriber Units assist the process.

Figure 2.12 illustrates a NAMPS system. NAMPS requires few changes from an AMPS system. The MTSO is essentially the same. The Base Station has one new 10 kHz NAMP channel, and the scanning receiver is modified to measure 10 kHz channels. The Subscriber Unit is dual mode, functional on either an AMPS channel or NAMPS channel. The NAMPS Subscriber Unit accesses the system through AMPS control channels (step 1). During the request for service, the Subscriber Unit indicates it has NAMPS capability. The AMPS control channel responds by assigning the NAMPS Subscriber Unit to either an 10 kHz NAMPS voice channel or a 30 kHz AMPS radio channel (step 2). Regardless of whether the radio channel is AMPS or NAMPS, each voice channel uses a single radio channel.

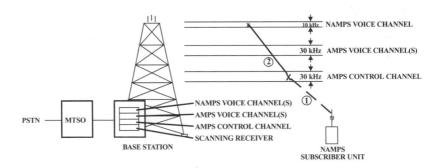

Figure 2.12, NAMPS Cellular System

The NAMPS handoff procedure differs from the AMPS procedure. Unlike the AMPS system, NAMPS Base Stations can command cellular radios to report information on received signal quality. This information is called Mobile Reported Interference (MRI) information. MRI information consists of the current operating signal strength and Bit Error Rate (BER). When the signal strength or BER fall below minimum levels, the Base Station can either increase its power or command the cellular radio to handoff to another channel. In some circumstances, MRI can provide voice quality superior to AMPS and increase radio channel efficiency.

One characteristic of FM radio signals used in AMPS and NAMPS is the degree to which they are allowed to modulate or shift the radio frequency as they transmit information. The more FM signals are allowed to modulate, the better they resist noise, fading, and interference. The modulation allowed for NAMPS radio channels is lower than that allowed for AMPS, so NAMPS is more susceptible to co-channel interference than AMPS. However, MRI more than compensates for the decreased modulation, and Motorola reports that overall, subscribers on NAMPS systems experience less co-channel interference than subscribers on AMPS systems [12].

2.11.2 IS-54 Time Division Multiple Access (TDMA)

IS-54 TDMA cellular service was commercially introduced in Calgary, Canada, in August 1992. It is now being deployed in countries around the world. Like AMPS, TDMA divides the radio spectrum into 30 kHz channels. The distinguishing feature of TDMA is that it employs digital techniques at the Base Station and in the cellular radio to subdivide the time on each channel into time slots. Each time slot can be assigned to a different subscriber. Voice sounds and access information are converted to digital information which is sent and

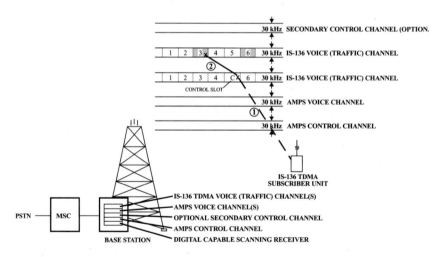

Figure 2.13, TDMA Cellular System

received in bursts during the time slots. The bursts of digital information can be encoded, transmitted, and decoded in a fraction of the time required to produce the sound. The result is that only a fraction of the air time is used, and other subscribers can use the remaining time on the radio channel. Current TDMA technology allows three subscribers to share one channel without interference. With improved compression techniques, six subscribers could share a single 30 kHz radio channel.

TDMA Subscriber Units are much more complex than AMPS or NAMPS phones. This added complexity, limited demand, and lower production numbers have resulted in higher costs for TDMA Subscriber Units. However, TDMA allows more subscribers to simultaneously use radio channels in the cell site, reducing the total number of required cell sites and radio channels, and ultimately reducing the cellular system infrastructure cost.

Figure 2.13 illustrates an IS-54 TDMA cellular system. Several changes are required to convert it from an AMPS system. The MTSO requires new software, and becomes a Mobile Switching Center (MSC). The Base Station requires two new types of channels, a 30 kHz optional secondary control channel, and a 30 kHz TDMA voice (traffic) channel. IS-54 Subscriber Units are dual mode, functioning on either AMPS or TDMA channels.

The TDMA Subscriber Unit accesses the system through an AMPS control channel (figure 2.13 step 1) or through an optional secondary control channel. The optional secondary control channel is identical to an AMPS control channel, except that it operates on a different frequency. The control channel responds to a service request by assigning the TDMA Subscriber Unit to either a 30 kHz AMPS voice channel or a 30 kHz TDMA radio channel (step 2).

2.11.3 IS-136 TDMA Digital Control Channel (DCC)

IS-136 is an evolution of the IS-54 TDMA standard which includes a digital control channel (DCC) and other new features. The changes to IS-54 were so significant that it was renamed IS-136, and is often referred to as TDMA DCC. In the United States, commercial deployment of IS-136 TDMA cellular service on a trial basis occurred in 1995 [13].

The physical construction of TDMA DCC Subscriber Units is not much more complex than that for IS-54. The primary difference is advanced software which allows access through the new digital control channel. Because IS-136

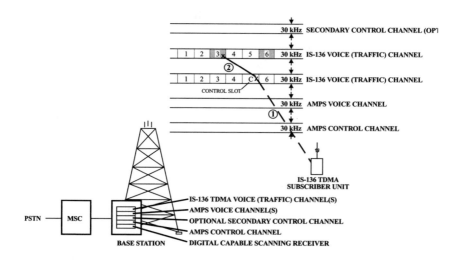

Figure 2.14, TDMA IS-136 Digital Control Channel System

Subscriber Units are so similar to IS-54 units, they are expected to be comparable in price. IS-136 systems use advanced messaging services and hierarchical cell structures, allowing co-existing private and public systems and a range of new services.

Figure 2.14 illustrates an IS-136 TDMA system. The MTSO is essentially the same. The Base Station has just one new 30 kHz IS-136 TDMA channel, and it uses the same IS-54 scanning receiver. The Subscriber Unit is tri-mode, operable on AMPS, IS-54 TDMA, or IS-136 TDMA channels.

The IS-136 TDMA Subscriber Unit accesses the system either through an AMPS control channel or a control slot on an IS-136 TDMA traffic channel (figure 2.13 step 1). The control slot responds by assigning the IS-136 TDMA Subscriber Unit to either an 30 kHz AMPS voice channel or a 30 kHz IS-136 TDMA radio channel (step 2).

2.11.4 IS-95 Code Division Multiple Access (CDMA)

Like TDMA, CDMA is a digital system. CDMA differs from TDMA in that it divides the radio spectrum into wide 1.23 MHz digital radio channels with each channel carrying up to 64 separate coded channels. Each coded channel is identified by a unique pseudo-random noise (PN) code. Digital receivers separate the channels by correlating signals with the proper PN sequence and enhancing the correlated one without enhancing the others. A CDMA RF channel uses some of its coded channels as control channels. The control channels include a pilot, synchronization, paging, and access channel. IS-95 CDMA cellular service began testing in the United States in San Diego California during 1991. In 1995, IS-95 CDMA commercial service began in Hong Kong and several test systems were operating in North America.

CDMA Subscriber Units are similar to TDMA phones in complexity. The CDMA system uses similar digital voice compression, but the channel coding process uses variable rate compression. Variable rate compression is an advantage because it decreases the average number of bits representing the voice so that more users share a radio channel.

Figure 2.15 illustrates an IS-95 CDMA system, and reveals several changes from the AMPS system. The modified CDMA MTSO allows simultaneous handoff between cell sites. The Base Station adds a 1.23 MHz CDMA RF channel. No new scanning receiver is needed because the CDMA radio channel measures signal strength. The first IS-95 Subscriber Units were dual mode, operable on either an AMPS or CDMA radio channels.

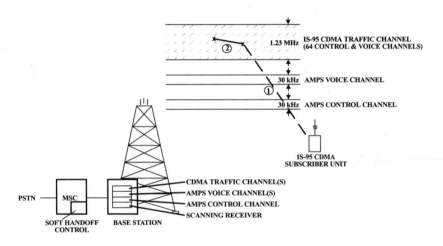

Figure 2.15, CDMA Cellular System

The CDMA Subscriber Unit accesses the system either through an AMPS control channel or coded channel on a CDMA RF channel (step 1). The CDMA control channel responds by assigning the CDMA Subscriber Unit to either a 30 kHz AMPS voice channel or a 1.23 MHz CDMA radio channel (step 2).
A Subscriber Unit requesting access to the cellular system competes on the access channel. This competition is a random process where the cellular radio continually increases its access power until the Base Station responds. If the Base Station does not respond within an allotted time, or the cellular radio exceeds a maximum power level, the access is aborted. This procedure avoids interference

between cellular radios during access attempts. When access is complete, the voice information is transmitted over separate coded voice channels.

The IS-95 CDMA system's "soft" handoff procedure is different from that of any other cellular technology. During soft handoff, a Subscriber Unit can simultaneously receive a channel from two cell sites (or radio channels) at the same time. The cellular radio measures and compares the received strength of pilot signals from Base Stations other than the one with which it is communicating. The strength of the received pilot signal indicates which cell can best communicate with the cellular radio, thereby determining the necessity for handoff. The advantage over other systems is that audio is not muted during handoff because the cellular radio is simultaneously receiving the same signal from two cell sites.

2.11.5 Global System for Mobile Communications (GSM)

The GSM system divides the radio spectrum into 200 kHz radio channels. This digital system subdivides each 200 kHz radio channel into 8 time slots per frame, allowing up to 8 subscribers per radio channel. Voice sounds and access information are converted to digital information that is sent and received in bursts during the time slots. The bursts are encoded, transmitted, and decoded in a fraction of the time required to produce the sound, so only a fraction of the air time on the radio channel is used. Other subscribers on the radio channel can then use the remaining time. The first generation of GSM technology allows eight subscribers to share one channel without interference. As speech compression techniques improve, sixteen subscribers will be able to share a single 200 kHz radio channel. GSM cellular service was first commercially introduced 1992, and is now being deployed in more than 30 countries around the world (see appendix 4).

GSM Subscriber Units are about as complex as IS-136 TDMA and IS-95 CDMA phones, and much more complex than the original analog Totally Accessible Communications System (TACS) or Nordic Mobile Telephone (NMT) cellular systems used in Europe. The increased complexity might have raised prices for Subscriber Units, but high global demand allowed economies of scale and cost reductions. GSM infrastructure costs are also reduced because more subscribers simultaneously use the radio channels in each cell site, requiring fewer radio channels.

The GSM system's form of digitized voice compression differs slightly from that IS-136 TDMA and CDMA. The GSM speech compression process is called Residual Excited Linear Prediction (RELP) coding. RELP coding uses more bits to represent a voice signal (13 kbps). In the future, a coding process at half the 13 kbps rate is planned. The increased efficiency will allow up to 16 subscribers per radio channel.

Figure 2.16 illustrates a GSM system, revealing several differences from early European analog systems. The GSM Base Station has only has one type of radio channel, a 200 kHz traffic channel. This channel provides both control and voice communication. GSM Subscriber Units access the cellular system through a time slot on one of the standard radio channels (step 1). After the traffic channel is assigned, the Subscriber Unit communicates its identity information to the system (step 2). In contrast, North American Subscriber Units communicate identity information on control channels. If the GSM identity information matches, conversation continues on the assigned traffic channel.

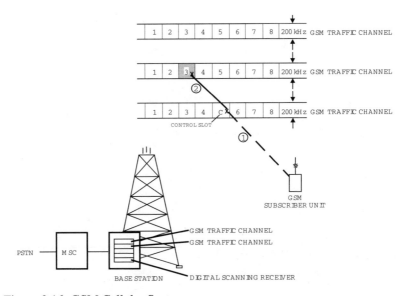

Figure 2.16, GSM Cellular System

The removable subscriber identity module (SIM) card is another unique feature of the GSM system. The SIM card contains the unique subscriber information such as the phone number, secret key, and the authentication algorithm which validates the identity of the subscriber. The Subscriber Unit does contain its own unique information, such as physical capabilities of the device (i.e. maximum power level, ability to receive short messages), and shares this information with the cellular system.

GSM Subscriber Units can synchronize to multiple control channels, allowing them to select cell site radio channels both when initiating a call and during the call. The Subscriber Unit assists handoff by measuring the signal strength and channel quality of other cell site radio channels and returning this information to the system.

2.12 Dual Mode Cellular

If a Subscriber Unit could receive only a digital technology, it could not ROAM into areas without its type of digital service. Users such as taxi cabs or local delivery services may find this acceptable, but others must ROAM large geographic regions.

In the Americas, the universal availability of AMPS cellular service has created the demand for dual mode Subscriber Units. Dual mode Subscriber Units can utilize a digital radio channel if available, and if not, they can have cellular service through the analog AMPS system.

In Europe and other parts of the world, analog cellular systems and frequencies vary significantly, and GSM service is readily available. There, the single digital standard allows subscribers to ROAM into other regions not previously accessible. Dual mode GSM phones have been planned, but these plans combine the Digital European Cordless Telephone (DECT) system and GSM [14].

References

1. CTIA, User Performance Requirements, Issue 1, September 8, 1988.

2. Electronic Industries Association, EIA Interim Standard IS-54.

3. Electronic Industries Association, EIA Interim Standard IS-136.

4. Electronic Industries Association, EIA Interim Standard IS-95.

5. Electronic Industries Association, EIA Interim Standard IS-88.

6. FCC Regulations, Part 22, Subpart K, "Domestic Public Cellular Radio Telecommunications Service," 22.903, June 1981.

7. Cellular system manager LA Cellular, personal Interview, industry expert.

8. Cellular service provider manager, personal interview, industry expert.

9. MA/COM manager, Personal Interview, 1995.

10. CTIA, Technology Conference, Washington DC, October 1991.

11. Michel Mouly and Marie-Bernadette Pautet, The GSM System for Mobile Communications, (M. Mouly et Marie-B. Pautet), pp 478,479.

12. Special Seminar, NAMPS Panel, CTIA Next Generation Cellular: Results of the Field Trials, Washington, DC, December 4-5, 1991.

13. Product manager for cellular telephone manufacturer , personal Interview, 1995.

14. Product manager, Ericsson, personal interview, February, 1996.

Chapter 3
TDMA (IS-54)

3. TDMA (IS-54)

The IS-54 Time Division Multiple Access (TDMA) standard combines digital TDMA and AMPS technologies. This chapter describes IS-54 TDMA dual mode technology at a semi-technical level. Chapter 1 describes the AMPS technology associated with the IS-54 TDMA system.

In 1990, the first TDMA digital standard, called IS-54 Rev 0, evolved from the EIA-553 AMPS specification. IS-54 Rev 0 identified the critical parameters (e.g. time slot structure, type of radio channel modulation, message formats) needed to begin designing TDMA cellular equipment. Unfortunately, IS-54 Rev 0 lacked some basic features that were introduced in the first commercial TDMA phones, and IS-54 Rev A was soon introduced to correct errors and add essential basic features (such as caller ID) to the TDMA standard. In 1991, IS-54 Rev B added features such as ESN authentication, voice privacy, and a more capable caller ID with greater benefit to the user. Figure 3.1 summarizes the evolution of the IS-54 specification.

Figure 3.1, Evolution of the IS-54 Specification

Digital TDMA technology has continued to evolve beyond the current IS-54 Revision B to the digital control channel system covered by the IS-136 specification. The IS-136 system is covered in chapter 4.

A primary feature of the IS-54 system is its ease of adaptation to the existing AMPS system. Much of this adaptability is due to the fact that IS-54 radio channels retain the same 30 kHz bandwidth as AMPS system channels. Most Base Stations can therefore place TDMA radio units in locations previously occupied by AMPS radio units. Another factor in favor of adaptability is that new Subscriber Units were developed to operate on either IS-54 digital traffic (voice and data) channels or the existing AMPS radio channels. Some of this dual mode capability is due to the fact that IS-54 systems use an analog control channel very similar to that of AMPS systems.

3.1 System Overview

The IS-54 dual mode cellular system has three types of channels: analog control channels (ACC), analog voice channels (AVC), and digital traffic channels (DTC). The system uses the same type of control channels and analog voice channels as AMPS. The new digital traffic channel is divided into time slots and uses digital phase modulation to transfer voice and control signaling information. Each user on the radio channel is assigned a distinct time slot for reception and transmission.

All IS-54 TDMA digital traffic channels are divided into frames with 6 time slots. Every communication channel consists of two 30 kHz wide channels, a forward channel (from the cell site to the Subscriber Unit) and a reverse channel (from the Subscriber Unit to the cell site). The time slots between forward and reverse channels are related so that the Subscriber Unit does not simultaneously transmit and receive.

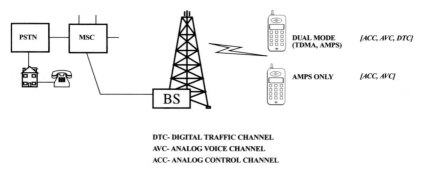

DTC- DIGITAL TRAFFIC CHANNEL
AVC- ANALOG VOICE CHANNEL
ACC- ANALOG CONTROL CHANNEL

Figure 3.2, Overview of an IS-54 System

The IS-54 system has defined a new type of radio channel called a traffic channel. The traffic channel carries both digitally coded voice and data. Although the IS-54 system continues to use the AMPS control channel format, a new set of optional control channels has been defined. Figure 3.3 shows the IS-54 radio channel types.

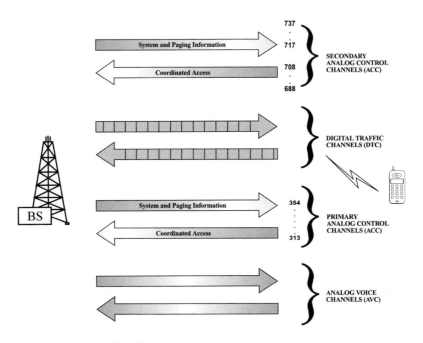

Figure 3.3, IS-54 Radio Channels

3.1.1 Speech Data Compression

Under the IS-54 system, three different Subscriber Units share time on a traffic channel. To split the time three ways, Subscriber Units must compress voice data to transmit in one third the speech time. To do this, IS-54 digital transmissions are divided into 20 ms frames, and each Subscriber Unit sends its speech data in a third of the 20 ms frame (6.67 ms). Figure 3.4 shows the 20 msec speech signals converted into 6.67 msec data bursts. To send the signal in a third of the time, the Subscriber Unit must digitize the audio signal, process the digitized signal to reduce the number of bits required, and compress the data bits in time. A speech coder compresses the digitized audio signal by analyzing the digital bits and creating coded information to represent the voice. The codes use 1/8th as many bits as used in the public "land line" switched telephone network. The coded data bits are then interleaved between adjacent 20 msec time slots prior to phase modulating the RF carrier.

The speech coder's performance is one of the greatest concerns in any digital voice transmission system. If the speech coder does accurately code and decode voice data, users will notice noise or errors. Fortunately, even in most situations of poor radio conditions, the IS-54B speech coder has been shown to provide voice quality superior to that of analog cellular. The high IS-54B voice quality is due to error correction that resists random errors, interleaving that reduces effects of burst errors, and a speech coder that replaces lost frames with similar previous frames. The replacement process takes advantage of the fact that speech information is periodic for short intervals, allowing the speech coder to fill in lost frames with previous frames. As the number of lost frames increases, the subscriber gradually hears more muting. The result is a more natural-sounding voice, even with errors.

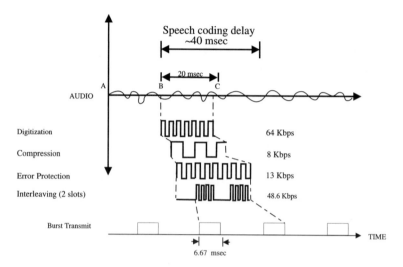

Figure 3.4, Speech Data Compression

3.1.2 Radio Channel Structure

IS-54B systems are either full rate or half rate. Full rate IS-54 systems allow three users to simultaneously share a radio channel. Half rate IS-54 systems allow 6 users to share a radio channel. Figure 3.5 shows how users share IS-54 full rate and half rate radio channels.

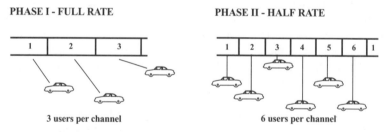

Figure 3.5, TDMA Radio Channel Sharing

Full Rate TDMA

The time on full rate TDMA channels is divided into frames containing 6 time slots. During a voice conversation, two time slots are used for transmitting, two are used for receiving, and two are idle. The Subscriber Unit typically uses the idle time to measure the signal strength of surrounding channels to assist in handoff. Subscriber Units transmit every third slot so that phone #1 uses slots 1 and 4, phone #2 uses slots 2 and 5, while phone #3 uses slots 3 and 6. This time sharing results in a user-available data rate of 13 kbps. However, some of the user data is used for error detection and correction, so only 8 kbps of data are available for compressed speech data.

Subscribers must be able to talk and listen at the same time, so the Subscriber Unit must function as if it is simultaneously sending and receiving (called full duplex). When operating on a digital radio channel, IS-54 Subscriber Units do not actually transmit and receive simultaneously. However, they appear to be full duplex because the speech data bursts alternate between transmitting and receiving, and when received, the compressed speech bursts are expanded in time to create a continuous audio signal.

Figure 3.6 shows how TDMA full duplex radio channels are divided in time. IS-54 digital radio channels transmit on one frequency and receive on another frequency 45 MHz higher, but not at the same time. The Subscriber Unit transmits a burst of data on one frequency, then receives a burst on another frequency, and is briefly idle before repeating the process.

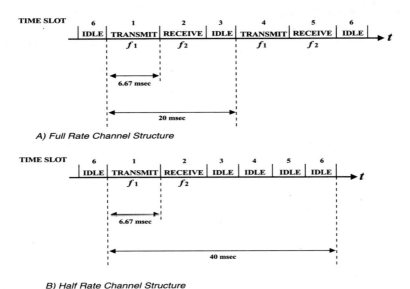

Figure 3.6, IS-54 Channel Structure

3.1.3 Half Rate TDMA

A radio channel's capacity can be doubled by dedicating only one slot per frame per subscriber, creating a half rate channel. Half rate channels use one of the six slots to transmit and one to receive, leaving four idle. Figure 3.6 illustrates half rate TDMA channel structure. Using only one of the six time slots results in a user-available data rate of less than 6.5 kbps. The decrease is due to the increased percentage of signaling data. A half rate system supports six simultaneous users per radio channel. Introduction of half rate service is planned for the near future, after the standards committee has approved a suitable digital speech coder.

3.2 System Attributes

Dual mode digital cellular systems meet the cellular telecommunications industry association (CTIA) user performance requirements (UPR) by maintaining the old RF control channel structure, improving voice quality, cost-effectively expanding capacity, and maintaining coordinated system control. The dual mode system maintains compatibility with the established access technology because both analog (EIA-553) and dual mode (IS-54) Subscriber Units use the same analog control channels. Even in varying radio conditions, voice quality remains reasonably stable. Some voice channels are converted to allow digital multiplexing of several users per RF channel, so systems can cost-effectively expand by sharing time on digital RF channels. They can also reduce the minimum cell size. A Mobile Switching Center (MSC) continues to provide the central control as Subscriber Units move through the system.

3.2.1 Frequency Reuse

The IS-54 TDMA system was originally believed to tolerate higher interference levels because of its error correction capability. Error correction does improve voice quality in high interference, but in real application, TDMA radio channels tolerate only about 17 dB of interference, allowing frequency reuse similar to that of AMPS [1].

Where error correction has not greatly improved frequency reuse, a feature called Mobile Assisted HandOff (MAHO), to be discussed later, has improved it. MAHO has assisted adaptive channel allocation, or the re-assignment of radio channels among cell sites. MAHO assists adaptive channel allocation by using both the Base Station and the Subscriber Unit to detect interference levels, thereby making reassignment of frequencies more efficient. The more efficient frequency reassignment system improves frequency reuse.

3.2.2 Capacity Expansion

Initially, TDMA can expand a system's capacity expansion by serving three subscribers on one 30 kHz radio channel. Next, half rate systems can serve six subscribers on one 30 kHz radio channel. Potentially, even more users may be able to share slots using digital speech interpolation, to be discussed in chapter 11.

3.2.3 Secondary Dedicated Control Channels

The IS-54 standard also specified a new optional band of secondary dedicated control channels. Secondary dedicated control channels are sometimes used when a new digital system must co-exist with an older AMPS-only cellular system. When the AMPS system cannot coordinate digital radio channel assignment, dual mode Subscriber Units can access the system through the secondary dedicated control channels. The dual mode Subscriber Unit searches for a digital protocol capability indicator (PCI) on AMPS control channels. If it cannot find digital capability on the primary dedicated control channels, it assumes that it has found an older AMPS only control channel that does not have digital capability. On existing AMPS control channels, the PCI bit is a reserved bit set to 0. If a dual mode Subscriber Unit finds PCI equal to 0, it searches for a secondary dedicated control channel (separate dual mode system). If finds no system message on the secondary control channel (perhaps no secondary channel is in operation or it may be a voice channel), the Subscriber Unit attempts to access the system as an analog unit on the primary dedicated control channel. A service provider's secondary dedicated control channels are 737 to 757; B service provider's secondary dedicated control channels are 688 to 708.

3.2.4 Optional Reversed Paging and Access Channels

Dual mode cellular systems serve more subscribers by efficiently sharing radio channels, but initially, dual mode systems' control channel capacity remained the same as for AMPS systems, delivering about 20-40 pages per second. An AMPS access channel can only handle 3-4 messages per second. Most pages are unanswered, either because the Subscriber Unit is turned off or is located in another cell. Eventually, to serve more subscribers, it would be necessary to increase control channel capacity. Therefore, an optional access method for digital subscribers was created.

The typical AMPS system allocates one control channel per cell or sector, even though the AMPS system allows two control channels per cell site, one for paging and one for access. A data field called CPA (combined paging and access) is used to define whether or not separate paging and access channels are used. Both control channels are rarely used, however, because most paging channel information is in the forward direction and most access channel information is in the reverse direction. In this circumstance, one control channel can carry almost as much information as two.

For many years to come, most cellular systems' subscriber populations will be a mixture of AMPS and dual mode subscribers. This mix of subscribers will make it practical to dedicate two control channels, one for paging and one for access. If dual mode Subscriber Units reverse their paging and access functions, they can use the excess forward capacity of the access channel and the excess reverse capacity of the paging channel. Figure 3.7 illustrates AMPS, dual mode, and combined paging capacities. However, even though the situation illustrated in (C) *Combined Analog & Dual Mode* increases system capacity, very few systems have ever used separate paging and access channels.

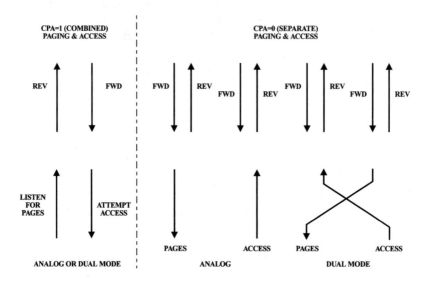

Figure 3.7, Paging and Access Channel Capacity

The reversal of paging and access channels is optional for digital Subscriber Units. The reversal occurs only when the primary dedicated control channel has digital capability and is designated as separate paging and access channels. Figure 3.8 illustrates how reversed paging and access channels work.

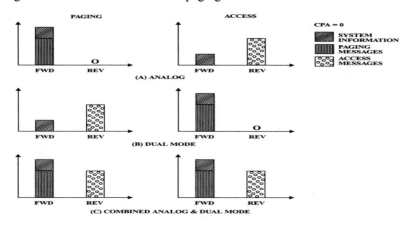

Figure 3.8, Reversed Paging and Access Channels

3.2.5 Dynamic Time Alignment

Dynamic time alignment is a technique that allows the Base Station to receive digital Subscriber Units' transmit bursts in an exact time slot. Time alignment keeps different digital subscribers' transmit bursts from colliding or overlapping. Dynamic time alignment is necessary because subscribers are moving, and their radio waves' arrival time at the Base Station depends on their changing distance from the Base Station. The greater the distance, the more delay in the signal's arrival time. Dynamic time alignment adjusts for differences in radio wave travel time according to each Subscriber Unit's changing distance from the Base Station.

The Base Station adjusts for the delay by commanding Subscriber Units to alter their relative transmit times based on their distance from the Base Station. The Base Station calculates the required offset from the Subscriber Unit's initial transmission of a shortened burst in its designated time slot (necessary only in large diameter cells where propagation time is long). To account for the combined receive and transmit delays, the required timing offset is twice the path delay. The Subscriber Unit uses a received burst to determine when its burst transmission should start. The Subscriber Unit's default delay between receive and transmit slots is 44 symbols, which can be reduced in 1/2 symbol increments to 15 symbols. This standard allows Subscriber Units to operate at a maximum of 72 miles from the Base Station without slot collisions. Figure 3.9 illustrates dynamic time alignment.

Path propagation delay = 5.3 usec/mi
Round trip delay = 10.6 usec/mi
One symbol = 41.1 usec : 1/2 symbol = 20.55 usec
2 miles require 1/2 symbol =>>12 miles = 3 symbols
The standard guard time is 3 symbols
Maximum distance = 12 mi + (4 mi/symbol * 15 symbols) = 72 mi

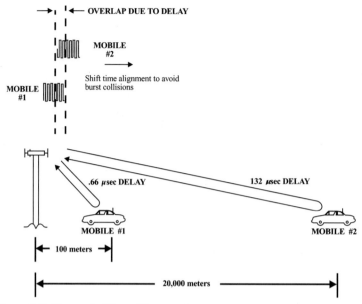

Figure 3.9, Dynamic Time Alignment

3.2.6 Mobile Assisted HandOff (MAHO)

Mobile Assisted Handoff (MAHO) is a system in which the Subscriber Unit assists the hand-off decision by sending radio channel quality information back to the Base Station. In existing analog systems, handoff decisions were based only on measurements of Subscriber Units' signal strength made by receivers at the Base Station. IS-54 systems use two types of radio channel quality information; signal strength of multiple channels and an estimated bit error rate of the operating channel. The bit error rate is estimated using the result of the forward error correction codes for speech data and call processing messages. Having the Subscriber Unit report quality information also allows for measurements of the downlink quality that are not possible from the Base Station.

Figure 3.10 illustrates the MAHO process. The system sends the Subscriber Unit a MAHO message containing a list of radio channels from up to 12 neighbor cells. During its idle time slots, the Subscriber Unit measures the signal strength of the channels on the list including the channel it is currently operating on. The Subscriber Unit averages the signal strength measurements over a second, then continuously sends MAHO channel strength reports back to the Base Station every second. The system combines the MAHO measurements with its own information to determine which radio channel will offer the best quality, and it initiates handoff to the best channel when required.

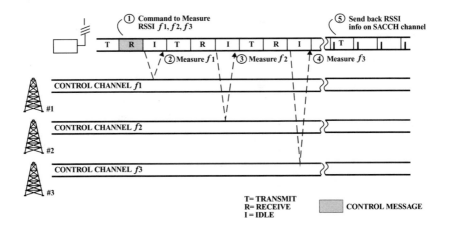

Figure 3.10, Mobile Assisted handoff

3.2.7 Multipath

An RF signal can be distorted when it bounces and propagates over different paths, staggering the signal's arrival times at the receiver. Such errors are called multipath signal distortion. Multipath causes symbol amplitude and phase distortion. It can distort TDMA transmissions if the delayed signal (called the 2nd ray) arrives more than 20 μsec after the original signal, and if the 2nd ray's amplitude is within 10 dB of the original signal. Figure 3.11 a. illustrates a significant multipath problem. Radio signals travel at 5.3 usec per mile, so to distort TDMA transmissions, the 2nd ray must travel an extra 4 miles and be received at approximately the same power level as the original signal. Figure 3.11 b. illustrates a circumstance in which reflected signals are only slightly delayed

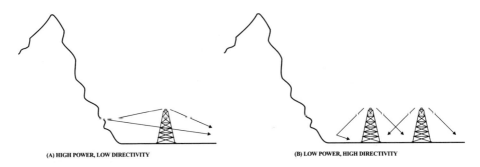

Figure 3.11, Delay Spread Consideration

Received signal equalization compensates for multipath amplitude and phase distortions. Equalization compensates for the varying radio channel conditions. Given a known bit stream input (such as synchronization words), it is possible to adjust the decision points as channel quality changes.

Selecting a synchronization word bit sequence allows equalizer training. The sync word is composed of 28 bits, the last 4 being unique to identify each slot. Digital voice color coded (DVCC) bits allow for additional equalization. In small cells with minimal multipath signals, equalization may unnecessary, and a command from the Base Station can turn off the Subscriber Units' equalizer to reduce processing and power consumption. When there are no significant multipath rays present, the received signal quality is better with the equalizer turned off.

3.2.8 Doppler Effect

When an operating transmitter and receiver are in motion relative to each other, the received frequency varies in proportion to the rate of motion toward or away. This effect, known as the Doppler effect, matters little to analog cellular but it significantly affects digital phase transmission and equalizer operations. Therefore, in addition to overcoming multipath distortion, the equalizer must overcome phase and amplitude distortions arising from the Doppler effect.

3.2.9 Slot Structure

The IS-54 standard describes four slot structures: forward speech slot, reverse speech slot, FACCH data message slot, and shortened burst slot. Each slot has dedicated fields. With the exception of its data bit encoding method, the FACCH data slot format is the same as forward and reverse speech slots. Each slot is composed of 324 bits (162 symbols).

Interleaving, or the continuous distribution of data bits between adjacent slots, is used to overcome effects of burst errors due to Rayleigh fading. Diagonal interleaving is used so that information is distributed between adjacent slots to reduce temporary fades. Since the number of consecutive bit errors is smaller, the error protection code works more accurately to correct non consecutive errors.

3.2.9.1 Forward Data Slot

The forward data slot transfers voice and data from the Base Station to the Subscriber Unit. It contains 324 data bits, of which 260 are available to the subscriber. The initial field in the slot contains the synchronization field which identifies the slot number, timing information which allows decoding. It is a standard pattern which also allows equalizer training. The equalizer adjusts the receiver to compensate for radio channel change (distortion). The SACCH field contains a set of dedicated bits for sending messages. The data fields carry the subscribers voice and data information. The digital voice color code (DVCC) is similar in function to SAT in analog cellular because it is a standard pattern set to a different value in each cell and can also be used for equalizer training.

28	12	130	12	130	12
S	SA	DATA	D	DATA	RSV

NOTE: S   -IS SYNCRONIZATION WORD
       SA -IS SACCH
       DATA -IS USER INFORMATION
       D   -IS CODED DIGITAL TRAFFIC COLOR CODE
       RSV -IS RESERVED

Figure 3.12, Forward Traffic Channel Slot Structure

3.2.9.2 Reverse Data Slot

The reverse data slot transfers voice and data from the Subscriber Unit to the Base Station. It differs from the forward data slot in that it includes guard and ramp time. During the guard time period (approximately 123 usec), the Subscriber Unit's transmitter is off. Guard time protects the system from bursts being received outside the allotted time slot interval due to the propagation time between the Subscriber Unit and cell site (see dynamic time alignment). Ramp time slowly turns on the transmitter to protect other Subscriber Units from interference (outside the allotted 30 kHz bandwidth) that occurs if a Subscriber Unit turns on instantaneously. The synchronization word, DVCC, and SACCH fields provide the same functions as described in the forward traffic channel slot.

6	6	16	28	122	12	12	122
G1	RA	DATA	S	DATA	SA	D	DATA

NOTE: S   -IS SYNCRONIZATION WORD
       SA -IS SACCH
       DATA -IS USER INFORMATION
       D   -IS CODED DIGITAL TRAFFIC COLOR CODE
       RSV -IS RESERVED
       G1 -IS STANDARD GUARD TIME
       RA -IS RAMP UP TIME

Figure 3.13, Reverse Traffic Channel Slot Structure

3.2.9.3 FACCH Data Slot

When urgent control messages such as a handoff command are sent, signaling information replaces speech information (260 data bits) similar to the blank and burst process used for control on the AMPS voice channel. A FACCH control message slot is identified by use of a different type of error correction coding. Initially, all slots are decoded as a speech data slots, but if a FACCH message is in a speech data slot, the CRC check sum and other error detection code outputs will fail, and the message will be decoded as a FACCH data slot. This process is used so that information bits need not be dedicated to indicate whether a data slot is for speech or control as is done in the GSM system. The FACCH Data Slot structure is identical to a speech slot, and only the data bits are FACCH data rather than digitally coded speech.

3.2.9.4 Shortened Burst

When a Subscriber Unit begins operating in a large diameter cell or following a handoff bewteen two adjacent cells of markedly different size, it sends shortened bursts to help determine the propagation time between the Subscriber Unit and Base Station. Propagation time in large cells could be so long (round trip in excess of 500 μsec) that overlapping bursts could cause significant problems. The shortened burst allocates another guard time which prevents received bursts from overlapping before a Subscriber Unit's dynamic time alignment has adjusted. Figure 3.14 illustrates the format of a shortened burst slot.

Notice that there are several synchronization fields. Any two sync fields are separated by different amounts of time so that a Base Station receiver can simply detect the relative time the burst is being received in comparison to other bursts. If bursts are received out of their expected time periods, the Base Station can command the Subscriber Unit to adjust its transmit time. After the shortened burst has been used to determine time alignment, the Subscriber Unit will begin to use the standard reverse traffic channel slot structure to send user information.

6	6	28	12	4	28	12	8	28	12	12	28	12	16	28	22
G1	RA	S	D	V	S	D	W	S	D	X	S	D	Y	S	G2

```
NOTE:   G1  -IS STANDARD GUARD TIME
        S   -IS SYNCRONIZATION WORD
        D   -IS DTCC
        V   -IS OH
        W   -IS OOH
        X   -IS OOOH
        Y   -IS OOOOH
        G2  -IS ADDITIONAL GUARD TIME
        RA  -IS RAMP UP TIME
```

V,W,X and Y fields are defined in hexadecimal, each hexadecimal 0 is equivalent to 0000 binary.

Figure 3.14, Shortened Burst Slot Structure

3.3 Signaling

Signaling is the physical process of transferring control information to and from the Subscriber Unit. As discussed in chapter 2, signaling on the voice channel has been divided into two channels: the Fast Associated Control Channel (FACCH) and the Slow Associated Control Channel (SACCH). The fast channel replaces speech with signal data. The slow channel uses dedicated sub frames within each burst. Messages sent via the SACCH channel have long delays of 440 msec compared with 40 msec required to send a FACCH message. To save time on time critical messages such as a handoff, SACCH messages may be sent as FACCH messages.

When a Subscriber Unit is operated in the DTX (discontinuous transmission) mode to conserve battery power, the SACCH channel is not operating continuously. Therefore, all messages are sent as FACCH rather than SACCH.

3.3.1 Control Channel Signaling

Like the AMPS system, the IS-54 system's control channel signaling continues to be FSK (Frequency Shift Keying). IS-54 control channel signaling differs from AMPS in four ways. The IS-54 system uses a dedicated bit to indicate that the Base Station has digital capability; it introduces a new band of secondary dedicated control channels; it adds parameters for assigning a digital traffic channel; and it allows a reversal of separate paging and access channel functions.

3.3.1.1 Digital Capability Indicator (PCI)

A Dual Mode Subscriber Unit is normally given first priority for obtaining a digital voice channel when it attempts to access the system. The Subscriber Unit verifies that the Base Station has digital capability by first checking the Protocol Capability Indicator (PCI) bit in the overhead message. The PCI bit lets the Subscriber Unit know that the control channel it is monitoring has digital traffic channel capability.

Additional parameters are needed to assign a digital traffic channel. These include a time slot and a type of digital channel indicator. Radio channel assignments could be a half rate, full rate, or other type of digital channel such as a data channel.

3.3.2 Analog Voice Channel Signaling

Analog voice channel signaling is identical to the original EIA-553, with the one exception of new parameters in the handoff command which allow assignment to a TDMA traffic channel.

3.3.3 Traffic Channel Signaling

The digital traffic channel offers several methods of transferring control information. These can be divided into in-band and out-of-band signaling. In-band signaling replaces voice data, and out-of-band is sent simultaneously with the voice data.

In- band signaling for the digital channel includes FACCH. Out-of-band signaling is called SACCH.

3.3.3.1 Slow Associated Control Channel (SACCH)

SACCH is a continuous data stream of signaling information sent beside speech data (out-of-band signaling). SACCH messages are sent by dedicated bits in each slot, so SACCH messages do not affect speech quality. However, the transmission rate for SACCH messages is slow. For rapid message delivery, SACCH messages are sent via the FACCH channel. The SACCH and FACCH system was designed to maximize the number of bits devoted to speech and minimize the number of bits devoted to continuous signaling.

The SACCH is allocated twelve bits per slot. A message is composed over 12 slots, resulting in a gross rate of 600 bps. The data is 1/2 rate convolutionally coded, reducing data transmission to 300 bps including a CF flag, a reserved bit, and CRC. Subtracting these fields yields transmission rates of:

- **218 bps throughput full rate**
- **109 bps throughput half rate**

Each SACCH message uses 50 bits allocated as follows:

```
|CF|RSV|    Message    |
  1   1       48
```
WHERE:
  CF = CONTINUATION FLAG
  RSV = RESERVED (SET TO '0')
  MESSAGE = DATA BITS

The CRC is generated by appending the DVCC to the message. The DVCC precedes the message field. This composite bit sequence is passed into the CRC generator. The output is a 16 bit CRC field. The resultant CRC field is appended to the original message without the DVCC field. This composite bit sequence is fed into the convolutional coder. The protected data that comes out of the convolutional coder is placed in an array for transmission. Like speech data, the data is interleaved to reduce the effects of burst errors. DVCC is included in the CRC calculation to prevent a receiver from responding to a false call processing message due to co-channel interference from the wrong cell. Figure 3.15 illustrates the SACCH signaling process.

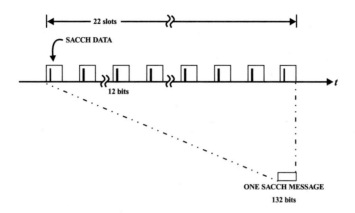

Figure 3.15, SACCH Signaling

3.3.3.2 Fast Associated Control Channel (FACCH)

FACCH control messages replace speech data with signal messages (in-band signaling). Even though speech quality is non-linearly degraded as more and more speech frames are replaced with signaling information, no limit has been established on how many speech frames may be replaced. Because it is important that control commands be successfully received even in adverse conditions (e.g., poor signal quality during handoff), FACCH data is error-protected by a 1/4 rate convolutional coder.

FACCH messages use the entire 260 data bits of the burst, providing a gross data rate of 13 kbps. However, the 1/4 rate convolutional coding reduces the data transmission rate to 3,250 bps. Subtracting signaling bits for the CF flag and CRC, FACCH signaling transmission rates are:

- 2.4 kbps throughput full rate
- 1.2 kbps net throughput half rate

Each FACCH message uses 49 bits. This is allocated according to the following format:

|CF| Message |
  1   48

WHERE:
      CF = CONTINUATION FLAG
      MESSAGE = DATA Bits

The CRC is generated by appending the DVCC to the message. The DVCC precedes the message field. This composite bit sequence is passed into the CRC generator. The output is a 16 bit CRC field. The resultant CRC field is appended to the original message without the DVCC field. This composite bit sequence is fed into the convolutional coder, and the protected data that comes out of the coder is placed in an array for transmission. Like speech data, FACCH data is interleaved to reduce the effects of burst errors. Figure 3.16 illustrates the FACCH signaling process. Notice that a FACCH message replaces speech slots, and each FACCH message is interleaved between two slots.

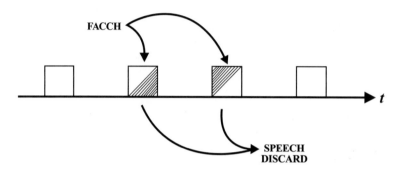

Figure 3.16, FACCH Signaling

No fields in a standard slot identify it as speech or signaling, so all slots are first decoded as speech. If decoding the speech slot indicates it is not a valid speech slot, the receiver will then attempt to decode it as a FACCH message. If the CRC calculates correctly, it is a FACCH message. If the CRC decodes incorrectly then it is probably data which has errors due to interference.

3.3.3.3 DVCC

Each cell site in a cellular system (or localized region of the system) has its own unique DVCC code used to detect interference from neighboring cells and to indicate that the Subscriber Unit is operating (off hook) on a channel. A unique DVCC for each cell site ensures that the correct Subscriber Unit is communicating with the proper Base Station since frequencies are reused in most cellular systems.

The DVCC is transponded back much like the SAT tone. The Base Station sends the DVCC and adds 4 parity bits to the 8 bit DVCC code. This is called the coded DVCC (CDVCC) The value 00h is not used as a valid CDVCC, so there are 255 unique codes. Figure 3.17 illustrates how the Subscriber Unit receives and transponds back the CDVCC to the Base Station.

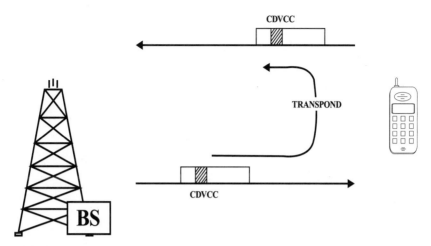

Figure 3.17, Transponding DVCC

3.3.3.4 DTMF Signaling

The digital voice coder used for IS-54 was not designed to handle non-speech audio such as DTMF tones, but testing has shown that it can pass them [2]. Unfortunately, the codec adds twist distortion which alters the amplitude relationship between one tone component and another within the DTMF signal. The situation can be further complicated if data compression devices are used between the cell site and MSC (which also use codecs such as ADPCM).

To avoid speech coder DTMF distortion, a DTMF On message can command the Base Station or MSC to create DTMF tones. Figure 3.18 shows a user pressing key number 2 (step 1) to create a FACCH message (step 2) that indicates digit #2 has been pressed. The receiver in the Base Station decodes this message (step 3) and commands a DTMF generator to create a number 2 DTMF touch tone (step 4). When the user releases the #2 key, a FACCH message is created which indicates the #2 key has been released and the DTMF touch tone is stopped. In addition to the PRESS and RELEASE messages the mobile station can also send a message which contains one or more DTMF digits. This message is used to "speed dial" service or similar purposes. The duration of each digit is preset acording to the industry standards in this case.

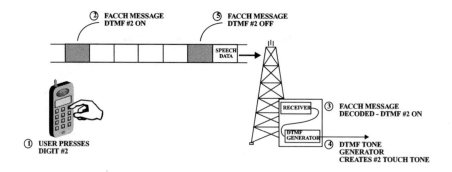

Figure 3.18, DTMF Signaling

3.4 System Parameters

The system parameters for an IS-54B system are very similar to the EIA-553 analog systems. The control channel and analog voice channel are basically the same as the analog system. However, the additional digital traffic channel requires new modulation methods.

3.4.1 Frequency Allocation

The 25 MHz x 2 cellular band allocation remains the same, although a secondary set of channels is authorized for use as dedicated control channels. These control channels are optional, and are an implementation issue left to the system operator. Figure 3.19 displays the IS-54 frequency allocation.

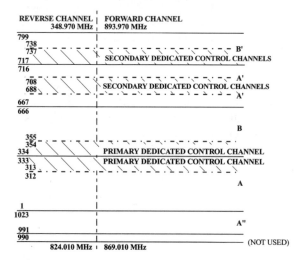

Figure 3.19, IS-54 Channel Frequency Allocation

3.4.2 Duplex Channels

The digital channel is frequency duplex, and receive and transmit frequencies are divided into time slots that also allow TDD (Time Division Duplex) operation. Forward and reverse channels are separated by 45 MHz. The transmit band for the Base Station is 869-894 MHz. The transmit band for the Subscriber Unit is 824-849 MHz. The 30 kHz AMPS channel bandwidth has been maintained. Figure 3.20 illustrates the 44 symbol offset between the forward and reverse channel that allows for dynamic time alignment.

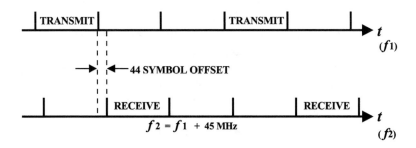

Figure 3.20, Time Division Duplex Radio Channels

3.4.3 Modulation Type

The IS-54 control channel continues to use FSK modulation (analog voice channels use both FM and FSK modulation). The new digital traffic channel (DTC) uses only pi/4 DQPSK modulation. Pi/4 DQPSK modulation was chosen to maintain spectral efficiency and to optimize the RF amplifier section. Spectral efficiency may be increased by multiplexing two signals in phase quadrature.

Multiplexing in phase quadrature may be explained using an I-Q pattern. When we combine two amplitude-modulated RF signals which are 90 degrees out of phase, depending on the signals' amplitudes, the resulting signal is at the same frequency and shifted in phase. This phenomena allows transfer of information, because different bit patterns input to the modulator cause a specific amounts of phase shift in the transmission. Therefore, if the received RF signal is sampled for phase transitions and amplitude at specific periods of time, it is possible to recreate the original bit pattern. The four allowed phase shifts (+45, +135, -45, and -135 degrees) represent the original binary information. The receiver looks for anticipated phase information, called a decision point.

Specific time periods are specified for transitions. The binary (digital) signals' phase changes vary the amplitude of each of the two RF signals. The I axis is the amplitude of the first signal source reference amplitude varied by binary signal X^k. The Q axis is the amplitude of the 90 degree phase shifted signal modulated by binary signal Y^k. The combination of the varied amplitudes results in amplitude and phase components shown on the I-Q pattern.

For quadrature modulation, information in all four quadrants is used to impose and determine modulation information. Figure 3.21 illustrates four points that are the result of combining the two signals with amplitudes determined by table A. By choosing relative 45 degree and 135 degree shifts (by having a different set of amplitude multipliers for the two RF signals), the second set of modulation information will result in a new set of four decision points which are different from the originals. For example, if the first shift is +45 degrees, and a second shift is +135 degrees, the end result is +180 degrees from the original phase. A 180 degree phase change was not part of the original set of decision points. These shift values were chosen to allow two sets of decision points. A total of eight decision points are displayed in figure 3.21. Valid decision points alternate. Note in figure 3.21 the x and o's mark alternating decision points. A transition from the 0 reference point (the top in this example) must result at one of the x decision points. If the resultant shift (phase change) does not reach a final value within those decision areas, the data is in error and must be discarded.

Each two-bit stimulus input has a corresponding phase shift. Figure 3.21 shows

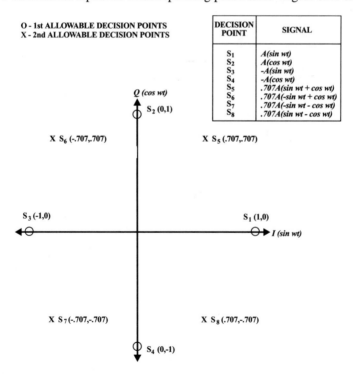

Figure 3.21, I Q Decision Points

a 01 bit input to the IS-54 standard quadrature modulator, and the corresponding phase-shifted output. The initial decision point is at s1. A 01 bit input requires a phase shift of + 135°. Notice that the shifts never require a transition path through the origin. This was specified to simplify the RF amplifier design. Gray coding is used between decision points so that only a one-bit transition point occurs between decision point shifts. The transition period between decision points is 41.15 μsec, resulting in a symbol rate of 24.3 thousand symbols per second (ksps). Each symbol represents 2 bits, so the input data rate is 48.6 kbps.

Modulations in the carrier offset the frequency above and below the reference, and power is distributed over a frequency range. The requirements for this spectral density are more stringent for than AMPS channels. The requirement for this spectral mask was derived from adjacent channel and co-channel interference levels. As modulation occurs, the out of channel spectral density interferes with Subscriber Units operating on adjacent channels.

FM signals maintain a constant amplitude envelope, but pi/4 DQPSK modulation does not. Figure 3.23 illustrates power level variations for Dual Mode transmission. Two types of intensity variations may cause interference and errors in measurement. First, the 33% duty cycle of RF energy bursts (on 6.67 msec, off 13.37 msec) creates a pulse repetition frequency (PRF) of 50 Hz in the electro-magnetic field. Harmonic components of this PRF may interfere with magnetically coupled devices. The 33% duty cycle also causes power meters which were designed for continuous signals to indicate about 1/3rd of the burst power level. A second amplitude variation results from DQPSK modulation at a symbol rate of 24.3 kHz. At this symbol rate, DQPSK modulation can interfere with electronic products that have nonlinear characteristics and lead lengths of about 3.5" (1/4 wavelength). The amplitude variation creates an average burst power level about 1.8 dB below the peak burst power.

3.4.4 RF Power Classification

A new class IV power class of Subscriber Unit has been added, with a maximum output power of -2 dBW, and a minimum of -4 dBm. The Class IV Subscriber Unit output power is identical to Class III, but its minimum power is 12 dB lower. The lower minimum power allows systems to reduce the minimum cell site radius. Table 3.1 shows the power classification types for the IS-54 radio system.

RF Power	Class I	Class II	Class III	Class IV	Class V-VI
Maximum Power	4 Watts	1.6 Watts	.6 Watts	.6 Watts	Reserved
Average Power (Full Rate TDMA)	1.333 Watts	.533 Watts	.2 Watts	.2 Watts	Reserved
Minimum Power	6 mW	6 mW	6 mW	.5 mW	Reserved

Table 3.1
IS-54 Power Classification

Dual mode Subscriber Units transmit in bursts, so their average output power is properly measured during the burst period. Figure 3.23 illustrates output power level variation during the transmitted burst. Unlike AMPS FM modulation, DQPSK modulation produces dips in the output signal power, so peak power level during a DQPSK burst is 1.8 dB above the average power level. Also, the 33% transmit duty cycle for a full rate IS-54 channel causes a typical power meter to measure about 30% of peak burst power.

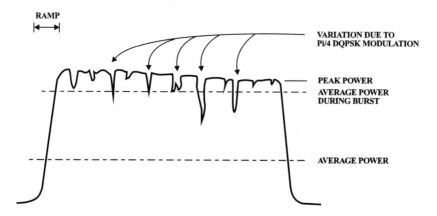

Figure 3.23, Digital Traffic Channel RF Power Level Variation

While the Subscriber Unit transmits in bursts, the Base Station transmits continuously. As long as one slot of the radio channel is in use, the Base Station radio transmitter remains on during all bursts.

3.5 System Operation

The IS-54 cellular system uses three types of channels, analog control channels (ACC), analog voice channels (AVC), and digital traffic channels (DTC). The analog control channels allow the Subscriber Unit to retrieve system control information and compete for access. The analog voice channels and digital traffic channels are primarily used to transfer voice information, but they also send and receive some digital control messages.

When it is first powered on, a Subscriber Unit initializes by scanning for a control channel and tuning to the strongest one it finds. During initialization, it also determines if the system is digital capable. If the system is not digital capable, the Subscriber Unit looks for an optional secondary control channel. After initialization, the Subscriber Unit enters idle mode and waits to be paged for an incoming call or for the user to place a call (access). When a call is to be received or placed, the Subscriber Unit enters system access mode to try to access the system via a control channel. When it gains access, the control channel sends commands to the Subscriber Unit to tune to an analog or digital traffic channel. The Subscriber Unit responds by tuning to the designated channel and entering conversation mode.

3.5.1 Access

Prior to accessing an IS-54 system, a Subscriber Unit listens for a protocol capability indicator (PCI) flag to determine if it is digital capable. When the Subscriber Unit attempts access to a system that is digital capable, the system indicates its dual mode capability with the mobile protocol capability indicator (MPCI). If the access attempt succeeds, the system assigns the Subscriber Unit an analog or digital voice channel. If it assigns an analog channel, the system sends an Initial Voice Channel Designation (IVCD) message which contains the voice channel number. If the system assigns a digital channel, it sends an Initial Traffic Channel Designator (ITCD) message containing the channel number and time slot. Figure 3.24 illustrates the TDMA system access process.

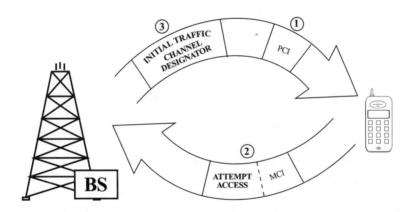

Figure 3.24, TDMA System Access

3.5.2 Paging

In response to a page message, but prior to accessing the IS-54 system, a dual mode Subscriber Unit determines if the system has digital capability. If the system has digital capability (PCI is set), the Subscriber Unit responds to the page message by attempting to access the cellular system. During the access attempt, the Subscriber Unit indicates that it is responding to a page message and that it has digital capability.

Dual mode cellular systems can also send caller number identification. After the page message is received and the voice channel or traffic channel assigned, the system can transfer the caller number ID (CNI). The CNI is transferred on a the voice or traffic channel prior to the subscriber answering the call. Figure 3.25 illustrates the IS-54 paging process.

3.5.3 Handoff

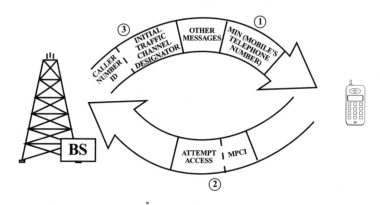

Figure 3.25, TDMA System Paging

When a dual mode Subscriber Unit is operating on a digital channel, the handoff process is different from AMPS. Figure 3.26 illustrates a TDMA system digital channel handoff. Long before a handoff, the Base Station typically commands the Subscriber Unit to measure nearby radio channels' signal strength (time 1). The Subscriber Unit continuously sends the radio channel quality information back to the Base Station using the SACCH channel (time 2). Use of the SACCH channel leaves voice quality unaffected. When the Base Station determines that the Subscriber Unit can be better served by another cell site, it commands the unit to tune to a new radio channel (time 3). After the Subscriber Unit receives the handoff message, it mutes its audio. When the Subscriber Unit has tuned to the new radio channel, it sends shortened bursts if the handoff was between two cells

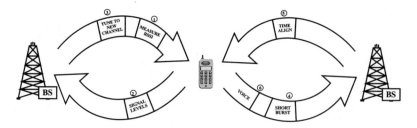

Figure 3.26, TDMA System Digital Channel Handoff

which differ by more than 2 miles (time 4) to avoid potential collisions with bursts sent from other Subscriber Units. When the Base Station determines the necessary time adjustment, it commands the Subscriber Unit to time align (time 5). The Subscriber Unit then unmutes the audio and begins voice communications (time 6). When the two cells are about the same size, no muting or shortened burst are used. This is also the case for a sector to sector handoff within the same cell. Then there is no gap in the speech, providing a so called "seamless" TDMA handoff.

3.6 Call Processing

IS-54 call processing has four major differences from analog call processing: new system control information, secondary control channels, reversed paging and access functions, and signaling on the digital traffic channel.

3.6.1 Initialization

Initialization is the process of obtaining information about the system type and the access procedures necessary to communicate with the system. Initialization for IS-54 Subscriber Units differs from AMPS Subscriber Units in the retrieval of a PCI bit indicating digital capability of the cell site. Figure 3.27 illustrates a

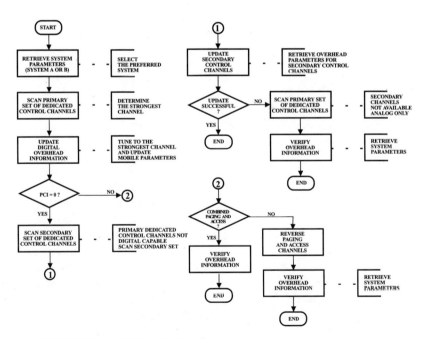

Figure 3.27, TDMA Initialization Mode

Subscriber Unit first powering on and searching for a digital-capable control channel. If it cannot find a digital-capable control channel, it initializes with the parameters sent on the analog control channel.

3.6.2 Idle

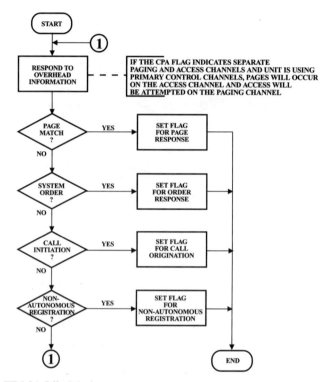

Figure 3.28, TDMA Idle Mode

Figure 3.28 illustrates operations carried out in the TDMA Idle Mode. While idle, the TDMA Subscriber Unit continuously monitors the system parameters to determine if it has been paged, if it has received an order, or if the operator is initiating a call. After obtaining the system parameters, the Subscriber Unit must determine if a call is ready to be received (a page) or if a call is to be initiated by the subscriber. To do so, it continually monitors the overhead messages for changes in parameters. Overhead messages include the system identifier, the number of access channels, and other parameters that affect how the Subscriber Unit will operate. To determine if an incoming call is to be received, the Subscriber Unit monitors page commands. To see if the user has initiated a call, the Subscriber Unit also monitors the status of its keypad.

3.6.3 System Access Task

The cellular system will assign the Subscriber Unit to a digital traffic channel if it is available. If none is available and the system cannot assign a digital channel, the Subscriber Unit accepts assignment to an analog radio channel.

Although the digital cellular system can serve many new subscribers by sharing radio channels, control channel capacity has remained the same. Anticipating likely future increases in control channel capacity, the IS-54 system changed digital Subscriber Units' page monitoring process. The typical AMPS system allocates one control channel per cell or sector, even though the AMPS system allows two control channels per cell site, one for paging and one for access. However, AMPS systems rarely use both control channels because most paging channel information is in the forward direction and most access channel information is in the reverse direction. In this circumstance, one control channel can carry almost as much information as two.

For many years to come, most cellular systems' subscriber populations will be a mixture of AMPS and dual mode subscribers. This mix of subscribers will make it practical to dedicate two control channels, one for paging and one for access. If dual mode Subscriber Units reverse their paging and access functions, they can use the excess forward capacity of the access channel and the excess reverse capacity of the paging channel.

When dual mode Subscriber Units are not using secondary dedicated control channels, they will check to see whether the Combined Paging and Access (CPA) field is set to 0. If the CPA field is 0, separate paging and access control channels are available, and dual mode Subscriber Units will listen to the access channels for pages.

Analog and dual mode system access functions are very similar. Figure 3.29 illustrates the necessary changes. Refer to figure 2.21 for a detailed system access flow diagram.

In addition to providing system and control information, the control channel provides an access point for the Subscriber Unit. Subscriber Units randomly attempt access to the cellular network. To prevent multiple Subscriber Units from initiating access simultaneously, a seizure collision avoidance procedure has been developed. Four elements make up the process: the busy status of the channel, system response time interval, random time delays, and maximum number of automatic access attempts.

Busy status of the channel: The forward control channel is interleaved with dedicated bits that indicate whether the Base Station is busy. The Subscriber Unit monitors these bits before attempting access to the system, and waits until they indicate that the Base Station is not busy. **System response time interval:**

When the Base Station is busy during an access attempt, the control channel indicates busy within a prescribed time period. If the control channel indicates busy before or after the time period, the Subscriber Unit assumes the Base Station is responding to a competing Subscriber Unit and inhibits transmission of the access attempt. **Random time delays:** Following an unsuccessful access attempt, the Subscriber Unit waits a random time before trying again. The random wait prevents two or more competing Subscriber Units from repeatedly attempting access again and again at identical intervals. **Maximum number of automatic access attempts:** The system limits the number of automatic attempts to prevent overloading when many Subscriber Units constantly attempt access.

The Subscriber Unit attempts access to the system when the operator initiates a call, or whenever the unit receives a page (incoming call), orders, or a registration request. The access attempt contains a message that indicates which of these types of access is required. Registration requests, which are sent by the Base Station, require the Subscriber Unit to indicate that it is operating in the system. Registering active Subscriber Units in each cell area limits paging requirements because paging messages are sent only on control channels of cells in the vicinity of the last registration.

Control channels may be divided into separate paging and access channels where

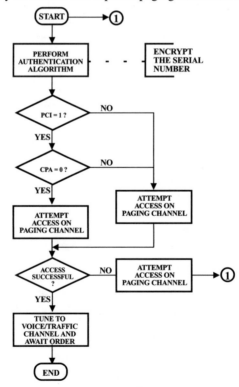

Figure 3.29, TDMA Access Mode

one control channel delivers pages and a different control channel coordinates access. Separating paging and access channels allows access channels to handle more service requests independent of paging requirements. However, assigning paging and access to different channels does not increase system capacity. The forward channel must carry all access signaling and the reverse channel carries all paging, leaving system capacity unaffected. As a result, paging and access are usually combined in one control channel. Separate paging and access channels was originally conceived for a proposed radio coverage of many adjacen cells using only one paging channel, with separate access chanels for each cell. In practice, no system operators have used this configuration, except during testing.

3.6.4 Conversation Mode

Figure 3.30 illustrates TDMA system control of the Subscriber Unit on the digital traffic channel in conversation mode. TDMA adds new messages to support, and it changes the way that signaling messages are physically transmitted and received. Refer to figure 2.22 for a more detailed functional flow diagram.

When access is granted, an Initial Voice Channel Designation (IVCD) message from the Base Station sends the Subscriber Unit a voice channel number to which it must tune to initiate conversation. While the conversation is in progress, the Base Station continues to control the Subscriber Unit's power level, handoff,

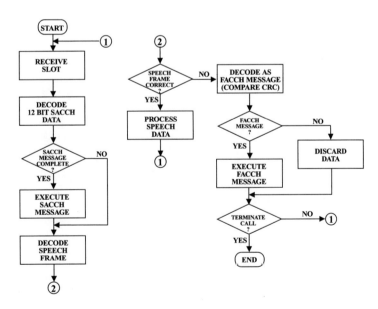

Figure 3.30, TDMA Conversation Mode

alerting, and other functions through blank and burst signals (voice information replaced with signaling commands).

To ensure a reliable radio link, a radio link timer monitors the time that the SAT (or CDVCC on a digital traffic channel) is interrupted to detect loss of radio continuity. After 5 seconds without the correct SAT or CDVCC (while the receive signal is below the level at which the Base Station can control it), the radio link is broken by both the Base Station and mobile station and the timer turns the Subscriber Unit off.

Discontinuous Transmission (DTX) is another feature that may be used in the conversation mode, particularly in portable Subscriber Units. To conserve power, the Subscriber Unit transmitter may be turned off during silent intervals in the conversation.

References

1. Dr. George Calhoun, "Digital Cellular Radio", Artech House, 1988.
2. Telecommunications Industry Association, "DTMF Tone Processing Via the Decoder, "TR45.3.2/90.07.26.10, Motorola Corporation, June 1990.

Chapter 4
TDMA (IS-136)

4. TDMA Digital Control Channel (IS-136)

The IS-136 specification evolved from IS-54 revision B, and was originally called IS-7X, then IS-54 revision C. IS-136 revisions include a new digital control channel, new system control methods, and advanced features resulting from the changes to the digital traffic channel that were defined under IS-54 revision B. The term "Digital Control Channel" (DCC or DCCH) is often used interchangeably with IS-136, but strictly speaking, DCC refers only to the control channel and not to the IS-136 specification as a whole. The IS-136 specification consists of two parts: IS-136.1, which covers digital control channel messaging and operation; and IS-136.2, which covers modifications to existing IS-54 systems.

The IS-136 specification concentrates on features that were not present in the earlier TDMA system. These include longer standby time, short message service functions, and support for small private or residential systems that can coexist with the public systems. In addition, IS-136 defines a digital control channel to accompany the Digital Traffic Channel (DTC). The digital control channel allows a Subscriber Unit to operate in a single digital-only mode.

Since the first IS-54 (revision 0) specification, many proposals and comments have been put forward on how to add features and improve the specification. Many of the proposed changes were incorporated into revisions of the IS-54 standards (Rev. A, B and the Technical Service Bulletins). However, recent changes such as the digital control channel advanced the overall structure of the specification enough to warrant its designation as a new standard, IS-136.

Until the IS-136 standard, revisions to IS-54 specified only features and parameters for a new digital voice channel (traffic channel), leaving the IS-54 control channels virtually identical to those in the older EIA-553 systems. IS-54 planners re-used the existing control channel structure to allow cellular systems to preserve their existing RF layout and to facilitate rapid deployment. Unfortunately, using the analog control channel structure restricted new IS-54 TDMA systems to the limited features available under that structure. Figure 4.1 summarizes the evolution of the IS-136 specification.

Revision A of the IS-136 specification now supports operation in two hyperbands. The specification allows for operation in the 800 MHz range for the existing AMPS and DAMPS systems as well as the newly allocated 1900 MHz bands for PCS systems. This permits dual band, dual mode phones (800 MHz and 1900 MHz for AMPS and DAMPS). The primary difference between the two bands is that there will not be ACC or AVC operation at 1900 MHz. All discussions in

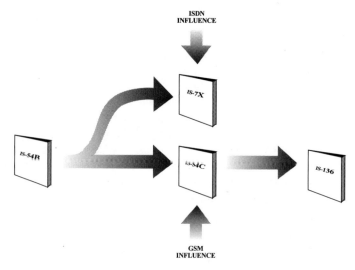

Figure 4.1, Evolution of the IS-136 Specification

this chapter apply equally to both bands.

As revision B of the IS-54 specification revision began, planners envisioned a new voice channel and a new data channel using ISDN signaling. They defined this as the IS-7X series (e.g. IS-70, IS-71, etc.) of specifications. Once work on the specification was under way, it became clear that the work required to develop the specification and the time required to design the system and Subscriber Units would delay market availability by several years. Planners compromised, creating a new digital control channel that used the existing IS-54 revision B digital traffic channel. This was originally known as IS-54 revision C. However, changes from IS-54 revision B were so substantial that IS-54 revision C was renamed IS-136.

During development of IS-136, many new features were influenced by or borrowed from the GSM (Global System for Mobile Communications) specification. GSM is the European digital cellular standard adopted by more than 25 countries (see Appendix 4). The DCC's overall architecture is very similar to GSM control channel structure. The DCC benefitted from the best of the GSM specification's features, adding to and improving upon them.

4.1 System Overview

The IS-136 system has four types of channels: analog control channels (ACC), analog voice channels (AVC), digital traffic channels (DTC), and digital control channels (DCC). The ACC, AVC, and DTC channels have been slightly modified to enhance system operation. A new digital control channel uses the slots of a DTC to page Subscriber Units and coordinate system access.

Figure 4.2 illustrates the structure of an IS-136 system. The IS-136 system contains the same basic subsystems as other cellular systems, including the mobile switching center, the base station, and the Subscriber Units (phones). The MSC for IS-136 systems is very similar to the MSC on existing IS-54 revision B systems.

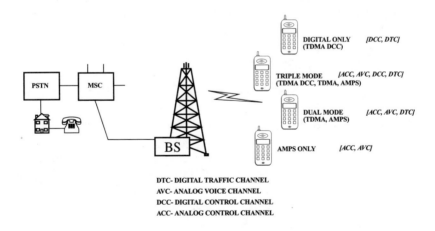

Figure 4.2, Overview of an IS-136 system

Because the new digital control channels use the IS-54 radio channels, frequency planning is similar to the AMPS system.

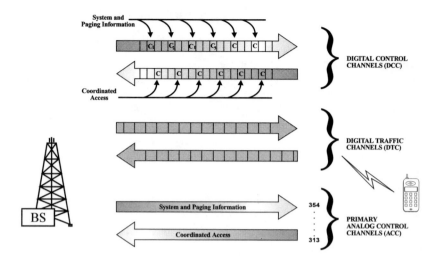

Figure 4.3, IS-136 Radio Channels

The IS-136 system has defined a new type of control channel called a digital control channel (DCC). The DCC carries system and paging information, and coordinates access similar to the analog control channel (ACC), however the DCC has many more capabilities. The DCC adds features such as extended sleep mode, short message service (SMS) private and public control channels, and others. The DCC uses the DTC slot structure, and may co-exist with DTC traffic channels used for voice. Figure 4.3 shows the IS-136 radio channel types.

4.1.1 New Digital Control Channel

IS-136 systems use the IS-54 30 KHz TDMA RF channel structure. The most significant change is in the data format on the control channel. IS-136 systems have a new digital control channel which co-exists with standard IS-54 TDMA slotted RF channel. Its radio frequency can be assigned to any one of the cellular radio channels, not limited to predefined frequencies as in previous revisions. The RF carrier spacing and designation, and the modulation characteristics are identical to the IS-54 Rev B systems. In many cases, the specification was copied directly from the appropriate paragraphs of the Rev B specification.

IS-136 TDMA uses the same frame structure as IS-54. Each frame is 40 ms long, and divided into six time slots of equal size. An IS-136 system is either half rate or full rate. A full rate system allocates two time slots per user, allowing only three different users per channel. For example, under a full rate system user A would use time slots 1 and 4, user B would occupy time slots 2 and 5 and user C would occupy time slots 3 and 6. On the other hand, in a half rate system, each user occupies only one of the six time slots. Full rate systems are being fielded first. Future upgrades will allow half rate operation.

Subscriber Units functioning in older radio systems scanned for available channels before attempting to access the system. The scanning process could be relatively time-consuming, so a dedicated control channel (ACC) was created. Similarly, in the IS-136 system, the DCC might be located anywhere, so a means was designed to quickly find the control channel. Figure 4.4 shows how the IS-136 system can locate the digital control channel. The AMPS control channel has been modified by adding a new message which points to the DCC. The IS-54 slot structure has also been modified to indicate the range of channels in which the DCC is located. The release message for analog voice channels and digital traffic channels has also been modified to include information about the location of nearby DCC's.

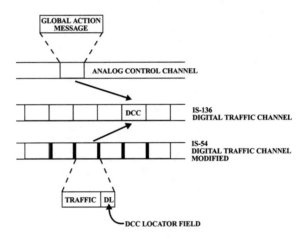

Figure 4.4, IS-136 DCC Locating Process

Much of the design of IS-136 systems is similar to IS-54 Rev B or EIA-553 systems. However, one major difference is the method for allocating control channels. IS-136 systems can be set up as small private systems within the existing public systems. Therefore, it is a much more complicated task for a Subscriber Unit to find the "best" channel. The small private systems are controlled by two new types of system identifiers: PSIDS (private system id's) and RSIDS (residential system ID's). The billing system also has the capability to charge users different rates depending on which system they access.

Multiple DCC's can exist on different RF channels in the same cell site, allowing public and private systems to co-exist (further discussed in Chapter 9).

4.1.2 Logical Control Channels

New messaging and control functions have been added to the control channel. Some of the bits in each control slot have been dedicated for this capability. These new logical control channels consist of bits combined from a period of 32 associated slots to form a single superframe. The superframe is divided into logical sub channels which are simply logical groupings of the time slots. The groupings are obtained by combining predefined data and by using information

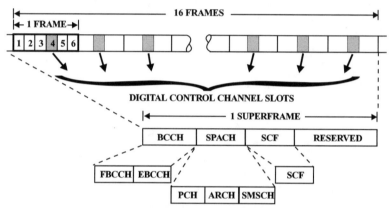

FBCCH - FAST BROADCAST CHANNEL-(CRITICAL SYSTEM OVERHEAD INFORMATION)
EBCCH - EXTENDED BROADCAST CHANNEL-(OPTIIONAL SYSTEM OVERHEAD)
SPACH - SHORT MESSAGE, PAGING, AND ACCESS RESPONSE CHANNEL
BCCH - BROADCAST CONTROL CHANNEL
SCF - SHARED CONTROL FEEDBACK
PCH - PAGING CHANNEL
ARCH - ACCESS RESPONSE CHANNEL
SMSCH - SHORT MESSAGE SERVICE CHANNEL

Figure 4.5, Logical Sub-Channel of IS-136

contained on the DCC. Figure 4.5 shows how the DCC is divided into its logical sub channels.

The broadcast channels (BCCH) contain general system information needed by all Subscriber Units. Two types of broadcast information are sent on fast and extended channels. The fast channel sends information that the mobile station needs to obtain quickly. An extended channel carries information that the mobile station uses after locking on (or camping on) to the DCC. The SMS, paging and access channel (SPACH) contains data for the Subscriber Unit about pages and short messages, as well as access responses. The shared control feedback (SCF) flags are part of the access channel, and provide information to control access to the channel by multiple Subscriber Units.

4.2 System Attributes

IS-136 systems have modified IS-54 channels to allow both voice and control channels to exist on the same RF carrier. System control has also been modified to allow private (office and home) and public systems to co-exist. The new paging capability extends battery life. A short message service has also been added. IS-136 also supports advanced data transmission services through the extensions provided by the IS-130 and IS-135 data specifications. Some call processing control has been shifted to the Subscriber Unit, simplifying system control.

4.2.1 Multi-Function Radio Channels

The IS-136 specification modifies the IS-54 traffic channel for use as both a control channel and a voice channel. As the number and density of subscribers grows, systems require more capacity to handle calls. One of the simplest (but maybe not the least costly) of ways to increase voice channel capacity is to add cell sites. TDMA adds voice channel capacity by allowing multiple users on each radio channel. AMPS systems use a single control channel per cell site. Because of the increased channel capacity, it was necessary to add control channel capacity. IS-136 has the advantage using any or multiple channels for control.

IS-136 radio channels support both voice and control channels in a single cell. Several different system ID's and voice channels (digital traffic channel) can exist on a single RF channel. Private and residential systems would allow access only for previously approved users, and the public system would be available for public cellular users.

4.2.2 Digital Control Channels

Each control channel can deliver a limited number of pages, limiting the number of users each cell can serve. As the density of users increased, especially in urban areas, additional cells and control channels were needed. IS-136 allows the control channels to be located anywhere in the set of valid channels. A control channel can be a different time slot on the same RF channel as a voice channel. In IS-136 systems, the Subscriber Unit performs a process called cell reselection. As soon as the phone "locks on" to a DCC, it begins determining the next channel or cell it should lock on to. Cell reselection is a complicated process in which the Subscriber Unit examines parameters such as system ID's and signal strengths to determine when to monitor a new DCC. This differs from hand-off on the voice channel because the Subscriber Unit has much greater control in selecting the new DCC.

Several features of DCC provide for more capacity expansion than existing TDMA systems. First is the use of paging classes. A Subscriber Unit is assigned to one of eight paging classes. Each class has a different rate at which it sends page messages. Because pages are usually repeated, being able to group them into classes helps to increase each individual base station's capacity. IS-136 also uses a temporary mobile station identifier (TMSI) to increase paging capacity. A TMSI is shorter than a MIN, so that up to five messages can be packed into a frame of data.

Another large capacity improvement in IS-136 systems is the ability to use any channel as a control channel. The increase in the number of available control channels solves many of the frequency reuse issues by simply using more control channels. In practice, however, control channels cannot be added indefinitely without running out of voice and data channels.

4.2.3 Public, Private, and Residential Systems

The overall system control for an IS-136 system is very similar to the existing IS-54 TDMA systems. However, Public System Identifiers (PSIDS) and Residential System Identifiers (RSIDS) have been added to allow small cellular systems to be installed and controlled within existing public systems. PSIDS, RSIDS, and the increased capacity for control channels will allow small private systems to exist under the umbrella of existing commercial systems. For example, an office complex could have its own private cellular system to be used only at work. When you left work, your phone would automatically lock on to the commercial system.

4.2.4 Paging Classes

As portables took over a larger share of the market, the length of a battery's operating time became a concern. To help increase the battery life, the system can allow the phone to "sleep" during times when messages are not expected. Paging channels (a logical channel, not a physical RF channel) are multiplexed to reduce power consumption during receive mode. Cellular telephones in 1994 had an average battery life of 15 to 20 hours with standard sized batteries [1]. The new specification allows for sleeping between paging groups, and IS-136 phones should double or even triple the average standby time.

If a mobile user can tolerate a slightly longer delay when a page is being received, that user can be assigned to a paging class which sends pages at a slower rate. The phone knows when the next frame containing page messages will

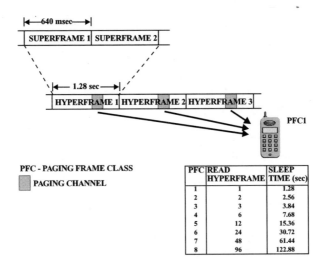

Figure 4.6, Paging Classes in an IS-136 System

occur, so it can sleep to conserve battery power until the next frame for its paging class. Figure 4.6 shows the range of paging classes for IS-136. Delays in receiving pages for classes 1 - 3 range from 1.2 to 3.84 seconds, a range that is probably acceptable for normal use. However, paging class 8 has a maximum delay of over 120 seconds, too long for a caller to wait for the subscriber to answer. This paging class is targeted for remote control applications where maximum battery life is needed and a quick response to a page is secondary.

4.2.5 Short Message Services

Short message service (SMS), or transmitting and receiving short text messages on the phone's alphanumeric display, has been added to the IS-136 specification. Related services, called broadcast services, send the same text message to many users. These may be advertisements, traffic reports, or weather reports [2]. Point-to-point messages are transmitted from one sender to one receiver. This messaging system is like a two-way pager. Initially the IS-136 specification defined only point-to-point messaging, but a broadcast SMS is very likely to be added as one of the first enhancements.

4.2.6 Digital Only Phones

IS-136 phones may be designed to operate in a digital only mode. Because an IS-54 revision B dual mode phone must have the ability to operate on the analog control channel to gain access to a digital traffic channel, a dual mode phone must have analog and digital functionality. This makes it more costly to produce than an analog only or digital only phone. IS-136 phones can operate as a multi-mode (analog or digital) or digital only. One of the biggest savings of building a digital only phone would be the elimination of the frequency duplex filter [3] resulting in some size and cost savings. Talk time would increase (longer battery life) because a frequency duplex filter only allows 25-50% of the radio energy to pass through [4].

4.2.7 Data Transmission

Data transmission on IS-136 systems is covered by an entirely new method of encoding and transmitting data: the IS-130 and IS-135 specifications. The IS-130/135 data transmission system has defined a new data encoding method and eliminated the need to send data on voice channels (as on IS-54 systems). A rate 5/6 encoder is used to send data on digital channels, and it operates at much higher data rates than the rate 1/2 encoder used for voice data.

4.2.8 SPACH ARQ (Automatic Retransmission Request)

Another new attribute of IS-136 systems is the automatic retransmission request (ARQ). To ensure that the Subscriber Unit receives messages, the base station puts the Subscriber Unit into ARQ mode. In ARQ mode, the Subscriber Unit tracks the frames that have been received, and sends a status list back to the base station. When the Subscriber Unit receives an "ARQ begin frame," it stores the partial echo assigned (PEA). The PEA is then used to identify subsequent "ARQ continue frames." In addition to storing the PEA, the Subscriber Unit initializes the frame number map to all zeros, indicating that no continue frames have been received. The frame number map can handle transactions up to 33 frames long (32 continue frames plus the begin frame).

After initializing the ARQ frame map, the Subscriber Unit sends a status frame to the base station. Then, as each frame of data is received, the Subscriber Unit updates the frame map corresponding to the frame number. If a frame is not received by the Subscriber Unit, the base station will retransmit it with its original frame number. If the Subscriber Unit does not receive a request to send another status frame within 26 frames, it automatically sends one to the base station. If the Subscriber Unit is ever unable to send a status frame, or it does not receive any ARQ continue frames within 26 frames, then it leaves ARQ mode. When sending ARQ status frames, the Subscriber Unit uses reservation-based access if the value of ARM (ARQ response mode) equals 1. Otherwise, it uses contention-based access.

4.2.9 Monitoring of Radio Link Quality (MRLQ)

An IS-136 subscriber can measure the signal quality of control channels. The Subscriber Unit is required to use this quality indicator to help it select an optimal control channel. Control channel signal quality is measured by the number of word errors during one slot of each paging frame. The CRC calculation is used to determine errors. If the CRC check fails, a counter is decremented by one, and if the CRC passes, the counter is incremented by one. The MRLQ counter is initialized to a value of ten, and is never incremented beyond a value of ten. Whenever the counter reaches zero, the Subscriber Unit enters the cell reelection state and reselects the best DCC channel to camp on.

4.2.10 Control Channel Reselection

Control channel reselection is the process of determining if there are better control channel candidates to monitor. After the control channel is initially selected, the reselection process continues as long as the Subscriber Unit is in idle mode. The reselection process consists of two steps. First, the Subscriber Unit measures the signal strength of the current DCC and the signal strength of the DCCs on a neighbor list (sent as part of the F-BCCH data). Second, the Subscriber Unit performs a detailed process of determining which (if any) DCC should be reselected, called the selection criteria.

To measure the DCCs' signal strengths, the Subscriber Unit keeps a running average of the last 5 measurements. Only the last 2 measurements are required for the currently selected DCC. Generally, signal strength measurements are required at least once per hyperframe (1.2 seconds), although criteria exist to allow the Subscriber Unit to modify the frequency at which it measures channels on the neighbor list.

After obtaining the first 5 signal strength measurements on the neighbor list channels, the Subscriber Unit can begin examining the reselection criteria. The three major reselection criteria steps each contain many sub-steps, so only an overview can be presented here. To reselect a new DCC, a Subscriber Unit must progress through three steps. First the trigger conditions must be passed. Trigger conditions consist of items such as a radio link failure based upon MRLQ measurements, receipt of a "go away" message, determining that the current DCC does not contain the service desired, or simply a periodic rescan. After trigger conditions are assessed, the Subscriber Unit enters the candidate eligibility filtering (CEF) process. CEF is used to determine which channels on the neighbor list are possible new DCCs. The CEF process uses different criteria to determine eligible channels depending on which trigger condition was used to enter the reselection process. For example, much more stringent conditions are used in a periodic rescan than when a radio link failure has occurred. The final step in reselection is applying the candidate reselection rules to determine if a reselection should take place.

4.2.11 Mobile Assisted Channel Allocation (MACA)

MACA is a set of information reported from the Subscriber Unit to the base station to assist in determining the best channel. When a Subscriber Unit is in DCC monitoring (camping state), it performs two MACA-related functions: long term MACA (LTM) and short term MACA (STM).

Long term MACA is a set of data containing the word error rate, bit error rate, and received signal strength for the current DCC. Short term MACA contains received signal strength for the current DCC, and possibly for other channels. A MACA list is sent from the base station with a list of other channels to be measured.

MACA reports are sent back to the base station during other activities such as originations, pages, and registrations. The MACA Status and the MACA type fields sent from the base station as part of the FBCCH data determine the time when a Subscriber Unit sends MACA reports, and the type of report to be sent.

4.2.12 Slot Structure

Several new types of slot structure exist in the IS-136 standard. Most of these are part of the new digital control channel architecture. Three different slot structures are used in an IS-136 control channel:
    1. Normal slot format for Subscriber Unit to base station
    2. Abbreviated slot format for Subscriber Unit to base station
    3. Slot format for base station to Subscriber Unit

In addition, the slot format of the forward digital traffic channel (base station to Subscriber Unit) is modified to include a new field called the digital control channel locator. This field provides information to help the Subscriber Unit search for DCCs. The reverse digital traffic channel (Subscriber Unit to base station) slot format remains unchanged from IS-54B systems. Figure 4.7 shows the IS-136 forward traffic channel slot structure.

28	12	130	12	130	1	11
S	SACCH	DATA	CDVCC	DATA	RSV	CDL

NOTE:
S- SYNCHRONIZATION WORD
SACCH- SLOW ASSOCIATED CONTROL CHANNEL
DATA- USER INFORMATION
CDVCC- CODED DIGITAL VOICE COLOR CODE
RSV- RESERVED
CDL- CODED DCC LOCATOR

Figure 4.7, IS-136 Forward Traffic Channel Slot Structure

The IS-54 forward traffic channel slot format was modified to create a CDL field by reducing the size of the reserved bits from 12 bits to 1 bit. The field coding is similar to the coding used for the CDVCC field. The field contains a value that indicates a range of 8 RF channels in which a digital control channel can be found. This field helps a Subscriber Unit find a DCC during its initial scan. It is very likely that a DCC Subscriber Unit will detect activity on an IS-54B traffic channel while looking for a DCC. The CDL field can then be used as a pointer to a block of channels to search.

The slot format for the reverse traffic channel is identical to the slot format used on IS-54B systems. It is shown in Chapter 3, Figure 3.13.

28	12	130	12	130	10	2
S	SCF	DATA	CSFP	DATA	SCF	RSV

NOTE:
- S -IS SYNCRONIZATION WORD
- SCF -IS SHARED CHANNEL FEEDBACK (FORMERLY SACCH)
- DATA -IS USER INFORMATION
- CSFP -IS CODED SUPER FRAME PHASE (FORMERLY DVCC)
- RSV -IS RESERVED

ONLY USED FOR DCC

Figure 4.8, IS-136 Forward Control Channel Slot Structure

The forward control channel slot format, shown in Figure 4.8, is similar to the slot format used on the forward traffic channel, but some significant differences exist. The SACCH field is replaced with a portion of the shared control feedback (SCF) field used to control the uplink access. The CDVCC field is replaced with the coded superframe phase information. The data fields hold the paging and SMS message information. The CSFP is an index number (O-31) that identifies the slot number within a superframe.

The reverse control channel slot format, shown in Figure 4.9, is also similar to the reverse slot format used on the traffic channel. It starts with GUARD and RAMP fields to allow transmitters to ramp up and ramp down. The guard field prevents a Subscriber Unit whose timing is not precisely aligned from transmitting on the next slot. A PREAMBLE field follows. The PREAMBLE field helps with symbol synchronization because it contains eight symbols, each with a phase change of -pi/4 radians. The PREAMBLE field contains data in the traffic channel. The SYNC field follows. The SYNC field allows the base station to

6	6	16	28	122	24	122
G	R	PREAM	S	DATA	SYNC+	DATA

NOTE:
- G -IS GUARD TIME (TRANSMITTER OFF)
- R -IS TRANSMITTER RAMP UP TIME
- PREAM -IS PREAMBLE TO AID AGC (FORMERLY DATA)
- DATA -IS USER INFORMATION
- SYNC+ -IS SECOND SYNCHRONIZATION WORD (FORMERLY SACCH + DVCC)

Figure 4.9, IS-136 Reverse Control Channel Normal Slot Structure synchronize to the transmitting Subscriber Unit. A SYNC+ field has also been added to the reverse control channel slot format. The SYNC+ field contains extra sync information if the base station needs it. The user information (voice / data) is carried in two data fields of 122 bits each.

The abbreviated slot structure shown in Figure 4.10 reduces the second DATA field to 78 bits, and adds another RAMP (R) and ABBREVIATED GUARD (AG) field. This large guard time assists large macrocells where time delays near cell boundaries are very large. During the access attempt, when time alignment is not known, data can be lost if initial transmissions are misaligned. The total of 44 bits in the R and AG fields corresponds to about 10km. After the base station receives the initial abbreviated slot, it can assign time alignment and use the normal slot structure. Most systems with small cell areas do not need the abbreviated slot format, and in these systems, Subscriber Units use the normal slot format.

6	6	16	28	122	24	78	6	38
G	R	PREAM	S	DATA	SYNC+	DATA	R	AG

NOTE:
G — IS GUARD TIME (TRANSMITTER OFF)
R — IS TRANSMITTER RAMP UP/DOWN TIME
PREAM — IS PREAMBLE TO AID AGC (FORMERLY DATA)
S — IS FIRST SYNCHRONIZATION WORD
DATA — IS USER INFORMATION
SYNC+ — IS SECOND SYNCHRONIZATION WORD (FORMERLY SACCH + DVCC)
AE — IS ADDED GUARD TIME (FORMERLY DATA)

Figure 4.10, IS-136 Reverse Control Channel Abbreviated Slot Structure

4.3 System Parameters

Most system parameters on IS-136 systems are similar to IS-54B systems. The RF carrier spacing and designations, power output, and modulation characteristics are identical to IS-54B. The major changes include a time division duplex control channel, DQPSK modulation, new control channel encoding, and a new channel structure.

4.3.1 Duplex Channels

To prevent a Subscriber Unit's receive and transmit messages from interfering, IS-136 systems separate transmit and receive channels by the same 45 MHz as IS-54 Rev B systems. In addition to separating transmit and receive frequencies, IS-136 voice (traffic) and control channels can separate transmission and reception in time. The time separation, called Time Division Duplex (TDD) operation, simplifies the design of transmitters and receivers. Figure 4.11 shows IS-136 system time and frequency separations between receive and transmit channels.

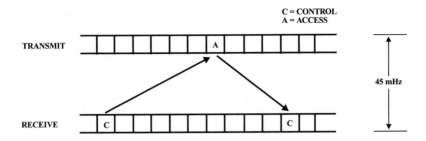

Figure 4.11, DCC Duplex Channel

4.3.2 Modulation Type

Some proposals received during development of IS-136 suggested changing the modulation characteristics at the physical layer, but these were not adopted [5]. In the interest of maintaining commonality and minimizing development time, the Pi/4-DQPSK modulation schemes for data were maintained as identical between IS-136 and IS-54.

4.3.3 Special Encoding Methods

The shared control feedback (SCF) fields are used to control access on the reverse access channel. The Subscriber Unit receives these fields before attempting access to the system. Like the busy idle bits on the AMPS control channel, the SCF fields provide system busy information. In addition, the SCF field also provides busy/reserved/idle indication, echoes back a response, and indicates if the access was received or not received.

Some of these fields, including the busy/reserved/idle (BRI), coded partial echo (CPE) and received/not received (R/N), are block coded to protect against errors. In addition to the special encoding, these fields are interleaved to protect against Rayleigh fading.

Convolutional Encoding

The logical sub channels on DCC (BCCH, SPACH and RACH) use a rate 1/2 convolutional encoding scheme. The encoding scheme uses the same polynomials as full rate speech from IS-54 Rev B. To detect and correct errors, a CRC field and 5 tail bits are appended to each message. The input data bits (length varies depending on the channel type) are first run through the CRC calculator. The seven CRC bits are added to the original data bits along with five tail bits (all zeros). The resulting bit stream is input to the convolutional encoder, and the output is used for transmission on the channel. A total of 260 data bits are transmitted from the base station to the Subscriber Unit. For normal slot lengths, the Subscriber Unit transmits a total of 244 bits to the base station. For abbreviated slot lengths, the Subscriber Unit transmits 200 bits.

Interleaving

The data bits for the IS-136 control channel slots are interleaved over only one slot. As a result, the Subscriber Unit reads only one slot to obtain paging information, thereby maximizing sleep time. If, for example, control channel data were interleaved over two slots, Subscriber Units would remain awake twice as long while decoding the paging channel, greatly reducing standby time. For IS-136 traffic channel slots (voice), data bits are interleaved over two slots.

4.3.4 Channel Structure

The DCC consists of several logical channels. The uplink channel (from Subscriber Unit to base station) is the Random Access Channel (RACH) used to coordinate random access attempts to the system. The downlink channel (from base station to Subscriber Unit) consists of three components: 1) short message, paging and access channels (SPACH), 2) a broadcast control channel (BCCH), and 3) system control field (SCF) information. The SPACH combines the short message, paging, and access channels. The BCCH broadcast channel provides system identification and access information. The SCF helps coordinate access to the system.

As Subscriber Units randomly attempt to access the cellular system, the RACH transfers the access request data from Subscriber Units to the base station. The SCF is a type of acknowledgment on the downlink channel.

The first downlink channel is the SMS point-to-point Paging and Access response Channel (SPACH). This channel contains broadcast information for specific Subscriber Units. As the name suggests, this channel contains three logical sub channels to carry point-to-point SMS messages, page messages and access responses.

The second downlink channel is the Broadcast Control Channel or BCCH. The BCCH contains general information intended for all Subscriber Units currently on the system. It is also further divided into two sub channels. The two sub channels are the Fast Broadcast Control Channel (F-BCCH) and the Extended Broadcast Control Channel (E-BCCH).

Built into every downlink burst are the Shared Control Feedback (SCF) flags. These flags are used as a type of acknowledgment of the uplink channel. The phone monitors these channels to determine if it is permitted to attempt access on the RACH, and to determine if the base station received the access attempt.

SMS Point-to-point, Paging and Access Response Channel (SPACH)
The SPACH contains three separate channels, the paging channel, the access response channel and the point-to-point SMS channel.

Paging Channel

Subscriber Units spend most of their time waiting for a page, so paging channel structure directly affects their "sleep" time. Most portable battery-powered phones use power conservation algorithms that turn off unused portions of the hardware, but periods of decreased power consumption are usually only tens of milliseconds. The IS-136 paging channel was designed to maximize the sleep time available to a portable phone, and it is a primary reason for increased stand-by times.

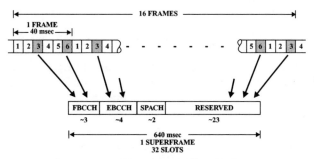

Figure 4.12, IS-136 Superframe Structure

The paging channel is a multiplexed channel comprised of many paging channel groups. Each group has a different period during which it is "active," or transmitting page information. A Subscriber Unit finds the proper paging channel through an algorithm based on its Subscriber Unit identification number and information found on the BCCH. Figure 4.12 shows the IS-136 superframe structure.

Access Response Channel (ARCH)

The access response channel is a shared channel that all Subscriber Units use for sending information to the base station to request service. The ARCH is a contention based access system, but some control flags on the downlink help to arbitrate the access. Contention based access is a system by which several Subscriber Units can talk to the base at the same time station without interference.

When two or more Subscriber Units contend for access at the same time, the base station eventually recognizes one of them and begins communicating with it. While monitoring the DCC, a Subscriber Unit attempts access by first listening to the SCF field to see if the channel is busy. If it is not busy, the Subscriber Unit sends its message. After starting to send its message, the Subscriber Unit looks for the base station to indicate that the channel status has changed to busy. If the channel becomes busy within a specific period of time, the phone assumes that the base station has received its message. The base station also sends other information, called the Partial Echo (PE) field, to indicate whether it has received the Subscriber Unit's message or a message from another Subscriber Unit. The PE field is a value obtained from the Subscriber Unit's message echoed back from the base station.

SMS Channel (SMSCH)

The SMS Channel transfers short messages to and from the Subscriber Unit. Unlike IS-54 messages, SMS messages can be sent on the control channel, eliminating the requirement for Subscriber Units to tune to a traffic channel to receive a message. An indicator in the SPACH channel identifies this information as an SMS message. SMS messages received on this channel are point-to-point messages (targeted for a specific Subscriber Unit).

A SMS message is received much like a normal page message, but the message type field indicates an SMS message rather than a normal page. The Subscriber Unit then waits for the proper number of bursts until the SMS message is complete.

Broadcast Control Channels (BCCH)

The broadcast channels consist of the fast and extended broadcast channels. Both channels contain general overhead information that applies to all Subscriber Units. The information sent on these channels is not addressed to a specific Subscriber Unit.

Fast Broadcast Control Channel (F-BCCH)

The F-BCCH constantly transmits information about the system from the base station to Subscriber Units. The information is very similar to the control information in an EIA-553 system. Before attempting access to the system, the Subscriber Unit must acquire and store the F-BCCH information, much of which, such system ID and other access data, is necessary for access.

After the Subscriber Unit finds a Digital Control Channel, the F-BCCH is the first channel it must decode. The information is repeated only every 32 frames, so it can take a significant amount of time to acquire the overhead data. The first frame of F-BCCH data contains information about the structure of the remaining frames. The number of each type of frame is included in the DCC structure message.

Extended BCCH (E-BCCH)

The Extended Broadcast Channel (E-BCCH) also transmits system information from the base station to the Subscriber Unit. Information sent on the E-BCCH is not critical to the operation of the Subscriber Unit, and is therefore updated much more slowly than F-BCCH data. It is updated so slowly that a flag indicating that information has changed is on the E-BCCH so that Subscriber Units need only wake up to read E-BCCH data when changes occur.

Earlier versions of the IS-136 specification contained another sub channel for the BCCH SMS messages called the S-BCCH (SMS broadcast channel). The S-BCCH was intended for general messages applying to all Subscriber Units, such as messages about the weather, news or advertisements. The IS-136 specification planners had difficulty reaching agreement on the format of this channel, and in the interest of getting the specification approved, it was removed from the final balloted version. The S-BCCH is likely to reappear in early IS-136 specification updates.

4.4 System Operation

The new IS-136 system digital control channel slightly alters operation from IS-54B systems, but once an IS-136 Subscriber Unit is in conversation mode, operation is very similar to IS-54B. The addition of the DCC locator information to help IS-136 Subscriber Units find a DCC is the greatest change, with few changes to the analog voice or digital traffic channels.

4.4.1 Access

A Subscriber Unit's access to the system is controlled through the SCF flags. The base station controls these flags, and uses them to allow or disallow Subscriber Units to access the system.

Shared Control Feedback (SCF)

When a Subscriber Unit attempts access to the system, it uses BRI flags, which the base station uses to control simultaneous access attempts by multiple Subscriber Units. A Subscriber Unit cannot access the system until it finds an idle slot. The base station sets slots as busy when communication with a

Subscriber Unit begins. The Subscriber Unit sends the access attempt to the base station during an idle slot, then looks for a response from the base station indicating that the message was received. Figure 4.13 illustrates IS-136 system access using SCF flags.

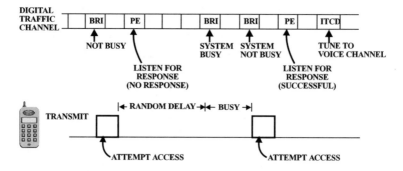

Figure 4.13, Access Using Shared Control Feedback

This type of access system works well, except when two Subscriber Units see the same idle slot and initiate transmissions simultaneously. If the base station correctly receives one of the messages, both Subscriber Units detect the received flag, but only one can be served. Such collisions are prevented with the partial echo (PE) field. The PE field is a value that allows Subscriber Units to identify which access attempt the base station is responding to. If a Subscriber Unit sees the received flag without the correct PE value, it stops the access attempt and retries after a random delay.

After a Subscriber Unit has gained access to the base station, it sends information about the call it is trying to place, and the base station responds with an initial traffic channel designation, just as in an IS-54B system.

4.4.2 Paging

Pages are received on the SPACH paging channel. A Subscriber Unit monitors its paging channel for any messages from the base station. An algorithm using the MIN of the Subscriber Unit determines which paging channel to monitor. Once a Subscriber Unit has locked on to its paging channel, it wakes up only once every two superframes (for PFC=1) to read a slot of information. If the Subscriber Unit detects a page message, it sends a page response to the base station. The base station responds with an initial traffic channel designation. Multiple paging channels can exist on the same cell, so the Subscriber Unit must use its MIN and information found on the BCCH channel to first find the correct paging channel. Figure 4.14 illustrates IS-136 system paging.

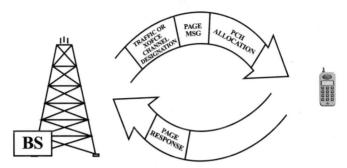

Figure 4.14, IS-136 System Paging

4.4.3 Hand-off

Hand-offs on an IS-136 system are very similar to those on IS-54B systems. Once the Subscriber Unit is on an analog voice channel or a digital traffic channel, it may hand off to another voice or traffic channel. The IS-136 specification alters very little of the on the voice and traffic channel messaging from IS-54B. Figure 4.16 illustrates the hand-off of a DCC.

IS-136 systems may hand off between different system ID's. For example, a Subscriber Unit could hand off from a private system to the public system. The Subscriber Unit experiences no difference when handing off between systems, but within the network, some additional information is required for billing. Rates on private systems are not necessarily the same as the public system, even when they occupy the same cell site and RF channel. Figure 4.15 shows the IS-136 handoff process.

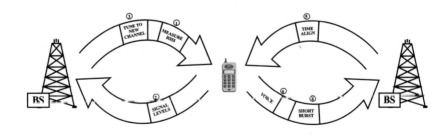

Figure 4.15, IS-136 Hand-off

4.5 Call Processing

The primary difference between call processing for an IS-54B system and an IS-136 system is the new control channel information and the new paging channel. Once a traffic or voice channel is assigned, processing is very similar to IS-54B systems.

4.5.1 Initialization

Control channel signaling for the DCC is considerably different from that of the existing IS-54 B control channel (an EIA-553 control channel with minor modifications). The initialization processing for an IS-136 Subscriber Unit is shown in Figure 4.16. One of the greatest differences is the additional messages for the Subscriber Unit to use in finding its next DCC channel. This process is called cell reselection.

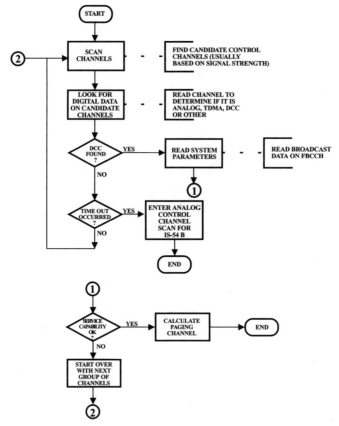

Figure 4.16, IS-136 Initialization Processing

After the difficult task of finding the DCC, the Subscriber Unit immediately begins a process to find its next DCC channel. The neighbor list received from the base station supplies information to assist in this task. Figure 4.17 shows cell reselection information on the DCC.

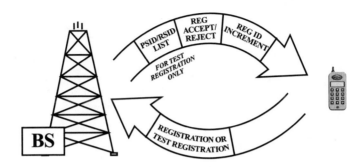

Figure 4.17, IS-16 Cell Reselection

Receiving a page or originating a call on a DCC is similar to the process used on the analog control channel. In the case of a received page message, the Subscriber Unit detects the page message while reading the PCH. Once the Subscriber Unit decodes the page message, it responds back to the base station. The base station acknowledges the response by sending a voice channel assignment for an analog or digital traffic channel.

Originations are even simpler. After the user requests a call to be placed, the Subscriber Unit sends an origination message to the base station, which responds with a voice channel assignment.

Another important part of control channel signaling is registration. Registration tells the base station that the Subscriber Unit is locked on its control channel, thereby notifying the base station where to send any messages, including pages. An important difference between IS-136 and IS-54B registration is the fact that IS-136 systems allow the base station to reject or accept a Subscriber Unit. If the phone should not be locked onto a particular control channel, for example if the user has not subscribed to a particular private system, then the system rejects its registration attempt. The system notifies the Subscriber Unit that pages will not be routed to this control channel for it to pick up, and that other messages will not be processed.

Several different types of registration are available: power up registration, periodic registration, and power down registration. As the names imply, power up registration is performed during the Subscriber Unit's power on sequence; periodic registration is performed periodically; and power down registration is performed when the phone is about to be turned off. Power down registration reduces the base station's message load by informing the base station that the

Subscriber Unit is no longer looking for a page. Figure 4.18 illustrates the IS-136 registration process.

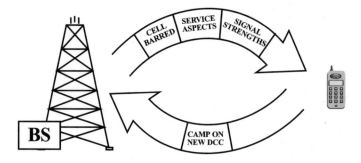

Figure 4.18, IS-136 Registration

4.5.2 Idle

An IS-136 Subscriber Unit in idle mode must monitor its paging channel to look for messages from the base station. It must also check to see if any information has changed on the broadcast channels. This check consists of a field on the paging channel that updates as it receives new broadcast information. The Subscriber Unit leaves its paging channel to read other slots in the DCC only when this field indicates new data. Figure 4.19 illustrates idle mode operation on an IS-136 system.

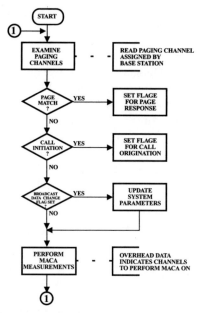

Figure 4.19, IS-136 Idle Mode

4.5.3 System Access Task

System access for IS-136 Subscriber Units is very similar to that for IS-54B Subscriber Units. After receiving a page or originating a call, the Subscriber Unit is assigned an analog voice channel or a digital traffic channel. This assignment is made using an initial traffic channel designation message just as in IS-54B systems. Figure 4.20 shows IS-136 access mode.

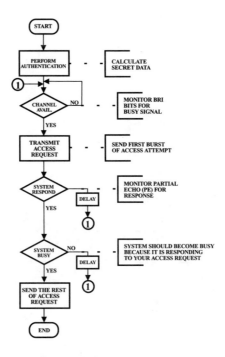

Figure 4.20, IS-136 System Access Mode

4.5.4 Conversation Mode

IS-136 and IS-54B systems have similar voice channel signaling. The differences are primarily in a few new fields to help locate DCCs. Digital control channels can be located virtually anywhere in the valid cellular frequency range, so it can be quite difficult for a Subscriber Unit to locate a DCC. Both the analog control channel and the digital voice channel of IS-136 systems have added fields to indicate the presence of a DCC in the same system, with information on how to find it. Figure 4.21 shows the processing flow for IS-136 Conversation Mode. Note that the new DCC locator field is not used during conversation, but during the search for a DCC during initialization.

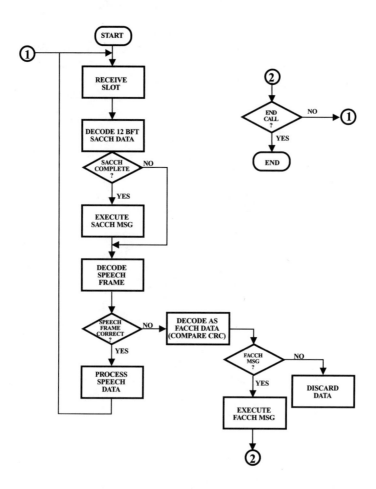

Figure 4.21, IS-136 Conversation Mode

References

1. Cellular handset manufacturer sales representative, personal interview, industry expert.
2. EMCI, "Digital Cellular Economics and Comparative Technologies", Washington DC, 1993.
3. ibid.
4. ibid.
5. TIA committee member, personal interview, industry expert.

Chapter 5
CDMA (1S-95)

5. CDMA (IS-95)

The IS-95 Code Division Multiple Access (CDMA) standard combines new digital CDMA and Advanced Mobile Phone Service (AMPS) functionality. This chapter contains semi-technical descriptions of IS-95 CDMA dual mode technology. For a description of the AMPS functionality associated with this IS-95 system, see chapter 1.

The Cellular Telecommunications Industry Association (CTIA) User Performance Requirements (UPR) specify the goals and objectives for the next generation of cellular technology. The Telecommunications Industry Association (TIA) is responsible for creating specifications to achieve these goals. As the IS-54 US TDMA specification process was nearing completion, the TIA realized that some requirements in the CTIA UPR were not mett [1]. In response, Qualcomm created a radio specification that satisfied the unmet requirements. Qualcomm proposed their proprietary specification to the TIA, which accepted it as the IS-95 CDMA specification. Figure 5.1 illustrates the evolution of CDMA specifications.

Figure 5.1, Evolution of the IS-95 Specification

5.1 System Overview

Figure 5.2 is an overview of an IS-95 system. Note that the IS-95 system includes many of the same basic subsystems as other cellular systems, such as the Mobile Switching Center (MSC), the Base Station (BS), and Subscriber Units (phones).

The MSC for IS-95 systems differs from existing EIA-553 system MSCs. The IS-95 MSC is enhanced to provide simultaneous communications among multiple Base Stations, simultaneously routing calls from one or more Base Stations to the Public Switched Telephone Network (PSTN). This simultaneous routing capacity makes handoffs more reliable and simplifies network signaling.

CDMA Base Stations use one or more CDMA radio channels to provide both control and voice functionality. The Base Station converts the radio channel to a signal that is transferred to and from the MSC. Like the MSC, the Base Station controller section can communicate simultaneously among different sectors in that cell, enhancing handoffs. Typically, each CDMA radio channel that has 64 independent channels replaces 2 AMPS radio channels.

IS-95 systems can serve Subscriber Units of three types: AMPS only, IS-95 Dual Mode, or IS-95 (digital only). AMPS phones use the AMPS Voice Channel (AVC) and AMPS Control Channel (ACC). IS-95 dual mode units use the ACC, AVC, Digital Control Channel (DCC), or Digital Traffic Channel (DTC). IS-95 digital-only units use the DCC and DTC. Figure 5.2 illustrates the IS-95 CDMA radio system.

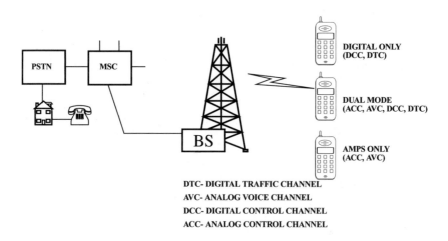

DTC- DIGITAL TRAFFIC CHANNEL
AVC- ANALOG VOICE CHANNEL
DCC- DIGITAL CONTROL CHANNEL
ACC- ANALOG CONTROL CHANNEL

Figure 5.2, Overview of an IS-95 CDMA Radio System

5.1.1 New Wide RF Channel

CDMA divides the radio spectrum into wide 1.23 MHz digital radio channels. CDMA differs from other technologies in that it multiplies (spreads) each signal with a unique pseudo-random noise (PN) code that identifies the channel. CDMA transmits digitized voice and control signals on the same frequency band. Each CDMA radio channel contains the signals of many ongoing calls (voice channels) together with pilot, synchronization, paging, and access (control) channels. Digital Subscriber Units select the signal they are receiving by correlating the received signal with the proper PN sequence. This correlation enhances the selected signal and leaves others unenhanced.

Each IS-95 CDMA radio channel is divided into 64 separate logical (PN coded) channels. A few of these channels are used for control, and the remainder carry voice information and data. Because CDMA transmits digital information combined with unique codes, each logical channel can transfer data at different rates (e.g., 4800 bps, 9600 bps).

CDMA systems use a maximum of 64 coded (logical) traffic channels, but they cannot always use all of these. A CDMA radio channel of 64 traffic channels can transmit at a maximum information throughput rate of 192 kbps [2], so the combined data throughput for all users cannot exceed 192 kbps. To obtain a maximum of 64 communication channels for each CDMA radio channel, the average data rate for each user should approximate 3 kbps. If the average data rate is higher, less than 64 traffic channels can be used. CDMA systems can vary the data rate for each user dependent on voice activity (variable rate speech coding), thereby decreasing the average number of bits per user to about 3.8 kbps [3]. Varying the data rate according to user requirement allows more users to share the radio channel, but with slightly reduced voice quality. This is called soft capacity limit.

Figure 5.3 shows how CDMA channels share each radio channel. Digital signals are coded to produce multiple chips (radio energy) for each bit of information to be transmitted. The receiver shifts the reference pattern in time until it matches the coded pattern (circles, squares, and diamonds in figure 5.3). Chips on the forward radio channel (from the Base Station to the Subscriber Unit) are selected to collide only infrequently with chips from other users. This is known as orthogonal coding. However, time chips being received from many subscribers on the reverse channel cannot be precisely controlled, so collisions on the reverse channel occur more often.

Several chips (134-536 chips) are created for each bit of user information (speech or data), so if some of them encounter interference such as the collision in figure 5.3, most of the remaining chips will still be received successfully. CDMA channels are designed to operate with interference among users (chip collisions), so

CDMA radio channels can tolerate a large amount of interference without significantly reducing voice quality.

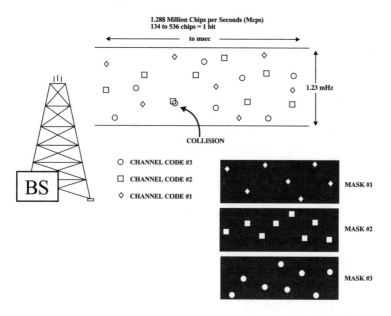

Figure 5.3, CDMA Radio Channel

The CDMA radio channel is constructed of coded signals on the same frequency, so adjacent cells use different codes and may reuse the same radio frequency. Interference from neighboring cells appears only as chip interference. The effect of chip interference is only to reduce system capacity.

The IS-95 system adds several types of information to the transmitted data. Error protection bits are added to protect the digitized speech from errors created during radio transmission. Control messages are inserted to coordinate operation between the Base Station and the Subscriber Unit, facilitating soft handoff and other channel maintenance functions.

The digitized voice signal is sent with control messages to a PN code sequence modulator, which adds the unique identifying PN code. The PN code sequence modulator produces a low-level RF signal at the desired radio channel frequency.

5.2 System Attributes

The IS-95 CDMA cellular system has several important advantages. It eliminates the need for frequency planning, provides high quality voice transmission under varying radio signal conditions, allows system capacity expansion, and maintains coordinated system control. The CDMA system is compatible with the established access technology, and it allows analog (EIA-553) and dual mode (IS-95) subscribers to use the same analog control channels. Some of the voice channels are converted to digital transmissions, allowing several users to be multiplexed on one RF channel. As with other digital technologies, CDMA reduces the cost of capacity expansion by allowing multiple users to share a single digital RF channel. However, in addition, the wide radio channel maintains stable voice quality even under poor radio conditions, and the CDMA system reduces network complexity by moving some of the system control to the Subscriber Unit. Like other cellular technologies, CDMA uses a Mobile Switching Center (MSC) to centralize control of Subscriber Units moving through the system.

5.2.1 Frequency Reuse

Figure 5.4 shows that CDMA allows the same frequency (f1) to be reused in all adjacent cells (N=1). In the shaded area where interference is significant, chip collisions from adjacent cells and other subscribers are more frequent, but this only reduces the number of users that can share the radio channel. Reusing the same frequency in every cell eliminates the need for frequency planning in CDMA systems.

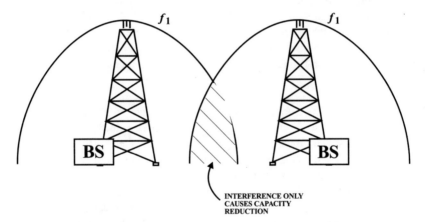

Figure 5.4, CDMA Frequency Reuse

Each CDMA radio channel occupies 1.23 MHz of spectrum (about 40 AMPS radio channels). However, one CDMA radio channel typically replaces only 2 AMPS radio channels in a single cell site or sector. This is because AMPS radio channels in each cell or sector are placed 21 channels apart (630 kHz) to allow for frequency planning.

5.2.2 Capacity Expansion

Because CDMA systems dedicate an average of less than 4 kbps [4] to end users, and they reuse the same frequencies in every cell, they can serve 20 [5] times more subscribers per cell site than an AMPS system [6]. CDMA has already been optimized with speech activity detection and variable rate speech coders, and no future enhancements are on the immediate horizon. However, a lower bit rate speech coder would increase system capacity still further.

5.2.3 System Control

While the MSC maintains overall control in CDMA systems, the Subscriber Unit does initiate some control processing. The Subscriber Unit monitors the signal strength of nearby cells, and when a neighboring cell's signal is strong enough, the Subscriber Unit requests handoff from that cell. This subscriber control reduces the MSC's Base Station coordination requirements.

5.2.4 Soft Handoff

In AMPS cellular systems, handoff occurs when the Base Station experiences a deterioration in signal strength from the Subscriber Unit. As AMPS subscribers approach handoff, signal strength varies abruptly and the voice is muted. In contrast, CDMA uses a unique "soft handoff," which is nearly undetectable and loses few if any information frames. CDMA's soft handoff is much less likely to lose a call during handoff.

Soft handoff allows the Subscriber Unit to communicate with two or more cell sites to continuously select the best signal quality until handoff is complete. The CDMA Subscriber Unit measures the pilot channel signal strength from adjacent cells and transmits the measurements to the serving Base Station. When an adjacent Base Station's pilot channel signal is strong enough, the Subscriber Unit requests the adjacent cell to transmit the call in progress. The serving Base Station also continues to transmit. Thus, prior to complete handoff, the Subscriber Unit is communicating with both Base Stations simultaneously. Using two radio channels simultaneously during handoff maintains a much higher average signal strength throughout the process. During soft handoff, the Subscriber Unit chooses the best frames from either Base Station. Soft handoff is nearly inaudible for voice communications, and is more reliable for modems, credit card machines, and other services transferring digital data. Figure 5.5 shows how CDMA systems use two channels simultaneously during handoff.

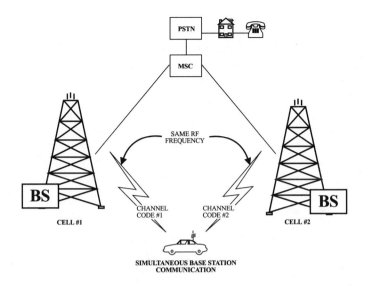

Figure 5.5, CDMA Soft Handoff

5.2.5 Variable Rate Speech Coding

CDMA speech coding occurs at a variable rate. The coding process begins with an analog-to-digital converter digitizing the user's voice at a fixed rate of 64 kbps. The digitized voice is supplied to a speech coder, which reduces the number of bits representing the speech by characterizing the speech into voice parameters. When voice activity is low, the variable-rate CDMA speech coder represents the speech signal with fewer bits. This added efficiency increases CDMA system capacity by about three times that provided by fixed-rate coders [7].

Figure 5.6 shows how the speech coder compression rate varies with speech activity. The 64 kbps speech signal is divided into 20 msec intervals. The speech coder characterizes and compresses this rate to 9600-1200 bps. As the speech activity decreases, the bit rate decreases.

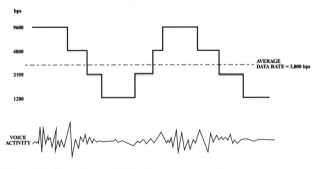

Figure 5.6, Variable Rate Speech Coding

131

5.2.6 Primary and Secondary Channels

A new feature of the IS-95 CDMA system is the Subscriber Unit's ability to simultaneously receive multiple channels. The single 9600 bps communication channel (higher data rates are possible per user) can be divided into separate primary and secondary channels. The primary channel can be used for communicating voice, and the secondary channel can be used for simultaneously transferring user data. The data rate, shared between the primary and secondary, is variable. If all of the channel capacity is used for the primary channel (e.g. voice), none is available for the secondary channel. If the voice activity is low, most of the bits are available for the secondary channel.

5.2.7 Discontinuous Reception (Sleep Mode)

Discontinuous Reception (DRX) enables Subscriber Units to power off non-essential circuitry during periods (sleep) when pages will not be received. To provide for this sleep mode, the paging channel is divided into paging sub-channel groups.

Figure 5.7 shows the DRX (sleep mode) process. When the Subscriber Unit registers on a CDMA radio channel, it informs the system of its sleep mode capability. CDMA paging channels are divided into 200 msec slots (paging groups) which allow the subscriber to sleep during unwanted groups. Each paging group is composed of 10 frames (200 msec). The system can dynamically assign up to 640 paging groups to allow a maximum sleep period of 2 minutes and 8 seconds. For normal operation, about 10 groups will be used for a maximum delay of about 2 seconds. Approximately 400 msec before the end of a sleep period, the Subscriber Unit wakes up to allow reacquisition with the control channel.

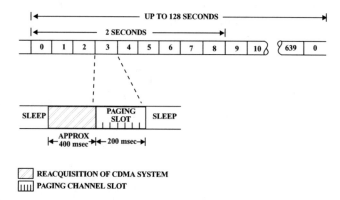

Figure 5.7, Discontinuous Reception (Sleep Mode)

5.2.8 Soft Capacity

A cellular system is in a condition of over capacity when more subscribers attempt to access the system than its radio interface can support at a desired quality level. CDMA allows the system to operate in a condition of over capacity by accepting a higher-than-average bit error rate, or reduced speech coding rate. As the number of subscribers increases beyond a threshold, voice quality begins to deteriorate, but subscribers can still gain access to the system.

Figure 5.8 shows that as more users are added to the system, voice quality deteriorates. When voice quality falls below the allowable minimum (usually determined by an acceptable bit error rate), the system is over capacity. Allowing more subscribers on the system by trading off voice quality (or the subscribers data rate) creates a soft capacity limit.

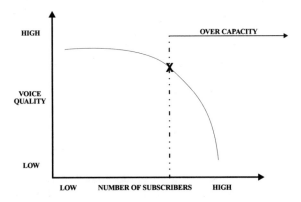

Figure 5.8, IS-95 Soft Capacity Limit

5.2.9 Precise RF Power Control

To effectively separate the coded channels, the received signals at the Base Station from all Subscriber Units must be at almost the same level. If one received signal were much more powerful than the others, the receiver could not effectively decode the weaker ones, making it much less sensitive to weaker channels. To accommodate this requirement for uniform signal levels, the CDMA system precisely controls Subscriber Unit power. The power control system has two parts; open loop and closed loop. The open loop is a coarse adjustment and the closed loop is a fine adjustment. The power control system maintains received signals within +/- 1 dB (33%) of each other. Demonstrations have shown that a strong interfering signal reduces the number of users per radio channel in a serving cell site. When interference is too great, Subscriber Units are handed off to another cell [8].

A CDMA Subscriber Unit's coarse (open loop) RF amplifier adjustment is controlled by feedback from its receiver section. The Subscriber Unit continuously measures the radio signal strength received from the Base Station to measure the signal strength loss between the Base Station and Subscriber Unit. Figure 5.9 shows that as the Subscriber Unit moves away from the Base Station, the received signal level decreases. The stronger the received signal, the less the Subscriber Unit amplifies its RF signal output; the weaker the received signal level, the more it amplifies RF signal output. The end result is that the signal received at the Base Station from the Subscriber Unit remains at about at the same power level regardless of the Subscriber Unit's distance.

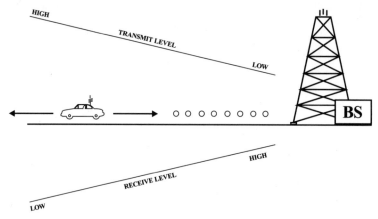

Figure 5.9, CDMA Open Loop RF Power Control

Because the open loop power adjustment does not adequately control received signal level, the Base Station also fine-adjusts the Subscriber Unit's RF amplifier gain by sending power level control commands to the Subscriber Unit during each 1.25 msec time slot. The commands are determined by the Base Station's received signal strength. The power control bit communicates the relative change from the previous transmit level, commanding the Subscriber Unit to increase or decrease power from the previous level.

Figure 5.10 illustrates closed loop power control. As the received signal power increases, the power control bit signals the Subscriber Unit to reduce transmit power level. When the received signal is lower than desired, the power control bit commands the Subscriber Unit to increase power. The closed loop adjustment range (relative to the open loop) is +/- 24 dB minimum. The combined open and closed loop adjustments precisely control the received signal power at the Base Station.

Chapter 5

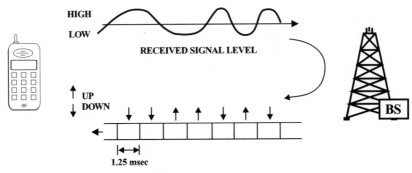

Figure 5.10, CDMA Closed Loop RF Power Control

Subscriber Units also reduce their average power by transmitting only in bursts when the channel data rate is reduced. Figure 5.11 shows how a 4800 bps channel transmits only 1/2 of the time.

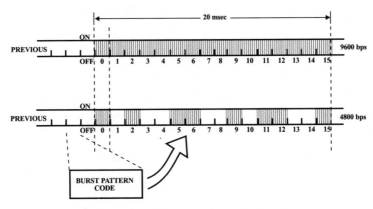

Figure 5.11, CDMA Subscriber Unit Transmit Power Bursts

5.2.10 Digital Control Channels

Each CDMA 1.23 MHz bandwidth radio channel pair contains its own digital control channel (DCC). A CDMA DCC is composed of four types of channels, each identified by a unique code: pilot, synchronization, paging, and access.

Each CDMA Base Station transmits a pilot signal on a unique pilot channel (shown in figure 5.12). The pilot signal is the reference for demodulating the signal and for estimating received signal strength to indicate which cell site can best communicate with the Subscriber Unit. CDMA Subscriber Units simultaneously measure the pilot signal strengths of all neighboring Base Stations in the system.

After the Subscriber Unit determines the strongest pilot channel, it demodulates the synchronization channel. The synchronization channel contains information parameters that allow the Subscriber Unit to synchronize to other CDMA channels (paging, access, and voice). These include system parameters, access parameters, channel list information, and a neighboring radio channel list.

The CDMA paging channel continuously sends system parameter and paging information intended for a single or group of Subscriber Units. After the Subscriber Unit has initialized, it continuously listens to the strongest paging channel to determine if a call is to be received.

When system attention is required, the Subscriber Unit competes for access on the access channel. This is a random process in which the Subscriber Unit continually increases its access power until the Base Station responds. If the Base Station does not respond within an allotted time, or the Subscriber Unit exceeds a maximum power level, the access attempt is aborted. This procedure avoids interference between Subscriber Units during access attempts. When access is complete, the voice information is transmitted over separate voice channels. Figure 5.12 shows that each CDMA radio channel contains a pilot channel, synchronization channel, paging channel, and several different traffic channels.

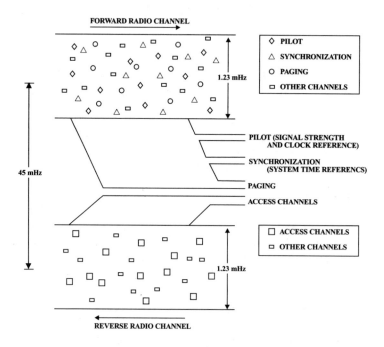

Figure 5.12, CDMA Digital Control Channels

5.2.11 Frequency Diversity

The CDMA radio channel spreads the signal over a wide 1.23 MHz frequency range, making it less susceptible to radio signal fading that occurs only over a specific narrow frequency range. As a result, radio signal fades affect only a portion of the CDMA chip codes, and most of the chip codes get through successfully. With only a small portion of information corrupted, digital information transmissions over a CDMA radio channel are relatively robust.

Figure 5.13 shows how a frequency fade in a CDMA signal only affects a small percentage of the chip codes. Each bit of information is represented by several chips (typically several hundred), so information bits are not usually affected by the radio signal fades.

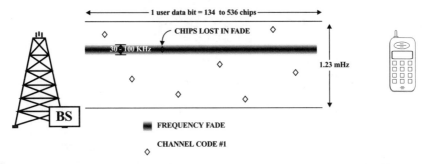

Figure 5.13, CDMA Frequency Diversity

5.2.12 Time Diversity

When multiple reflected (multipath) signals are received at slightly different times, CDMA systems are able to select only the strongest signal, whether reflected or direct. The gains from this process are similar to those obtained from antenna diversity. CDMA can also combine multipath signals, adding several weak multipath signals to construct a stronger one. This process is called rake reception. The result is better voice quality and fewer dropped calls than would otherwise be available.

Figure 5.14 shows how a multipath signal can be added to the direct signal. The radio channel shows two code sequences. The shaded codes are time delayed because the original signal was reflected and received a few microseconds later. The original signal is decoded by mask #1. The mask is shifted in time until it matches the delayed signal. The output of each decoded channel is combined to produce a better quality signal.

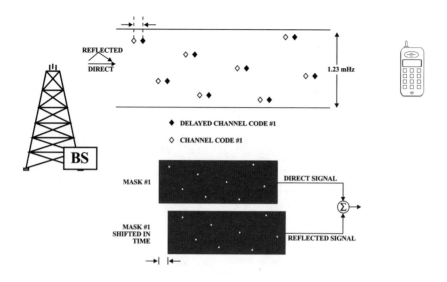

Figure 5.14, CDMA Rake Reception

5.3 Signaling

Signaling is the physical process of transferring control information to and from the Subscriber Unit. When operating on the AMPS channel, signaling continues to be blank-and-burst. When operating on the CDMA channel, signaling is sent by blank-and-burst or by dim-and-burst. Dim-and-burst signaling sends control information in unused bit locations during periods of low speech activity. Variable rate speech coding varies the coding rate so that both voice and control messages may be sent during each 20 msec frame, thereby allowing for fast or slow dim-and-burst signaling.

5.3.1 AMPS Control Channel Signaling

The AMPS control channel signaling is unchanged in dual mode CDMA systems. The AMPS control channel was not modified to allow assignment to a CDMA radio channel because each CDMA radio channel has its own control channel which can assign either an AMPS or CDMA radio channel.

5.3.2 CDMA Control Channel Signaling

The CDMA system adds a new control channel composed of a pilot, synchronization, paging, and access channel.

5.3.2.1 Pilot Channel

The pilot channel provides a reference clock for channel demodulation and is used as a reference signal level for handoff decisions. The Subscriber Unit must initialize to the pilot channel before accessing any other control channel.

5.3.2.2 Synchronization Channel

The synchronization channel provides additional system parameters such as system identification and standard offset time of the cell site. The Subscriber Unit must acquire these parameters before monitoring the paging or access channels.

5.3.2.3 Paging Channel

The CDMA paging channel offers new features that the AMPS paging channel does not. The CDMA paging channel supports sleep mode. The sleep mode is made possible by combining pages into groups. Each paging channel is divided into slots. The Subscriber Unit can operate in the slotted mode (using sleep intervals) or non-slotted mode.

Because the CDMA Subscriber Unit can simultaneously decode more than one coded channel, it can monitor multiple paging channels. When the Subscriber Unit determines that a neighboring cell site has a higher quality paging channel, it initiates handoff to the new paging channel.

5.3.2.4 Access Channel

The CDMA access channel is on the reverse radio link (Subscriber Unit to Base Station). Each CDMA radio channel can have up to 32 separate (coded) access channels. Access channel messages are grouped into 20 msec frames of 88 information bits. The gross channel rate for the access channel is 9600 bps. However, access channel messages are repeated twice, reducing the effective channel rate to 4800 bps.

The Subscriber Unit accesses the CDMA channel by increasing its power level stepwise until the Base Station acknowledges the request. The power level increases continue up to a maximum power limit received from the control channel before the access attempts begins. If the first access attempt (a complete sequence of power level increases) is unsuccessful, the Subscriber Unit waits for a random period before attempting access again.

5.3.3 AMPS Voice Channel Signaling

The handoff command sent on the AMPS voice channel does not support AMPS-to-CDMA handoff. It is technically possible to handoff from AMPS to CDMA through a hard handoff (mute period) like the standard AMPS handoff, but this was deemed unnecessary [9]. Because there are no AMPS to CDMA handoffs, the signaling format of the AMPS voice channel is unchanged.

5.3.4 CDMA Digital Traffic Channel Signaling

CDMA control messages can be sent using either one of two alternative techniques. One technique, called blank-and-burst, momentarily blanks the audio and replaces it with a burst of control information. An alternative technique, called dim-and-burst, reduces the number of digitized speech bits available to characterize the voice, and adds a short burst of control information bits. Any control message can be sent by either blank-and-burst or dim-and-burst signaling.

Blank-and-burst signaling replaces speech data with signal messages. This is called "in-band signaling". No limit has been established on how many speech frames may be replaced. Blank-and-burst message transmissions degrade speech quality because they replace speech frames with signaling information. Blank-and-burst messages degrade speech quality non-linearly as the number of frames stolen increases. Figure 5.15 illustrates blank-and-burst signaling.

Dim-and-burst inserts control messages when speech activity is slow. As the

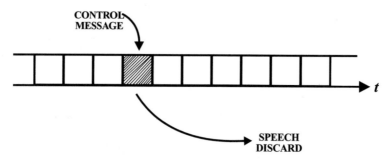

Figure 5.15, CDMA Blank-and-Burst Signaling

speech coder changes speed, some of the unused bits are reassigned as control message bits. Because all of the speech bits are not available for the control message, several frames are needed to send the message. The number of required frames varies according to speech activity.

The gross user data rate available is 9600 bps. The data rate is reduced by cycling the transmitter off for several 1.25 msec bursts (called power groups) during each 20 msec frame. The user data rate is determined by checking the frame quality bits (CRC). If the frame quality bits do not check for one data rate, decoding at another data rate will be attempted.

The mix between voice data and signaling data is determined by a mixed mode flag bit sent at the beginning of each frame. Additional flag bits include a burst format bit which indicates if the message is being sent via blank-and-burst or dim-and-burst, a traffic type which identifies primary or secondary channel, and a traffic mode bit which sets the proportional mix between voice and signaling data. Figure 5.16 illustrates a dim-and-burst message.

5.4 System Parameters

IS-95 system parameters are significantly different from the EIA-553 AMPS sys-

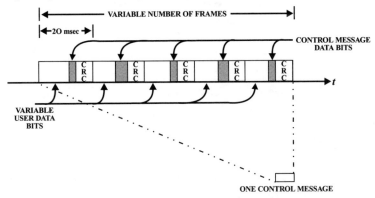

Figure 5.16, CDMA Dim-and-burst Signaling

tem. The IS-95 system parameters include a new wide RF radio channel, 45 MHz duplex channel separation, digital phase modulation, and new RF output power levels.

5.4.1 Frequency Allocation

While the 25 MHz x 2 cellular band allocation remains the same, the frequency allocation has been divided so that 9 or 10 1.23 MHz CDMA channels can be allocated in the A or B frequency bands. Because CDMA channels are wider than analog channels and require a guard band at the ends of the spectrum, the allowable channel assignments for CDMA channels are 1013 through 1023 and 1 through 777. The center frequency is used for the CDMA channel assignment. Figure 5.17 shows the CDMA channel assignment range.

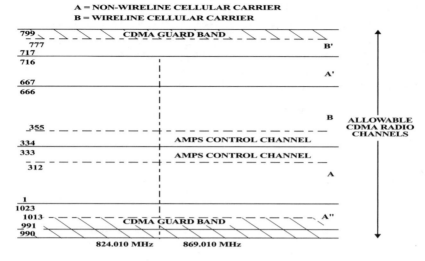

Figure 5.17, CDMA Frequency Allocation

5.4.2 Duplex Channels

Although the CDMA radio channel is digitized, the transmission continues to be frequency duplex. The separation of forward and reverse channels is 45 MHz. The transmit band for the Base Station is 869-894 MHz. The transmit band for the mobile station is 824-849 MHz.

Figure 5.18 illustrates the time offset between the forward and reverse channel, which is called time alignment. Time alignment advances or retards transmit bursts in 1.25 msec steps. Stepping the transmit time up or back effectively reduces interference to nearby cells. Unlike other technologies, CDMA systems do not need to receive frames at a precise time relative to transmit frames. This ability to shift Subscriber Units' time alignment allows CDMA systems to prevent all the Subscriber Units from transmitting at the same time, thus reducing the average interference level received by the Base Station.

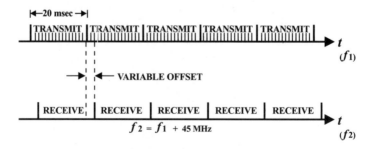

Figure 5.18, CDMA Duplex Channel

5.4.3 Modulation Type

IS-95 CDMA uses different types of modulation for the forward and reverse channels. The forward channel uses quadrature phase shift keying (QPSK) and the reverse channel uses offset quadrature phase shift keying (O-QPSK). Quadrature phase shift keying (discussed in chapter 2) transfers information by varying the phase of the transmit signal. Each 2 bits of information supplied to the modulator (called the chips) shift the signal's phase 0, +90, +/- 180, or -90 degrees. O-QPSK differs from QPSK in that it does not require the transmitter to pass the signal through the 0 signal level when both I and Q signals are zero. The Subscriber Unit's RF amplifier operates more efficiently with O-QPSK because it need not operate as linearly as it must with QPSK.

Figure 5.19 is a CDMA modulator block diagram. The PN multiplier multiplies the user data bits (1,200 - 9,600 bps) by a unique channel code. The output of the PN multiplier is a signal of 1.228 million chips per second (Mcps) which uniquely identifies the Subscriber Unit. The 1.228 Mcps signal is divided into I and Q bit channels which are supplied to a balanced modulator via a pulse shaper. The pulse shaper smoothes the edges of the digital signal to minimize rapid signal transitions which would result in radio frequency emissions outside the allocated bandwidth. The balanced modulator multiplies the shaped I and Q signals by two signals which are 90 degrees phase shifted (each two bits input to the modulator results in a single phase-shifted signal). Because of the PN multiplier, 32-256 phase shifts or 128-1024 chips represent a single user data bit. The number of symbols or chips per bit depends on the user data rate supplied to the PN multiplier. The output of the modulator is then supplied to the RF amplifier.
The pilot channel is always available on the forward radio channel for the

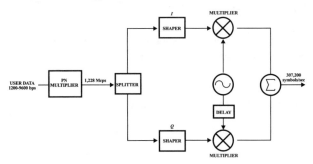

Figure 5.19, IS-95 CDMA Modulator Block Diagram

Subscriber Unit to use to demodulate the traffic and control channels. The pilot channel acts as a timing reference. This is called coherent demodulation. The pilot channel timing reference helps the demodulation process by indicating when the phase transition is complete. Figure 5.20 shows the modulation characteristics for the O-QPSK signal. A single phase shift (symbol) occurs every 3.25 usec. Each symbol represents two bits of information.

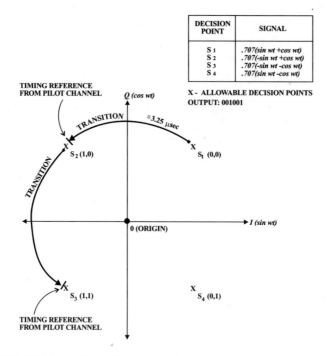

Figure 5.20, CDMA Coherent Demodulation

5.4.4 RF Power Classification

Maximum RF output power for AMPS and CDMA Subscriber Units is the same, but each CDMA 20 msec transmit frame is divided into sixteen 1.25 msec power control groups. These power control groups are controlled by the code assignment and user data rate requirement. The requirements for precise power control demand that CDMA Subscriber Units transmit at an average power 90% below analog FM Subscriber Units [10].

RF Power	Class 1	Class 2	Class 3	Class 4	Class 5-6
Maximum Power	4 Watts	1.6 Watts	.6 Watts	-	-
Minimum Power	6 mW	6 mW	6 mW	-	Reserved

Table 5.1
IS-95 Power Classification

5.5 System Operation

The IS-95 cellular system uses three types of channels: analog control channels (ACC), analog voice channels (AVC), and digital traffic channels (DTC). Each DTC provides voice and control channels. The control channels allow the Subscriber Unit to retrieve system control information and compete for access. The analog voice channels are primarily used to transfer voice information, but also to send and receive some digital control messages. Figure 5.21 shows the three types of IS-95 radio channels.

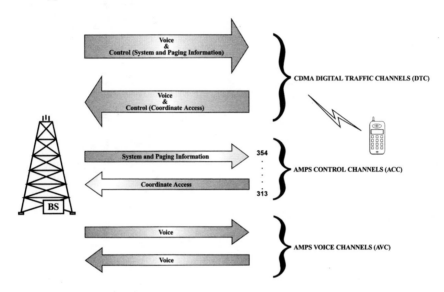

Figure 5.21, IS-95 Radio Channels

When a Subscriber Unit is first powered on, it initializes by scanning for control channels and tuning to the strongest one. In this initialization mode, it determines if the system is CDMA digital capable. If the system is not CDMA capable, the Subscriber Unit locks on to an AMPS control channel. After initialization, the Subscriber Unit enters idle mode and waits to be paged for an incoming call or for the user to place a call (access). When a call is to be received or placed, the Subscriber Unit enters system access mode to try to access the system via a control channel. When it gains access, the control channel commands the Subscriber Unit to tune to an analog or digital traffic channel. The Subscriber Unit tunes to the designated channel and enters conversation mode.

5.5.1 Access

Prior to accessing an IS-95 system, a Subscriber Unit listens for a CDMA pilot channel to determine if it is digital capable. If the system is digital capable, the Subscriber Unit attempts access via the CDMA radio channel. If the access attempt succeeds, the system assigns the Subscriber Unit an analog or digital voice channel. If it assigns an analog channel, the system sends an Initial Voice Channel Designation (IVCD) message which contains the voice channel number. If the system assigns a digital channel, it sends an Initial Traffic Channel Designator (ITCD) message with the channel number and channel code. Figure 5.22 illustrates the CDMA system access.

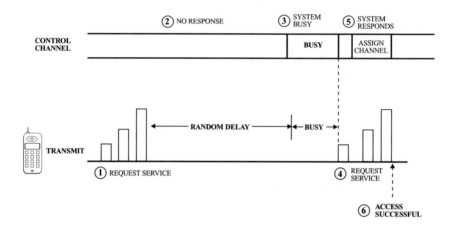

Figure 5.22, CDMA System Access

5.5.2 Paging

Paging is the process of sending a page message to the Subscriber Unit to indicate that a call is to be received. The IS-95 Subscriber Unit listens for pages on either the CDMA paging control channel or AMPS control channel (ACC), the preferred choice being the CDMA paging channel. CDMA paging channels can page in groups to allow sleep modes.

Page messages are sent on the paging channel. Figure 5.23 illustrates the paging process. Initially (step 1), the Subscriber Unit monitors the paging channel for pages. When the Subscriber Unit is paged, it requests service from the cellular system (step 2) indicating that it is responding to a page message. The cellular system then assigns it a new channel code (step 3) where it will be authenticated (step 4) and the conversation may begin.

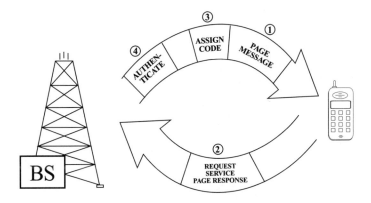

Figure 5.23, System Paging

The IS-95 cellular system can send Caller Line Identification (CLI). After the page message has been received and the voice channel or traffic channel has been assigned, the system can transfer the caller number ID. The caller number ID can be transferred on a voice or traffic channel before the subscriber answers.

5.5.3 Handoff

The handoff process for dual mode subscribers operating on a digital channel is different from that of AMPS. Figure 5.24 illustrates CDMA system digital channel handoff. Before handoff, the Subscriber Unit has received a list of neighboring cells' pilot channels that are candidates for handoff from the serving cell (#1, time 1). The Subscriber Unit continuously measures the signal strength of the candidate radio channels (time 2). When the pilot channel of a neighboring cell #2 is sufficient for handoff, the Subscriber Unit requests simultaneous transmission from that channel (time 3). The system then assigns the new channel to transmit simultaneously from cell #1 and cell #2 (time 4). The Subscriber Unit continues to decode both channels (different codes on the same frequency) using the channel with the best received quality. When the signal strength of the original channel falls below a threshold, the Subscriber Unit requests release of the original channel (time 5) and voice transmissions from cell #1 ends (time 6). The illustration shows the Subscriber Unit simultaneously communicating with only two Base Stations, but simultaneous communications with more than two Base Stations is possible.

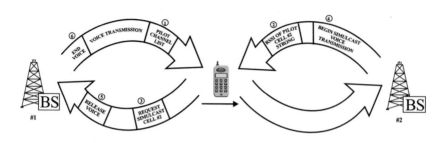

Figure 5.24, CDMA Handoff

Call processing is the group of tasks performed to 1) initialize information when the Subscriber Unit is turned on, 2) monitor the control channels, 3) attempt access, and 4) coordinate transmissions of voice or data. Call processing differs when operating in the AMPS and CDMA modes (see chapter 1 for a detailed description of AMPS call processing).

5.6.1   Initialization

Initialization is the process of obtaining system parameters to permit communication with the system. Figure 5.25 shows that the Subscriber Unit first scans to determine if there is a CDMA radio channel by looking for the pilot channel. If no CDMA pilot channel is available on a CDMA control channel, the subscriber initializes using the analog control channel and the parameters it sends. After acquiring the pilot channel, the Subscriber Unit synchronizes with the system and updates its time parameters for channel decoding.

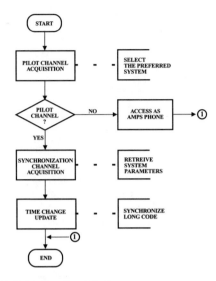

Figure 5.25, CDMA Initialization Mode

5.6.2 Idle

In idle mode, the Subscriber Unit monitors several different control channels to acquire system access parameters, and to determine if it has been paged, if it has received an order, or if the operator is initiating a call. Figure 5.26 illustrates idle mode call processing.

CDMA Subscriber Units monitor the paging channel for paging messages and system access information. If the Subscriber Unit is capable of discontinuous reception (sleep mode) and the system supports sleep mode, the Subscriber Unit turns off its receiver and other non-essential circuitry for a fixed number of burst periods. The system knows it has commanded the Subscriber Unit to sleep, and will not send pages designated for that Subscriber Unit during the sleep period. CDMA control channels on neighboring cells use different channel codes on the same radio channel frequency, so the Subscriber Unit will also simultaneously monitor neighboring radio control channels. If the Subscriber Unit finds a better control channel (higher signal strength or better bit error rate), it tunes to it.

The Subscriber Unit monitors the paging control channel to determine if it has received a page. If a call is to be received, a flag is set to indicate that the Subscriber Unit will enter access mode to respond. If the system sends the Subscriber Unit an order (such as a registration message), a flag will be set to indicate that the Subscriber Unit will perform an access attempt responding to

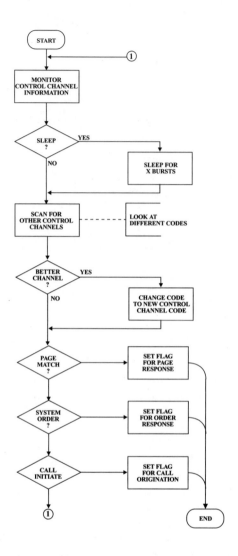

Figure 5.26, CDMA Idle Mode

the order. When a user initiates a call, a flag is set to indicate the access attempt is a call origination and that dialed digits will follow.

5.6.3 System Access Task

Subscriber Units randomly attempt access to the cellular network. To prevent multiple Subscriber Units from initiating accesses simultaneously, a seizure collision avoidance procedure has been developed. The contention resolution process consists of access class groups, gradual increase in access request power levels, random time delays, and maximum number of automatic access attempts.

Figure 5.27 illustrates a Subscriber Unit's access attempt. The dual mode Subscriber Unit first attempts access to the CDMA control channel. If the no access is available on the CDMA control channel, the Subscriber Unit attempts access on the AMPS control channel.

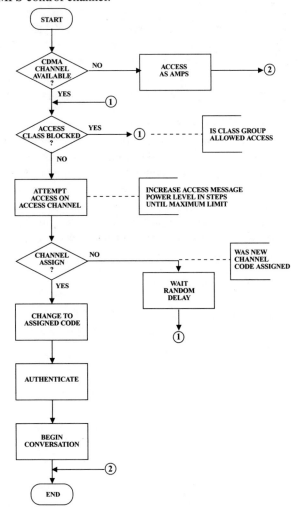

Figure 5.27, CDMA Access Mode

To limit the number of simultaneous access attempts, cellular systems can restrict access to specific groups of customers. Every Subscriber Unit is assigned one of 16 possible access classifications. The first ten classifications are random, and the remaining six are for emergency or high priority customers. Before an access attempt, the Subscriber Unit monitors the control channel to determine if its access class group has been restricted. If the access class is not blocked, the Subscriber Unit transmits an access burst on the access channel.

After transmitting an access attempt message, the Subscriber Unit listens to the control channel for a response message. If the access attempt is unsuccessful, the Subscriber Unit waits a random time before trying again. The random delay prevents repeated simultaneous access attempts from competing Subscriber Units. CDMA Subscriber Units vary their RF power during access attempts, beginning at a low power and increasing stepwise during sequential attempts until reaching a maximum. System parameters determine the initial and maximum access attempt power levels. To prevent overloading when many Subscriber Units constantly attempt access, the number of automatic attempts is limited.

The Subscriber Unit's initial request for service includes its identity information and the type of access requested (e.g. page response or call origination). If the system receives the information and has a radio channel available, the BTS sends an Initial Traffic Channel Designator (ITCD) message. If the Subscriber Unit does not receive the ITCD in a specified period, it delays a random amount of time, then attempts access again. The random delay prevents competing Subscriber Units from repeating simultaneous access attempts. Again, the number of automatic attempts is limited to prevent overloading when many competing Subscriber Units constantly attempt access.

5.6.4 Conversation Mode

Conversation mode call processing includes soft handoff selection, insertion and extraction of control messages, sharing between primary and secondary traffic channels, and monitoring the signal strength of other radio channels.

Figure 5.28 illustrates call processing for conversation mode. The Subscriber Unit continually receives bursts of data from the BS. If the Subscriber Unit is in soft handoff, it receives the same burst of data also identified by a different channel code. The Subscriber Unit compares the two (or more) bursts and selects the best one.

Chapter 5

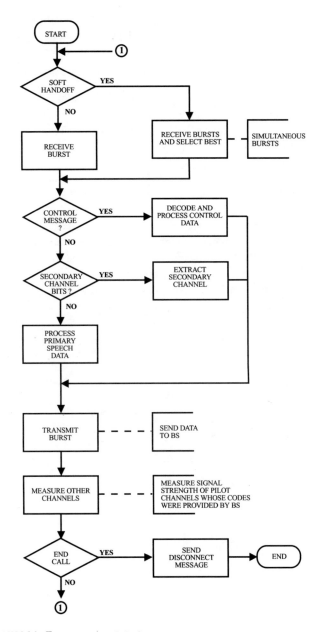

Figure 5.28, CDMA Conversation Mode

If the message is a control message, data will be extracted, decoded, and processed. Because control messages can be distributed over more than one slot (dim-and-burst signaling), several slots may be required to receive and decode the whole message.

If a secondary channel is in use (possibly a data channel), the bits for the secondary channel will be extracted, and the remaining bits will be sent to the speech coder for processing. The flag bits in the previous slot will indicate the rate of speech decoding.

The Subscriber Unit transmits its data bits at the same time it receives data, but the BS commands a slot period for transmitting relative to the received bursts. The BS also provides a pilot code list for neighboring cell sites. During conversation, the Subscriber Unit monitors the pilot channels of neighboring cells. When the signal strength of a neighboring cell's pilot channel is high enough to start soft handoff, the Subscriber Unit sends a message to the BS requesting soft handoff to begin. Conversation mode continues until the user or called party ends the call. The Subscriber Unit sends a disconnect message to the BTS confirming the end of the call.

References

1. CTIA, "User Performance Requirements," Issue 1, September 8, 1988.
2. Engineering manager, Qualcomm, personal interview, industry expert.
3. ibid.
4. Special Seminar, CDMA Panel, "CTIA Next Generation Cellular: Results of the Field Trials," Washington, DC, December 4-5, 1991.
5. ibid.
6. ibid
7. Engineer, Hughes Network Systems, personal interview, industry expert.
8. Special Seminar, CDMA Panel, "CTIA Next Generation Cellular: Results of the Field Trials," Washington, DC, December 4-5, 1991.
9. Engineering manager, Qualcomm, personal interview, industry expert.
10. Special Seminar, CDMA Panel, "CTIA Next Generation Cellular: Results of the Field Trials," Washington, DC, December 4-5, 1991.

Chapter 6
NAMPS (IS-88)

6. NAMPS (IS-88)

The IS-88 Narrowband Advanced Mobile Phone Service (NAMPS) standard combines Advanced Mobile Phone Service (AMPS) and Narrowband AMPS functionality. This chapter contains semi-technical descriptions of IS-88 NAMPS dual mode cellular technology. A description of the AMPS functionality associated with this IS-88 NAMPS system can be found in chapter 1.

NAMPS is an FDMA system commercially introduced by Motorola in late 1991. It is currently being deployed worldwide. Like the existing AMPS technology, NAMPS uses analog FM radio for voice transmissions. The distinguishing feature of NAMPS is its use of a "narrow" 10 kHz bandwidth for radio channels, about a third the size of AMPS channels. More of these narrow radio channels can be installed in each cell site, allowing NAMPS systems to serve more subscribers without adding cell sites. NAMPS also shifts some control commands to the sub-audible frequency range to facilitate simultaneous voice and data transmissions.

In 1991, the first NAMPS standard, named IS-88, evolved from the EIA-553 AMPS specification. The IS-88 standard identified parameters needed to begin designing NAMPS radios, such as radio channel bandwidth, type of modulation, and message format. During development, the NAMPS specification benefited from the narrowband JTACS radio system specifications. During the following years, advanced features such as ESN authentication, caller ID, and short messaging were added to the NAMPS specification. Figure 6.1 summarizes the evolution of the IS-88 specification.

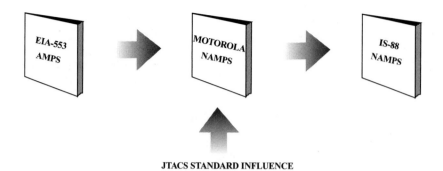

Figure 6.1, Evolution of the IS-88 Specification

One of the IS-88 system's greatest advantages arises from its use of AMPS system FM technology. Using this existing technology minimized changes in Subscriber Units and Base Stations, where other standards using new technologies have required extensive changes. The new Subscriber Units have dual mode capability, and operate either on existing AMPS radio channels or new NAMPS voice channels. NAMPS systems use an analog control channel very similar to that of existing AMPS systems.

6.1 System Overview

The IS-88 system has defined a new Narrowband Analog Voice Channel (NAVC) radio channel. The 10 kHz wide radio channel offers better sensitivity, noise rejection, and radio channel efficiency, and it can transfer simultaneous voice and data information.

NAMPS systems contain many of the same basic subsystems as AMPS cellular systems; the Mobile Switching Center (MSC), Home Location Register (HLR), Visitor Location Register (VLR), Base Station (BS), and Subscriber Units (called Mobile Stations (MS)). The NAMPS network also contains an Authentication Center (AuC) and Short Message Center (SMC). The MSC for NAMPS system connects Subscriber Units with other Subscriber Units or with networks such as the PSTN. The HLR and VLR are subscriber databases which list authorized services and create billing records. The AuC authenticates the Subscriber Unit's identity. The BS is the sum of all radio resources dedicated to a particular cell site. The Subscriber Unit (MS) is typically a mobile phone, but advanced paging and messaging services may allow unique types of Subscriber Units such as pagers.

In an IS-88 system, Subscriber Units may be one of two types: AMPS only or NAMPS IS-88 Dual Mode. AMPS phones use the AMPS Voice Channel (AVC) and AMPS Control Channel (ACC). IS-88 dual mode units use the ACC, AVC, and Narrowband Analog Voice Channel(NAVC). Figure 6.2 illustrates the IS-88 NAMPS radio system components.

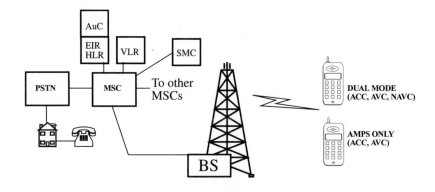

Figure 6.2, Overview of an IS-88 NAMPS Radio System

The IS-88 Dual Mode cellular system has three types of channels: Analog Control Channels (ACC), Analog Voice Channels (AVC), and Narrowband Analog Voice Channels (NAVC). The system uses the same ACC for AVC and NAVC channel assignment. The existing 30 kHz AMPS control channels are used for paging messages and access control. When a Base Station with NAMPS capability has established communications with a NAMPS Subscriber Unit, a NAMPS 10 kHz voice channel can be assigned if available. If no NAMPS channels are available, a 30 kHz AMPS channel will be assigned. Figure 6.3 illustrates IS-88 radio channel types.

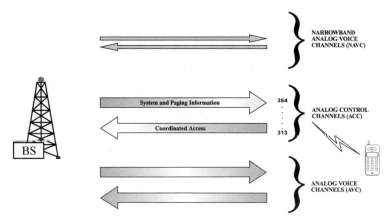

Figure 6.3, IS-88 Radio Channels

159

6.2 System Attributes

The NAMPS dual mode cellular system meets most of the CTIA UPR goals discussed in chapter 2 by maintaining the AMPS RF control channel structure, increasing the quality of voice transmission, providing cost effective capacity expansion, and maintaining coordinated system control. The dual mode system maintains compatibility with the established access technology by allowing existing analog (EIA-553) and new NAMPS dual mode Subscriber Units to use the same Analog Control Channels. To provide more advanced features and increase system capacity, some voice channels are converted to NAMPS. The new Mobile Reported Interference (MRI) feature maintains high voice quality by allocating radio channels based on radio channel quality. Cost effective capacity expansion is achieved by allowing more radio channels in a single cell site. Central control is provided by the Mobile Switching Center (MSC) as Subscriber Units move through the system.

6.2.1 Frequency Reuse

NAMPS systems allow frequencies to be reused much as AMPS systems do. Although NAMPS systems' narrow radio channels cannot tolerate as much interference as AMPS channels (see co-channel interference), adaptive channel allocation allows NAMPS radio frequency plans to be similar to AMPS frequency plans.

Using MRI information (discussed later), adaptive channel allocation can be assisted (sometimes called dynamic channel allocation) to increase the number of radio channel frequencies that can be reused in nearby cells. Adaptive channel allocation allows radio channel frequencies to be dynamically assigned as needed to cell sites. MRI can increase system capacity by assigning frequencies based on the interference levels detected by both the Base Station and the Subscriber Unit.

6.2.2 Sub-Band Digital Audio Signaling

A unique feature of NAMPS radio channels is sub-band digital audio signaling. In AMPS phones, an audio bandpass filter blocks the audio channel's lower range, but in NAMPS phones, a sub-band digital audio channel replaces the lower audio range (below 300 Hz) with digital information. Figure 6.4 shows how NAMPS sub-band digital and audio signals are combined.

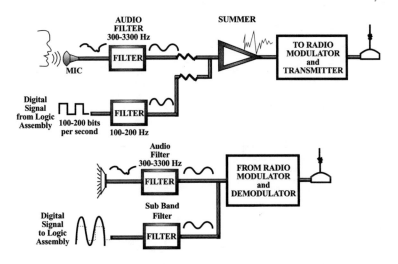

Figure 6.4, Digital Sub-band Signaling

6.2.3 Mobile Reported Interference (MRI)

NAMPS Subscriber Units can transmit radio channel quality information back to the Base Station to assist handoff decisions. This process, called MRI, sends channel quality information via the sub-band digital audio channel. Radio channel quality is measured using message parity bits to count the bits on the sub-band digital audio channel that are received in error. The Base Station sets a channel quality threshold level so that Subscriber Units can determine when signal strength and interference levels are unacceptable. When signal levels fall below the threshold, Subscriber Units inform the cellular system of poor radio channel conditions, and a handoff request is processed.

Figure 6.5 illustrates the MRI process. The process begins as the serving Base Station sends the minimum acceptable signal strength level to the Subscriber Unit. The information sets a minimum signal strength threshold in the Subscriber Unit (step 1). The Subscriber Unit continues to monitor the received signal strength indicator (RSSI) and the bit error rate (BER) of the sub-band digital signaling channel until the threshold is reached (step 2). The BER is an indicator of co-channel interference. When the threshold is crossed, the Subscriber Unit sends a single message to the Base Station indicating that received signal strength and BER rates are beyond tolerance. The Base Station then uses the information to assist in the handoff decision. If the Base Station wants another measurement from the Subscriber Unit, it sends a new message indicating a new threshold level.

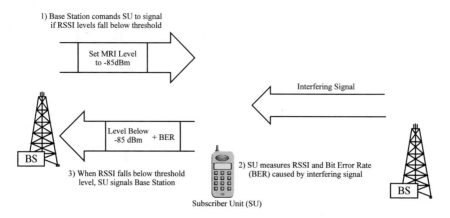

Figure 6.5, Mobile Reported Interference (MRI) Measurement

6.2.4 Capacity Expansion

Using NAMPS, the initial capacity expansion roughly triples the number of radio channels per cell site. For a typical cellular system, three NAMPS radio channels occupy the same frequency band as one AMPS radio channel. AMPS systems have 790 radio channels for voice communication (42 are used for control) while NAMPS has 2370 radio channels available for voice communication (3 times 790).

NAMPS further increases system capacity through the MRI feature's adaptive channel allocation. As demand for cellular service shifts from one area to another, MRI reassigns Base Station radio channel frequencies, deactivating a frequency in one Base Station and reactivated it in another. MRI determines interference caused by nearby Base Stations, and if interference is above a desired level, radio channel frequencies are reassigned.

6.2.5 RF Sensitivity

Narrow bandwidth reduces the noise entering the receiver and thereby increases the RF sensitivity to allow better signal reception in fringe coverage areas. The RF sensitivity increase is estimated between 1 to 4.8 dB [1,2]. Figure 6.6 shows how RF sensitivity is increased. The 30 kHz AMPS radio channel produces three times the thermal noise of a NAMPS channel. The receivers' intermediate filters (IF) reject noise outside the channel bandwidth and average RF signal energy remains the same, so in areas where signal strength is low, sensitivity is enhanced.

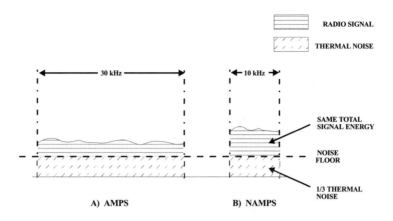

Figure 6.6, Increased NAMPS RF Sensitivity

6.2.6 Co-Channel Interference Rejection

Receivers that can reject co-channel interference can correctly receive a desired signal even when another radio on the same frequency is interfering. This ability to reject co-channel interference determines the distance between cells using the same frequency. Cell sites in most systems are typically surrounded by many other cell sites, so the ability to reuse frequencies determines a system's maximum capacity. The tracking of a desired signal in the presence of an unwanted signal is called capture effect, and it is important in rejecting co-channel interference. Narrowing radio channel bandwidth reduces the capture effect, and hence reduces the ability to reject unwanted signals. The reduced ability to reject unwanted signals results in decreased frequency re-use.

Figure 6.7 illustrates why AMPS receivers can reject a slightly higher level of co-channel interference than NAMPS receivers. Two radio channels are shown: channel A is an AMPS channel and channel B is a NAMPS channel. The receiver phase locks onto a radio signal. As the radio signal frequency changes, the receiver phase lock creates a correction voltage (the analog signal) which tracks the frequency. When an unwanted and sufficiently strong signal interferes, and the desired and undesired frequencies are close, the receiver can phase lock onto the unwanted signal. The unwanted signal might be co-channel interference from a nearby cell operating on the same radio frequency. A NAMPS radio channel is narrower than an AMPS channel, so for NAMPS channels, smaller changes in phase lock correction voltage are required, and the potential to lock onto an unwanted signal is increased.

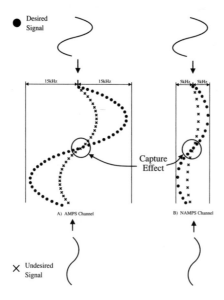

Figure 6.7, Co-Channel Interference Rejection

Modulation index is the ratio between audio signal and the amount of frequency change (deviation). NAMPS channels occupy 1/3 the AMPS channel bandwidth, but the NAMPS channel modulation index is 1/2 that of NAMPS. This comparatively small reduction of the modulation index is the result of two changes: eliminating the SAT and ST tones, and using the sub-band digital audio channel to transfer both types of control messages. The sub-band digital audio channel requires much less bandwidth than the SAT and ST signals.

6.2.7 Subscriber Unit Identity

The cellular system must view each Subscriber Unit separately and uniquely, and serve each subscriber with only the options specified. NAMPS systems uniquely identify Subscriber Units with a Number Assignment Module (NAM) and an Electronic Serial Number (ESN).

The NAM contains unique Subscriber Unit information such as its telephone number, home system identifier, access classification, and other subscriber features. The Subscriber Unit's 10-digit phone number is its Mobile Identification Number (MIN). The Subscriber Unit listens for its MIN on the paging channel to receive calls, and sends its MIN to transmit its identity. To help determine whether the Subscriber Unit is operating in its home system, the NAM contains a Home System Identifier (HSID) to compare with the one broadcast on the control channel. If the HSIDs do not match, the Subscriber Unit is not operating in its home system, and the control head displays ROAM to indicate changed billing rates (visited systems often charge a premium for service). The NAM also stores an Access Overload Class (ACCOLC) code (0-15) to selectively inhibit groups of Subscriber Units from transmitting when the system is busy. Higher level access classes are reserved for emergencies. Early Subscriber Units programmed NAM information into a standard Programmable Read Only Memory (PROM) chip. However, manufacturers soon made NAM information programmable via the handset keypad due to high costs for PROM chips and special programming devices.

Each NAMPS Subscriber Unit is created with a unique eleven-digit Electronic Serial Number (ESN). The first three digits represent the manufacturer, and the last eight are a serial number. The combined MIN and ESN uniquely identify a valid subscriber. However, duplicate MINs and ESNs are technically possible, so the NAM also stores information which is used during the authentication process (discussed in chapter 2).

6.3 Signaling

Signaling is the transfer of control information to and from the Subscriber Unit. All NAMPS voice channel signaling occurs on the sub-band digital audio channel.

6.3.1 Control Channel Signaling

Control channel signaling on a IS-88 system continues to be FSK (Frequency Shift Keying). IS-88 control channel signaling differs from that of EIA-553 primarily in its new short message capability and modified channel assignment command.

The NAMPS control channel can transmit signaling messages directly on the control channel without commanding the Subscriber Unit to tune to a voice channel. AMPS Subscriber Units cannot decode these signaling messages, so NAMPS messages do not affect their normal operations. The IS-88 system's initial voice channel designation command (channel assignment) contains two new bits of information about the assigned channel: one indicating if it is AMPS or NAMPS, and one indicating if it is above or below the AMPS channel center frequency.

6.3.2 AMPS Voice Channel Signaling

Operation on the Analog voice channel is identical to the original EIA-553 standard. The only differences in analog voice channel signaling are new handoff command parameters allowing a NAMPS voice channel assignment.

6.3.3 NAMPS Voice Channel Signaling

NAMPS voice channel signaling includes a Digital Supervisory Audio Tone (DSAT) to determine interference, a Digital Signaling Tone (DST) to indicate change in call status, and sub-band digital audio messaging to transfer control messages.

6.3.3.1 Digital Supervisory Audio Tone (DSAT)

Each cell site in a cellular system (or localized region of the system) has its own 24-bit unique digital supervisory audio tone (DSAT) code. DSAT codes are used to detect interference from neighboring cells and to indicate whether the Subscriber Unit is operating (off hook) on the radio channel. The DSAT is transponded back much like the SAT tone. Digital SAT has 7 unique codes which are transferred at a message rate of 200 bps. Figure 6.8 illustrates the Digital SAT Signaling Process. The Base Station is sending a 24 bit unique code (e.g. DAA4D4) on the forward channel (step 1). The Subscriber Unit receives

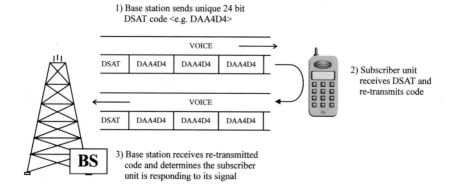

Figure 6.8, Digital SAT Signaling

the code and retransmits the same code (step 2). The Base Station receives the DSAT code to determine whether the Subscriber Unit is responding. If either the Subscriber Unit or Base Station do not receive the DSAT, or if they receive different DSATs, the audio will be muted. If the correct DSAT is not received after 5 seconds, communication will be terminated.

6.3.3.2 Digital Signaling Tone (DST)

The IS-88 standard replaces the AMPS Signaling Tone (ST) with a Digital Signaling Tone (DST) on the NAMPS channel. The DST indicates a change in call processing status. When a call is ended, a DST message is sent. To send a DST message, one 24 bit DSAT code is inverted. Figure 6.9 illustrates the DST

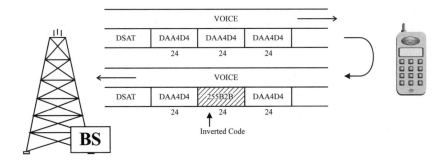

Figure 6.9, Digital Signaling Tone

signaling process. When the Subscriber Unit determines it must send a status change via DST, it inverts the DST received from the BS.

6.3.3.3 Sub-Band Digital Audio Messaging

Control messages on a NAMPS voice channel are sent via the sub-band digital signaling channel, eliminating blank and burst messages which momentarily mute the audio to send control messages.

Sub-band digital audio messages are composed of a synchronization word followed by the message word. A 30 bit synchronization word is sent at 200 bps to indicate that a message will follow, and a message word immediately follows the synchronization word at 100 bps. The message word is structured identically to messages sent on the wideband AMPS channel. Forward messages are 40 bits long with 12 BCH error check bits. Reverse messages are 48 bits long with 12 BCH error check bits. Figure 6.10 illustrates how the Subscriber Unit receives digital messages on the forward and reverse sub-band digital signaling channels.

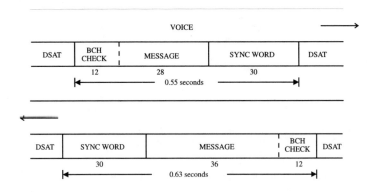

Figure 6.10, Sending Messages on the Sub-band digital audio channel

6.4 System Parameters

NAMPS systems allocate frequencies much as AMPS systems do, dividing AMPS radio channels into three NAMPS channels, using the same 45 MHz duplex channel separation, and maintaining the same output power levels. NAMPS radio channel modulation for voice continues to be FM. NAMPS signaling replaces SAT and ST with a sub-band digital audio channel. The amount of radio channel interference that NAMPS creates, and its tolerance to interference is different from AMPS.

6.4.1 Frequency Allocation

The Federal Communications Commission (FCC) regulates the frequency bands that US cellular carriers use. The 50 MHz cellular frequency band available in the US is shared between two cellular companies, called A and B carriers. Today, the US is divided into 734 cellular service areas, each with an A and a B carrier. The A carrier does not have a controlling interest in the local telephone company; the B carrier (often a Bell operating company) can have a controlling interest in a local telephone company. The A and B carriers are each licensed to use 25 MHz of radio spectrum. Each cellular carrier divides their 25 MHz into multiple frequency bands depending on the type of technology. For AMPS radio technology, each cellular carrier has 416 radio channels. Twenty-one of these channels are for control and paging, and 395 are for voice. For NAMPS systems, the 30 kHz AMPS radio channels can be divided into three 10 kHz narrowband

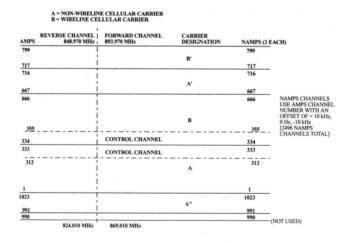

Figure 6.11, NAMPS Channel Frequency Allocation

channels which can provide as many as 2,496 10 kHz channels. The AMPS radio channel designation with an added frequency offset identifier identifies each NAMPS radio channel. The frequency offset can be 10 kHz above, at the center frequency (0 Hz offset), and ten kHz below. Figure 6.11 illustrates cellular channel allocations.

6.4.2 Duplex Channels

To support simultaneous transmission and reception (no need for push to talk), the Base Stations transmit on one set of radio channels (869 - 894 MHz), called forward channels, and receive on another set (824 - 849 MHz), called reverse channels. Forward and reverse channels in each cell are always separated by 45 MHz.

6.4.3 Modulation Type

RF Modulation converts an audio signal into a radio signal. The audio electrical signal containing the information to be transmitted is called the baseband signal. The RF carrier signal, called the broadband signal, transmits the information through the air.

The modulation method is FM (Frequency Modulation) for voice, and FSK (Frequency Shift Keying) for digital signaling. The mobile environment operates in the 800 MHz region, so audio processing is needed to enhance high frequency audio and minimize transients caused by 800 MHz signal fades [3].

The amount of NAMPS audio signal frequency deviation must differ from that of AMPS to allow NAMPS radio channels to be 1/3 the size of AMPS channels. The SAT and ST wideband data signals are not used for NAMPS radio channels.

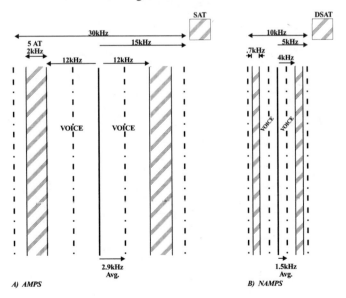

Figure 6.12, NAMPS Radio Channel Modulation Bandwidth

The functions of the SAT and ST have been incorporated into a new sub-band digital audio channel which transfers information at 100 to 200 bps. Figure 6.12 compares the bandwidths of AMPS and NAMPS radio channels.

6.4.4 Adjacent Channel Interference Guardbands

Three types of radio interference limit the maximum number of radio channels that can be used in a single cell: co-channel, adjacent channel, and alternate channel interference. Co-channel interference occurs when two nearby Subscriber Units operate on the same radio channel. Adjacent channel interference occurs when one radio channel interferes with a channel next to it (e.g. channel 412 interferes with 413).

Each NAMPS radio channel is only 10 kHz wide, but a low level of radio energy is transmitted outside the 10 kHz band. A Subscriber Unit or Base Station operating at high power can produce enough low-level radio energy outside the channel bandwidth to interfere with Subscriber Units operating on adjacent channels. A 30 kHz channel separation would protect NAMPS channels from adjacent channel interference, but AMPS channels may interfere with NAMPS channels that are only 30 kHz away. To prevent this interference, a 60 kHz guard band separates AMPS radio channels and NAMPS radio channels. Figure 6.13 illustrates the need for the AMPS-to-NAMPS guardband.

As with AMPS systems (discussed in chapter 9), NAMPS system radio channels are typically allocated among Base Stations or sectors to maintain a 21-channel separation from other radio channels in the same Base Station or sector.

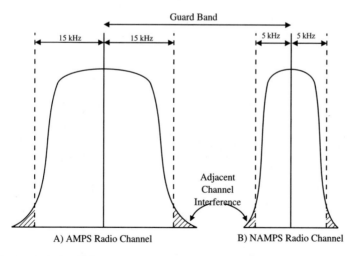

Figure 6.13, NAMPS Adjacent Channel Interference

6.4.5 Alternate Channel Interference

In a phenomenon called alternate channel interference, radio frequencies separated by two channels can interfere with each other. Figure 6.14 is a chart of allowable alternate and adjacent channel interference levels for NAMPS channels, called the bandpass spectral mask.

6.4.6 RF Power Classification

Subscriber Units are classified by maximum RF power output. NAMPS

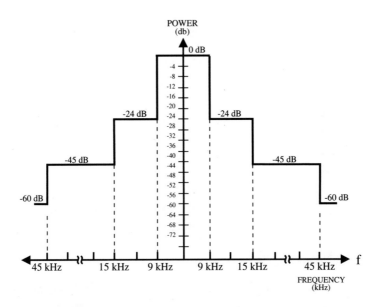

Figure 6.14, NAMPS RF Bandpass Spectral Mask

Subscriber Units have three classes of maximum output power. Class 1 maximum power output is 6 dBW (4 Watts), Class 2 is 2 dBW (1.6 Watts), and Class

RF Power	Class 1	Class 2	Class 3
Maximum Power	4 Watts	1.6 Watts	.6 Watts
Minimum Power	6 mW	6 mW	6 mW

Table 6.1
IS-88 Power Classification

3 is -2 dBW (.6 Watts). However, actual Subscriber Units' RF power outputs vary because Base Station commands adjust output in increments of 4 dB down to the minimum for all NAMPS Subscriber Units of -22 dBW (6 milliwatts). Table 6.1 lists the power classification types for the IS-88 radio system.

6.5 System Operation

NAMPS radio channels are divided into three groups: control channels, Analog Voice Channels (AVCs), and Narrowband Analog Voice Channels (NAVC). Control channels allow the Subscriber Unit to retrieve system control information and compete for access. Voice channels primarily transfer voice information, but also can send and receive digital control messages specific to the Subscriber Unit operating on the voice channel.

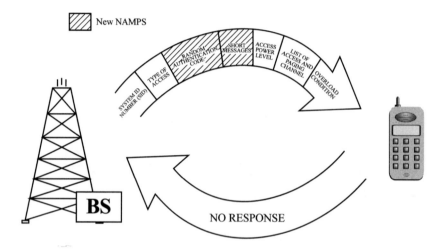

Figure 6.15, NAMPS Cellular System Broadcast Information

To initialize when it is first powered on, Subscriber Units scan a predetermined set of control channels and tune to the strongest one. In Initialization Mode the Subscriber Unit retrieves system identification and setup information. NAMPS systems can also send short messages and authentication information on the control channel. Figure 6.15 illustrates NAMPS system overhead system messages. After initialization, the Subscriber Unit enters idle mode and wait to receive its phone number (paging). In idle mode, it also listens for short messages (new) and waits for the user to originate a call (access). When a call is to be received or originated, the Subscriber Unit enters system access mode and attempts to access the system via a control channel. If access is successful, the control channel sends an Initial Voice Channel Designation (IVCD) message and sends a voice channel number to assign an AMPS or NAMPS channel. The Subscriber Unit tunes to the voice channel number in the message, and enters the conversation mode. As the Subscriber Unit operates on a voice channel, the system uses Frequency Modulation (FM) similar to the radio modulation used by commercial FM radio broadcast stations. When control messages are sent on the Narrowband Analog Voice Channel, the sub-band digital signaling channel (out of band signaling) transmits the digital message. As the Subscriber Unit moves out of range of one cell site radio coverage area, service is handed off to a radio channel at another nearby cell site.

6.5.1 Access

When a Subscriber Unit attempts to obtain service from a cellular system, the attempt is referred to as an "access" attempt. Subscriber Units compete to obtain access. An access attempt may begin either when the Subscriber Unit receives a command indicating that the system needs to serve it (a call to be received) or when the user places a call. As Subscriber Units randomly attempt access to the cellular network, the control channel serves as the access point to the system. To prevent multiple Subscriber Units from initiating access simultaneously, a seizure collision avoidance procedure has been developed. Four elements make up the process: the busy status of the channel, system response time interval, random time delays, and maximum number of automatic access attempts. For more detail on the collision avoidance procedure, see chapter 1.

Figure 6.16 illustrates the access process for originating a call. Initially, the user dials digits and presses the SEND button to originate the call. The Subscriber Unit sends the dialed digits and identification information in an access attempt message to the cellular system (step 1). If access is granted, the BS sends an Initial Voice Channel Designation (IVCD) message assigning the Subscriber Unit to an AMPS or a NAMPS voice channel, if available (step 2). If the attempt fails (the system may be busy), the Subscriber Unit waits a random time before sending another access attempt message. When access is granted, the Subscriber Unit tunes to the designated voice channel and authenticates with the cellular

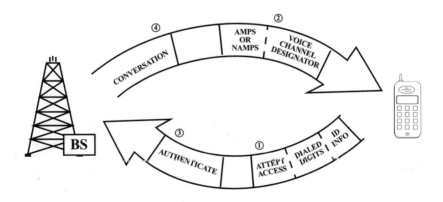

Figure 6.16, NAMPS Cellular System Radio Channel Access

system on that channel (step 3). When authentication succeeds, the conversation may begin (step 4).

6.5.2 Paging

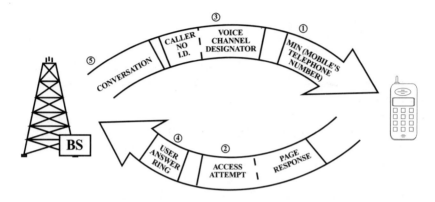

Figure 6.17, NAMPS Cellular System Paging

A call reaches the Subscriber Unit through the paging process. Figure 6.17 shows the NAMPS paging process. A page message is sent on the control channel with the Subscriber Unit's identification number (phone number) indicating that a call is to be received (step 1). If the Subscriber Unit is active in the system, it responds to the page by attempting access and indicating that the access request is in response to a page message (step 2). The cellular system then assigns an AMPS or NAMPS voice channel (step 3) with a message that may also contain a Calling Line Identifier (CLI) which displays the calling phone number to allow the user to decide if they want to answer. An alert message to activate the Subscriber Unit's ringer is also sent with the channel assignment command. If the subscriber answers the call (presses the SEND button), the Subscriber Unit transmits an acknowledgment message (step 4) to indicate that conversation can begin (step 5).

6.5.3 Handoff

When a Subscriber Unit moves far enough away from its serving cell, service must be transferred to a nearer cell in a process called handoff. For Subscriber Units using AMPS radio channels, the serving Base Station and adjacent Base Stations continuously monitor the Subscriber Unit's signal strength to determine when handoff is necessary. NAMPS Subscriber Units improve on this process by monitoring the serving radio channel's signal quality and returning that information to the Base Station on the sub-band digital signaling channel. The additional information assists in the handoff decision.

Figure 6.18 illustrates the NAMPS handoff process. In step 1, the Base Station sends a command containing a minimum acceptable signal strength level to the Subscriber Unit. The Subscriber Unit responds by beginning to continuously measure its received signal strength indicator (RSSI), called MRI measurements. When MRI measurements fall below the minimum acceptable signal strength level specified, the Subscriber Unit notifies the serving Base Station (step 2), which requests one or more adjacent Base Stations to tune to the Subscriber Unit's current operating channel and measure the signal strength. When an adjacent Base Station measures sufficient signal strength (step 3), the serving Base Station commands the Subscriber Unit to switch to (change frequency) that Base Station (step 4). After the Subscriber Unit starts communicating with the new Base Station, the communication link carrying the landline voice path is switched to the new serving Base Station to complete the handoff (step 5).

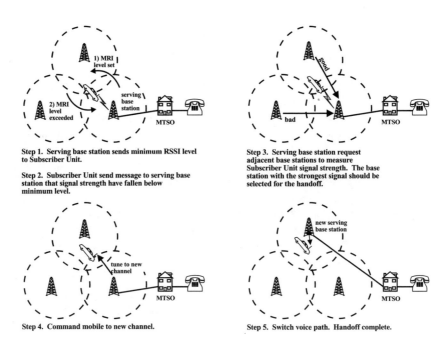

Figure 6.18, NAMPS Handoff Messaging

6.6 Call Processing

Call processing is the group of tasks performed to 1) initialize information when the Subscriber Unit is turned on, 2) monitor the control channels, 3) attempt access, and 4) coordinate transmissions of voice or data. A given sequence of processing for signaling messages must exist for a Subscriber Unit to operate correctly within a system. Functional operations are divided into four modes: initialization, idle, access, and conversation.

175

Cellular and PCS/PCN Telephones and Systems

In addition to these sequential tasks, other call processing functions occur in parallel. Timers are set and continuously monitored to permit Subscriber Unit operation. If a timer fails, the Subscriber Unit is not allowed to transmit.

NAMPS call processing is very similar to AMPS call processing. The method of message transfer differs (blank and burst for AMPS, sub-band digital audio for NAMPS), but the messages are the same. NAMPS also has a new simultaneous voice and data capability, MRI messaging, and some unique call processing steps.

6.6.1 Initialization

Initialization is the process of obtaining system parameters to permit communication with the system. System overhead information is continuously sent on control channels so that Subscriber Units can obtain parameters to establish communication and inform subscribers of system status. When the Subscriber Unit is first turned on, it scans a group of dedicated control channels and locks onto the strongest signal. From overhead messages on the control channel, the Subscriber Unit receives system parameters, such as System Identification (SID) and number of paging channels, and transfers them to its memory. If the received SID does not compare with the Home System Identifier (HSID) stored in the Subscriber Unit's Number Assignment Module (NAM), a ROAM status indicator is activated.

After obtaining system parameters, the Subscriber Unit monitors the control channel for overhead messages indicating changes in system parameters (e.g. system access power level), to obtain pages, and to determine if a call is to be initiated. Figure 6.19 illustrates idle mode call processing. The Subscriber Unit first reviews its stored information to select which system (A or B carrier) it will monitor. It then scans the control channels assigned to that system. The Subscriber Unit temporarily stores information from the system's overhead messages, including the System ID (SID) to update ROAM status, the number of access channels, the RANDom (RAND) access number for authentication, and other parameters. After initialization, the Subscriber Unit enters idle mode.

6.6.2 Idle Mode

In idle mode, the Subscriber Unit continually monitors overhead messages on the control channel for changes in system information. Such changes might include pages, a call is to be initiated, or a short message received. Figure 6.20 illustrates idle mode call processing. Overhead messages contain system information needed to access the system, which the Subscriber Unit temporarily stores. However, system information on the control channel can change, so the Subscriber Unit must continually receive and store it.

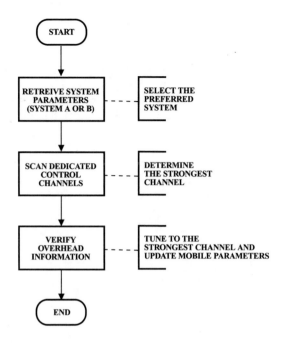

Figure 6.19, NAMPS Call Processing Initialization Mode

The paging channel is continually monitored to determine if an incoming call is to be received. Page messages contain the Subscriber Unit's MIN. If the Subscriber Unit has been paged, it sets up its response message (a flag) to indicate that it will attempt access as a response to a page message, and it enters access mode.

The Base Station may also command a Subscriber Unit to respond to system orders without the user's knowledge. Such commands include system registration and maintenance commands. When the Subscriber Unit receives an order, it sets up its response message (a flag) to indicate that it will attempt access as a response to a system order, and it enters access mode.

The Subscriber Unit also continually monitors the status of its control head (keypad) to determine if the user is initiating a call. If the user initiates a call, the Subscriber Unit sets up its response message (a flag) to indicate that it will attempt access to originate a call, and it enters access mode.

The NAMPS control channel can also send messages directly to a Subscriber Unit. The Subscriber Unit monitors the control channel short messages which contain its MIN. If a short message has been sent, the Subscriber Unit stores the message and alerts the user that a short message has been received. No access to the voice channel is required when short messages are sent on the control channel.

While monitoring a control channel during idle mode, the Subscriber Unit may move away from the serving cell site until the signal level falls below an acceptable level. It then scans for other control channels, and tunes to the strongest signal. During scanning (sometimes up to 3 seconds), the Subscriber Unit cannot receive pages or messages.

As cellular technology evolves, new messages (such as short message commands) may be created, and older Subscriber Units may not be able to understand them. If a Subscriber Unit does not understand a message, it disregards the

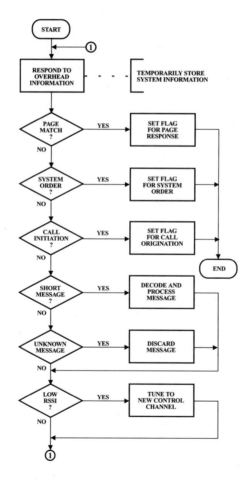

Figure 6.20, NAMPS Idle Call Processing Mode

message. This ability to respond only to messages that are understood will allow cellular systems to offer new features and services without causing older Subscriber Units to malfunction.

6.6.3 System Access

A Subscriber Unit attempts access to the system under four different conditions: when the user initiates a call; when a page (incoming call) is received; when a system order is received; and when the Subscriber Unit automatically registers with the system. The access attempt message indicates which of these types of access is required. Registration requests may be autonomous, such as when a Subscriber Unit enters a new system, or they may be required, such as when the Base Station requests a Subscriber Unit to indicate that it is operating in the system. Registering active Subscriber Units in each cell area limits paging requirements because paging messages are sent only on control channels of cells in the vicinity of the last registration.

Control channels may be divided into separate paging and access channels, where one control channel delivers pages and another coordinates access. When paging and access channels are separate, access channels can handle more service requests independent of paging requirements. However, assigning paging and access to different channels does not increase system capacity. The forward channel must carry all access signaling and the reverse channel carries all paging, leaving system capacity unaffected. As a result, paging and access are usually combined in one control channel. Figure 6.21 illustrates the steps in processing a call when a Subscriber Unit accesses the system.

The first step in system access is sending an access request message to indicate the type of access required. If the cellular system receives the access message, it returns an Initial Voice Channel Designation (IVCD) message to set the Subscriber Unit's new transmit channel. The Subscriber Unit confirms that the initial voice channel assignment is within the cellular system channel band, and that it can tune to that channel. If it cannot confirm, it aborts the access attempt.

If the access attempt is in response to a short message sent on the voice channel, the Subscriber Unit tunes to the voice channel, receives the message, alerts the user that a message has been received, and ends the access.

If the access attempt is to originate a call, the Subscriber Unit opens the audio path to allow the user to hear the called person's phone ring, and conversation can follow. If the access is in response to a page message, an alert message commands the Subscriber Unit to ring. If the user answers the ring (typically by pressing SEND), the ringer shuts off and the audio path opens to allow conversation.

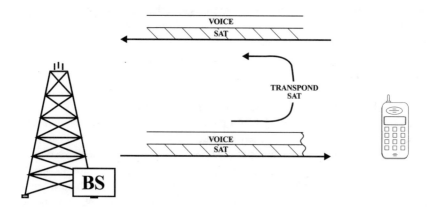

Figure 6.21, NAMPS System Access Call Processing Mode

When an access attempt is in response to a system order or a registration request, the Subscriber Unit accesses the system and indicates the type of order to process. After the order is processed, access ends.

6.6.4 Conversation

Conversation call processing occurs after an Initial Voice Channel Designation (IVCD) assigns a voice channel. The Base Station continues controlling the Subscriber Unit during conversation, including control of power level control, frequency handoff, and short messages. Control is accomplished either through blank and burst signaling on the AMPS channel or through sub-band digital audio signaling on the NAMPS channel.

Figure 6.22 illustrates the steps in the basic conversation mode call process. During conversation mode, the NAMPS Subscriber Unit continually extracts the digital sub-band audio data from the audio signal. The data includes a digital SAT (DSAT) code which uniquely identifies the signal to separate it from other nearby radio channels, and from other control messages.

To maintain a quality radio link during calls, a radio link timer monitors the DSAT code to detect interference and continuity loss. Initially, when the DSAT code is mismatched, an interfering signal may be present, and the Subscriber Unit mutes the audio so that the user will not hear the interfering audio. If DSAT code is continuously absent for 5 seconds, the DSAT timer expires, and conversation mode ends. Expiration of the timer indicates that the channel is unreliable and the Base Station can no longer control the Subscriber Unit.

During conversation, the Subscriber Unit continually monitors the radio channel quality, including measurement of the Radio Signal Strength Indicator (RSSI) and Bit Error Rate (BER). The Base Station may send the Subscriber Unit a message indicating a minimum channel quality level at which to operate. When signal quality falls below this level, the Subscriber Unit returns the signal quality measurement to the Base Station to indicate that handoff may be necessary.

Other commands such as power control, frequency handoff, and short messages may be received on the digital sub-band audio channel. When the Subscriber Unit processes these commands, it continues in conversation mode.

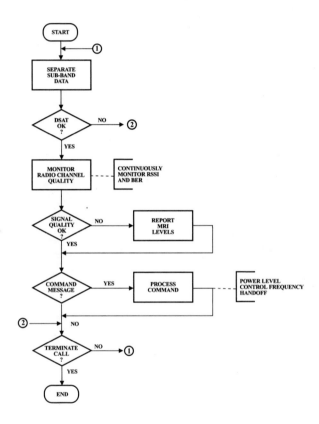

Figure 6.22, NAMPS Conversation Call Processing

Calls may be ended either by the user ending the call (pressing the END button) or by the other party hanging up. In either case, the Subscriber Unit sends a release channel signal Signaling Tone (ST) or Digital Signaling Tone (DST) to indicate that the call is finished.

References

1. Ericsson RF engineer, personal interview, industry expert.
2. Special Seminar, NAMPS Panel, <u>CTIA Next Generation Cellular: Results of the Field Trials</u>, Washington, DC, December 4-5, 1991.
3. The Bell System Technical Journal, p.110-114, January 1979, Vol. 58, No. 1, American Telephone and Telegraph Company, Murray Hill, New Jersey.

Chapter 7

Global System for Mobile Communications (GSM)

7. Global System for Mobile Communications (GSM)

Global System for Mobile Communications (GSM) is a digital cellular specification which was initially created to provide a single-standard pan-European cellular system. GSM began development in 1982, and the first commercial GSM digital cellular system was activated in 1991. Because it was created by representatives from many countries, it is accepted as the digital standard in more than 60 countries, including many countries outside Europe (see Appendix 4). This chapter contains semi-technical descriptions of GSM technology and its basic services.

Before GSM, most countries throughout Europe used their own unique cellular systems. Most subscriber equipment operated only on a single type of cellular system, so most customers could not roam to neighboring countries. With unique types of systems serving small groups of people, the mass production required to produce low-cost subscriber equipment was not feasible, so Subscriber Unit equipment, called a Mobile Station (MS) in a GSM system, costs remained high and early cellular systems enjoyed little success in the marketplace.

At the 1982 Conference of European Posts and Telecommunications (CEPT), the standardization body, Groupe Spéciale Mobile (GSM), was formed to begin work on a single European digital standard. The standard was later named Global System for Mobile Communications (GSM). This standard incorporated compatibility with Integrated Services Digital Network (ISDN) [1] and Signaling System Number 7 (SS7). In 1990, phase 1 of the GSM specifications were completed, including basic voice and data services. At that time, work began to adapt the GSM specification to provide service at 1800 MHz. This 1800 MHz standard, called DCS 1800, is used for the Personal Communications Network (PCN). Some GSM systems began operating in 1991, and by 1992 most major European GSM commercial systems were operating. Phase 2 of the GSM and DCS 1800 specifications, which added advanced short messaging, microcell support services, and enhanced data transfer capability, are now complete. Figure 7.1 shows how the GSM and DCS 1800 specifications evolved.

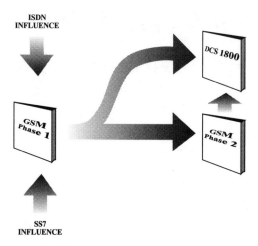

Figure 7.1, Evolution of the GSM and DCS 1800 Specifications

GSM standards development criteria included high spectral efficiency, subjective voice quality, reasonable Subscriber Unit and system equipment costs, the possibility of hand portables, new advanced services, and coexistence with existing systems [2]. The GSM digital cellular system meets these goals by providing a universal (global) digital standard, consistent quality of voice transmission, new services, cost effective capacity expansion, and a coordinated system of control. The digital voice quality is maintained even in hostile radio conditions through digital transmission. New services such as short message transmission have been added to provide for future services. GSM has also achieved its cost effective capacity expansion goal by allowing multiple subscribers to simultaneously use a single 200 kHz radio channel (carrier) and a single Base Transceiver Station. The goal of a coordinated system control was achieved through a subscriber identity module and coordinated SS7 connections.

Distinguishing features of GSM include a combined digital radio control and voice channel, slow frequency hopping, advanced data features, portable subscriber identity modules, and mobile assisted handover. GSM uses a single 200 kHz wide radio channel (called a traffic channel) as both control channel and voice channel. The radio channel is divided into groups of 8 Time Division Multiple Access (TDMA) time-slots per frame. Slots can be used either as control slots or as voice/data slots, thus simplifying Subscriber Unit and BTS radio design. Slow frequency hopping is the changing of frequencies during a data transmission or voice conversation. Slow frequency hopping adds frequency diversity to reduce the effects of signal fading and average interference from neighboring cells. GSM can also distinguish between different types of services such as voice, data, and facsimile. By supporting multiple types of information transfer, the radio channel and cellular system network equipment can be configured to enhance the information transfer rate. A portable subscriber identity module (SIM) card can be used in other GSM Subscriber Units, thereby identifying the subscriber's information to the system, not the Subscriber Unit's information. Mobile assisted handover in the GSM system assists the handover process by furnishing radio signal strength information back to the cellular system from the Subscriber Unit. The handover process also has rapid channel synchronization which further facilitates handover.

7.1 System Overview

Figure 7.2 illustrates GSM system components. The GSM system contains many subsystems, including a Mobile Switching Center (MSC), Home Location Register (HLR), Visitor Location Register (VLR), Base Station Controller (BSC), Base Transceiver Station (BTS), and Subscriber Units. The GSM network also contains an Equipment Identity Register (EIR) and an Authentication Center (AuC). The MSC for GSM systems connects Subscriber Units and other Subscriber Units or networks such as the PSTN. The HLR and VLR are subscriber databases which list authorized services and create billing records. The EIR checks the status of the mobile equipment for lost or stolen units. The AuC validates the subscribers identity via the SIM card (discussed later). The BSC coordinates radio resources in its geographic area. The BTS is a radio resource dedicated to a particular cell site. The combination of the BSC and BTSs is called the Base Station Subsystem (BSS). The Subscriber Unit is typically a mobile phone, but advanced paging and messaging services may allow other types of Subscriber Units such as pagers.

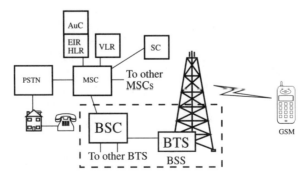

Figure 7.2, Overview of a GSM System

The GSM cellular system has one type of radio channel called a traffic channel (TCH) which is divided into frames and time slot periods. The traffic channel carries voice, data, and control information. From each frame, each user is assigned to a particular time slot for reception, and a particular time slot for transmission. Some slots in the TCH transfer control channel information, and some transfer voice channel information. Figure 7.3 illustrates the GSM radio channel and its division into voice channels and control channels.

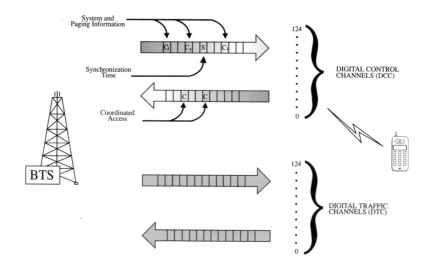

Figure 7.3, GSM Radio Channels

7.1.1 Speech Coding

To allow several users to share a single 200 kHz wide radio channel, voice signals are digitized, analyzed, and compressed before transmission. The conversion and compression process is called speech coding. Figure 7.4 illustrates the GSM speech coding steps. The analog voice signal is sampled 8,000 times each second and digitized into a 64 kbps digital signal. This 64 kbps digital audio burst is divided into 20 msec groups which are supplied to a GSM speech coder for compression[1]. The GSM speech compression process uses Regular Pulse Excitation-Long Term Prediction (RPE-LTP) to convert the 64 kbps PCM audio signal into a 13 kbps compressed voice signal [3]. Because radio channels can significantly distort information, error protection bits are added to the most important infor-

notes:
[1] A Subscriber Unit manufacturer may choose to directly digitize the audio at 13 kbps which eliminates the need to digitize at 64 kbps.

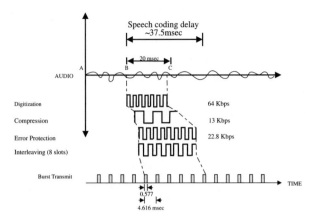

Figure 7.4, Speech Data Compression

mation of the compressed digital audio. The error protection bits increase the amount of data to 22.8 kbps. To help protect the data from short periods when the radio signal is poor (called rapid fading), the 22.8 kbps error protected data is interleaved (distributed) over 8 adjacent slot periods. These slot periods are assigned to 8 radio transmit bursts. While the basic processing delay of the speech coding process is approximately 37.5 msec, the total system delay is longer.

In a process is called Discontinuous Transmission (DTX), the GSM speech coder can stop transmitting the digital voice signals when speech activity is low. When the speech coder senses no speech activity, it changes the speech coding process to compress only background noise. The speech coder uses Voice Activity Detection (VAD) to determine if it is to transmit background noise only. When the VAD indicates low voice activity, the speech coder compresses background noise only at a data rate of about 500 bps. The compressed blocks, which contain the background noise characteristics, are called Silence Descriptor (SID) frames. When the SID frame blocks are received, they are expanded to re-create the background "comfort noise." Comfort noise prevents sudden disturbing changes in perceived sound characteristics when the caller stops talking. The speech coder sends a block of background noise at the beginning of low voice activity, then updates the background noise generator each 480 msec afterwards.

One of the greatest concerns in any digital voice transmission system is the performance of the speech coder. If the speech coder does not accurately code and decode voice data, subscribers notice the inaccuracies as noise or errors. Performance of the GSM speech coder performs well in poor radio conditions.

The mathematical operation of the GSM speech coder completely standardized in every detail, and it is therefore identical in each phone and system. This uniformity eliminates requirements for conformance testing of different manufacturers' equipment.

7.1.2 Radio Channel Structure

A GSM radio channel (carrier) is divided into logical channels and time periods. Time slots are the smallest individual time period available to each Subscriber Unit. The smallest frame is a 4.615 msec TDMA frame. Each time slot normally contains 148 bits of information. Of the 148 bits, some are dedicated as data, and others are dedicated as control.

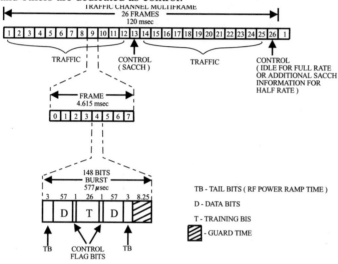

Figure 7.5, GSM Radio Channel Structure

Each radio channel contains several different multiframe time periods. 26 TDMA frames are combined to form traffic channel multiframes. The traffic channel multiframe allows individual call control information (e.g. power level control) to be shared with voice information. Frames 13 and 26 of each traffic channel multiframe are dedicated as control frames. Frame 13 is dedicated for slow control messages and frame 26 is unused for full rate speech coding applications. Figure 7.5 shows the basic GSM radio channel structure.

Multiframes are combined to form superframes and hyperframes. A superframe is composed of 51 multiframes (6.12 seconds). The hyperframe is the largest time frame in the GSM system, and is composed of 2048 superframes (approximately 3 1/2 hours). During a hyperframe period, every time slot has a unique sequential number composed of frame number and time slot number.

GSM systems allow several users to share each radio channel carrier frequency by dedicating a specific time slot from each frame to individual users. Voice channels can be either full rate or half rate. Full rate GSM systems assign one time slot per frame to each user, allowing 8 users to simultaneously share a radio channel. Half rate GSM systems assign one time slot every other frame to allow up to 16 users to share a radio channel.

Full Rate GSM

Subscribers talk and listen at the same time, so the Subscriber Unit must function as if it is simultaneously sending and receiving (called full duplex). When in conversation mode (called dedicated mode), GSM Subscriber Units do not transmit and receive simultaneously, but only appear to do so. To enable full duplex transmission, GSM voice channels are composed of different radio frequencies. One frequency is for transmitting from the Subscriber Unit; the other is for receiving to the Subscriber Unit. Speech data bursts alternate between transmitting and receiving, and when received, the compressed speech bursts are expanded in time to create a continuous audio signal.

During a full rate voice conversation, one time slot is dedicated for transmitting, one for receiving, and six remain idle. The Subscriber Unit uses the idle time slots to measure the signal strength of surrounding channels. These measurements assist in channel selection and hand-off. This time sharing results in a user-

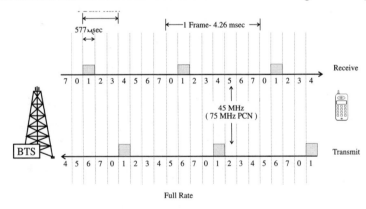

Figure 7.6, GSM Full Rate Radio Channel Structure

available data rate of 22.8 kbps. Because some of the 22.8 kbps are used for error detection and correction, so only 13 kbps of data are available for compressed speech data, or 12 kbps for data information.

Figure 7.6 shows how GSM full duplex radio channels are divided in time. GSM digital radio channels transmit on one frequency and receive on another frequency 45 MHz higher (75 MHz for PCN), but not at the same time.

189

Half rate GSM

The number of simultaneous users a radio channel can serve can be doubled by dedicating only one slot every other frame per subscriber, creating a half rate channel. Half rate channels use 1 of sixteen slots to transmit and 1 to receive, leaving 14 idle. Figure 7.7 illustrates the half rate GSM channel structure. Using

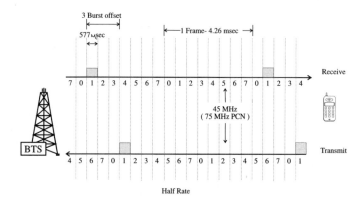

Figure 7.7, GSM Half Rate Radio Channel Structure

only one of the sixteen time slots results in a user-available data rate of about 11.4 kbps. A half rate system supports up to sixteen simultaneous users per radio channel.

Control Channels

GSM systems use digital control channels which co-exist with standard GSM digital voice channels on the same RF channel carrier. Digital control channels use one or more time slots on any of the available radio frequencies. Multiple types of control channels exist, and several of these can co-exist in the same cell site. Control channels send paging messages, coordinate access, and provide for the overall system control.

7.2 System Attributes

Some of the GSM system's unique attributes include enhanced frequency reuse, time division multiple access, dynamic time alignment, slow frequency hopping, multi-function traffic channels, discontinuous reception (sleep mode), RF power control, mobile assisted handover, multipath equalization, multiple slot structures, short message services, and subscriber identity modules (SIMs).

7.2.1 Frequency Reuse

GSM uses a wider radio channel than its analog predecessors, and it uses digital modulation to better reject interfering radio signals. These factors allow the radio channel frequencies to be reused more often than with earlier narrowband analog systems. In real application, the GSM radio channel can tolerate an interfering signal up to 20% of (7 dB below) the desired signal. By comparison, analog signals can only tolerate interfering signals from 1.6% to 6.3 % of (18-12 dB below) the desired signal [4].

Slow frequency hopping and mobile assisted handover information (discussed later) provide interference averaging and dynamic channel allocation. These additional factors further increase GSM system frequency reuse.

7.2.2 Time Division Multiple Access

Each GSM radio channel allows up to 8 simultaneous full rate users to share a single radio channel and up to 16 users for half rate systems for voice communications. Many more users could simultaneously share a single radio channel when low speed data communication rates are used. The TDMA access technology also allows the sharing of control, voice, and data communications on a single type of radio channel.

7.2.3 Dynamic Time Alignment

A finite amount of time is required for radio waves to propagate from the Subscriber Unit to the Base Transceiver Station. Subscriber Units transmit in bursts, so the BTS must receive all bursts in proper time sequence without any overlap. Without time alignment, a Subscriber Unit operating close to the BTS could overlap bursts with a Subscriber Unit operating far from the BTS. To compensate for this effect, commands sent from the BTS adjust Subscriber Units' relative transmit time to align the times their signals arrive at the BTS.

For dynamic time alignment to function properly, the BTS must determine how much offset time to use. Initially, the Subscriber Unit transmits a shortened burst (access burst) until the Base Transceiver Station can calculate the required offset time (timing advance). The required timing offset is twice the path delay, combining the downlink delay (Subscriber Unit receive) and uplink delay (Subscriber Unit transmit). The Subscriber Unit uses it's received time slot to determine when its burst transmission should start. The amount of dedicated guard time between adjacent time slots is 8.25 bits (30 usec). This allows a distance of only 4.5 km before bursts overlap. When the distance from the cell site exceeds 4.5 km, timing must advance to ensure the transmit burst does not overlap with the adjacent time slot. The timing can be advanced in 1/2 bit steps up to 237 usec.

This 237 usec maximum without slot collisions limits the distance the Subscriber Unit can operate from the cell site to about 40 km. Figure 7.8 illustrates the dynamic time alignment process.

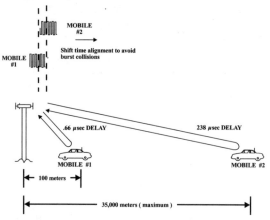

Figure 7.8, Dynamic Time Alignment

Path propagation delay = 3.33 usec/km
  Round trip delay = 6.66 usec/km
  One bit = 3.7 usec
  1 km requires 1.8 bits
  The standard guard time is 8.25 bits = 4.5 km
 Maximum distance = 4.5 km + (0.56 km/bit * 63 bits) = 40 km

Because the GSM system continuously transmits timing information on the control channels, Subscriber Units can perform some time alignment independent of the system time alignment commands sent by the BTS. While operating on a traffic channel (dedicated mode), the Subscriber Unit can monitor other radio channels between receive and transmit burst periods to capture the system time information from the Frequency Correction Channel (FCCH) and Synchronization CHannels (SCH) channels. This information provides a relative time shift between other cell site radio channels, which the Subscriber Unit can temporarily store in memory. Time alignment to other radio channels allows the Subscriber Unit to self-synchronize to a new BTS radio channel before receiving a handover command, greatly assisting handover.

If the Subscriber Unit is farther than about 40 km from the cell site, bursts received at the BTS will begin to overlap with the next adjacent time slot, even if the time advance is at it's maximum setting. If the adjacent time slot is not assigned to another user, it is possible for the BTS to receive the Subscriber Unit transmit burst in the adjacent time slot. This procedure increases the radius of the

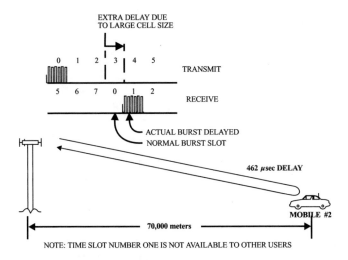

Figure 7.9, Extended Dynamic Time Alignment

cell site to more than 75 km. If additional time slots are unused (delayed relative to the assigned time slot) and the BTS can receive the delayed signal, a Subscriber Unit could operate even farther from the cell site. Figure 7.9 illustrates the extended cell dynamic time alignment process. The Subscriber Unit is assigned to Time slot Number (TN) 0. Due to the excessive delay, even with the timing advance set to a maximum of 237 usec, the burst transmitted from the Subscriber Unit is received at the BTS in TN 1. Because TN 1 is not assigned to another subscriber, the BTS can receive the Subscriber Unit's transmitted burst partially in the adjacent slot (TN 1).

7.2.4 Slow Frequency Hopping

Instantaneous radio signal fading and co-channel interference is often limited to particular narrowband radio channels at a specific time. GSM slow frequency hopping provides for the frequency diversity and interference averaging needed to overcome some of the narrowband channel signal distortion.

Frequency diversity is created by allocating bursts of data over time slots on different radio channel frequencies. Because radio signal fades (called a Rayleigh fades) are usually limited to a narrow radio band of less than 1 MHz, radio channels separated by more than 1 MHz are typically not faded and can still be received [5]. Frequency hopping with bit interleaving distributes the effects of the signal fading over many bursts.

Cellular system frequency plans can be designed for a worst-case situation, placing nearby cells using the same radio channel frequencies at a distance that eliminates interference. The acceptable Carrier to Interference Ratio (C/I) occurs when the desired signal is more than 5 times (7 dB) all interfering signals. Interference from nearby cells varies with Subscriber Unit location, activity, and power levels which are dynamically changing, so in many situations, there is very little interference from other cells. With interference averaging, frequency plans need not be designed for the worst case, and frequencies can be re-used in nearby cell sites more often than they could be without frequency hopping. The greater frequency re-use increases the system capacity.

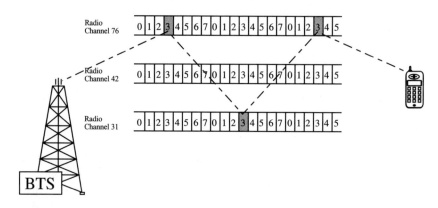

Figure 7.10, Slow Frequency Hopping Example (Symbolic)

GSM's 125 radio channel possibilities (375 for PCN) and 8 slots per channel, create a nearly infinite number of hopping possibilities. To create the hopping sequence pattern, the BSS assigns two parameters: Mobile Allocation Index Offset (MAIO) and Hopping Sequence Number (HSN). These two variables combine to form the hopping pattern. The result is a hopping rate of 217 hops per second for a specific Subscriber Unit, the same as the TDMA frame rate. Figure 7.10 illustrates a sample of frequency hopping.

7.2.5 Multi-Function Traffic Channels

GSM uses a single type of radio channel to multiplex (time share) control, voice, and data channels. The radio channels are divided into logical channels, and these channels' information format (e.g. fax or data) is defined. Control channels transfer broadcast, paging, and access control. Traffic channels transfer voice and data (e.g. fax) information.

Each GSM Subscriber Unit can measure the quality of nearby cell sites' control channels. Control channel selection is the process of finding neighboring control channels to determine if they are candidates for monitoring. The selection process consists of two steps. First, the Subscriber Unit measures the signal

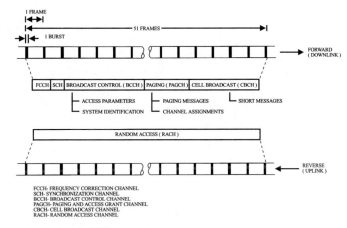

Figure 7.11, GSM Control Channel Structure

strength of the current control channel and the control channels of the neighboring cells. Then, the Subscriber Unit uses this quality indication to select an optimal control channel. This is accomplished by measuring the signal strength and number of errors received while decoding channel burst information.

The control channels have a 51 frame multiframe structure. Several types of control channels share the multiframe structure. Of the 51 bursts, some are used to broadcast system parameters, some are used to page and assign radio channels, and optionally, some are used to broadcast short messages.

The traffic channels (voice and data) have a 26 frame multiframe structure. In a

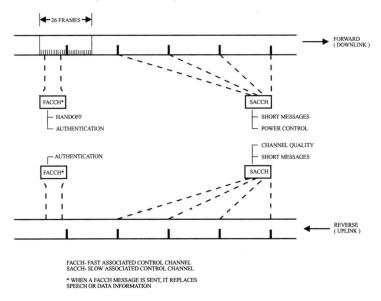

Figure 7.12, GSM Traffic Channel Structure

single traffic channel multiframe, time slot numbers 13 and 26 are dedicated as control bursts. Slot number 13 transfers control messages, and slot 26 remains unassigned for full rate traffic channels. The control bursts carry the Slow Associated Control CHannel (SACCH) used to transfer information that is sent continuously to the BTS, such as channel quality information. To send control messages quickly, some of the user data (speech bursts) can be replaced with control commands. This unscheduled replacement of the user speech bursts is called the Fast Associated Control CHannel (FACCH). Figure 7.12 illustrates traffic channel structure.

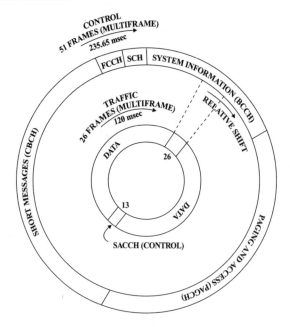

Figure 7.13, GSM Control and Traffic Channel Offset Cycle

Because control channel information time periods (large multiframe periods) are different from speech channel multiframe periods, eventually, other SCH can be monitored during the idle periods while operating in dedicated (conversation) mode. Storing the relative synchronization information of other radio channels allows the Subscriber Unit to pre-synchronize to those channels prior to handover. Figure 7.13 illustrates the variation in control channel and traffic channel cycles.

7.2.6   Discontinuous Reception (Sleep Mode)

Discontinuous Reception (DRX) allows the Subscriber Unit to power off non-essential circuitry (sleep) during periods when pages will not be received. To provide for sleep mode, the paging channel is divided into paging sub-channel groups. The number of the paging sub-channel is determined by the last digits of the IMSI. The system parameter information sent on the BCCH identifies the grouping of paging sub-channels

Figure 7.14 shows the DRX (sleep mode) process. Initially, the BCCH channel indicates to the Subscriber Unit which multiframes contain paging and access blocks, and which contain sub-paging classes. Subscriber Units can sleep during

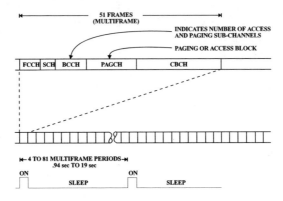

Figure 7.14, Discontinuous Reception (Sleep Mode)

multiframes which are not part of its paging sub-channel. Paging messages are contained in a PAGCH block of the control channel multiframe. The PAGCH block contains either all page messages or channel assignment messages. The paging blocks can be divided into 9 sub-paging channel groups. Alternating between paging and access channel blocks, and dividing into paging sub-channels allows sleep intervals to range between 4 and 81 multiframes. These intervals translate to sleep periods ranging from .94 second to 19 seconds maximum.

7.2.7 Power Control

Power control provides four important advantages: 1) it minimizes changes in base transceiver receiver RF signal strength from slot to slot; 2) it minimizes interference to nearby cell sites operating on the same radio channel 3) it increases battery life for portable Subscriber Units; and 4) it reduces out-of-band radiation [6]. Subscriber Units and BTSs require only enough radio energy to maintain a quality radio link. Minimizing the transmitted energy allows the same radio frequencies to be re-used in nearby cell sites with less interference. Base transceivers and Subscriber Units have several maximum power level classifications, and a minimum receive level of -104 dBm (-102 dBm acceptable for hand portables). Both the Subscriber Unit and the BTS can adjust their transmit output power level in 2 dB steps. Commands received from the BTS can adjust the

Subscriber Unit up to 15 steps below its maximum transmit power. Subscriber Units can only change power level 2 dB every 60 msec (13 bursts for full rate). Base station transmitters can adjust their output power level down to minimum of 13 dBm (20 mWatts). Because the RF transmitter in a Subscriber Unit typically consumes the most power, transmitting at the lowest output power to maintain a quality radio link also extends battery life.

GSM radio channels must co-exist with other GSM channels and radio systems (such as TACS and NMT) operating on nearby channel frequencies. Although the GSM transmitted signal uses Gaussian Minimum Shift Keying (GMSK) modulation, which does not produce significant radio energy outside the designated bandwidth, the cycling (rapid changes) of the RF energy between a high transmit level and an off period results in unwanted radio signals. The radio signal ramp up and ramp down specification control (during the tail bits in figure 7.5) limits the amount of spurious emissions by gradually increasing and decreasing the radio signal as the burst is transmitted.

7.2.8 Mobile Assisted Handover

Mobile assisted handover is a feature that helps the system determine when a handover to another radio channel is needed. Existing analog systems relied entirely upon receivers in the base stations to measure the signal strength of mobiles and determine when a handover was needed. GSM Subscriber Units continually return radio channel quality information to the BTS during conversation (dedicated mode).

Figure 7.15 illustrates mobile assisted handover. The Subscriber Unit initially receives a radio channel list of nearby cell sites from the BCCH channel (step 1). During idle periods (after the transmit burst and before the receive burst), the Subscriber Unit monitors up to 6 neighboring BTS control channels for signal

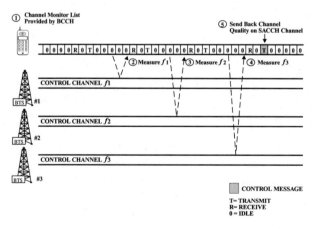

Figure 7.15, Mobile Assisted Handover

strength on all channels and bit error rate (steps 2-4). The Subscriber Unit returns the channel quality information to the BTS on the SACCH channel (step 5). The system uses the channel quality measurements provided by the Subscriber Unit and other signal quality information provided by the BTS to determine when the system will initiate a handover.

7.2.9 Multipath and Equalization

Multi-path signals occur when the same source RF signal takes different propagation paths to arrive at the receiver at a different times. This results in the reception of one or more delayed copies of the same signal which can cause symbol amplitude and phase distortion.

Multipath can be a significant problem or an asset for GSM, depending on whether the delayed signal can be extracted from the composite (combined)

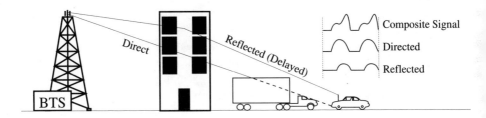

Figure 7.16, Multipath and Equalization

received signal. A radio signal travels a km in 3.33 usec, so a reflected signal (ray) traveling only 500 additional meters is received a half bit delayed. Figure 7.16 shows how a radio signal can travel two different paths to a Subscriber Unit. In this example, the direct path is attenuated while the indirect (reflected) is not attenuated. The resulting signal is the combination of the two rays.

GSM has relatively short bit intervals, requiring adaptive equalization. Adaptive equalization uses the characteristics of a known received signal to separate multipath signals from an unknown received signal. The GSM equalizer must be able to extract multipath signals with delays of up to 16 usec [7].

Equalization must also compensate for the doppler shifting of the burst's frequency as a result of the Subscriber Unit's movement. The maximum amount of doppler shift expected for the GSM system was calculated using a maximum speed of 250 km/hour which was based on the speed of the French train "A Grande Vitesse [8]."

7.2.10 Slot Structure

Four types of slot (burst) structures exist in the GSM standard; normal burst, access burst, S burst, and F burst. Each burst has dedicated fields which include data bits, training sequence bits, control bits, and guard bit intervals.

Blocks of data (such as a speech block) are interleaved (distributed) over several bursts. Interleaving overcomes the difficulty of short groups of consecutive bit errors due to Rayleigh fading. Speech data is continuously interleaved over 8 slots. Diagonal interleaving (odd and even bits) distributes information between adjacent slots to reduce the impact of temporary fades.

Normal Burst

The forward data slot transfers voice and data from the BTS to the Subscriber Unit. It contains a total of 148 data bits, of which 114 are available to transfer data. Two of the bits in the data block (one bit for each 57 bits of data) are dedicated as a flags to distinguish user data from FACCH signaling messages. The middle field in the slot is the synchronization (training) field which provides a standard pattern "training" to accommodate equalizer adaptive calibration. The equalizer adjusts the receiver to compensate for radio channel change (distortion). The data fields carry the subscribers voice and data information. Figure 7.17 shows the bit structure of the normal burst.

T - TAIL BITS
DATA - USER INFORMATION BITS
TRAINING - TRAINING BITS
C - FLAG BITS (DATA vs. FACCH)

Figure 7.17, Normal Burst Slot Structure

Access Burst

Access bursts are primarily used on the access channel allows Subscriber Units to randomly request service from the BTS. When a Subscriber Unit begins operating in cell, the propagation time between the Subscriber Unit and BTS is typically not known. In large cell areas, the propagation time could be so long that bursts could overlap, causing a significant problem. The access burst contains fewer bits, which allows for additional guard time and prevents received bursts from overlapping until dynamic time alignment adjusts delays.

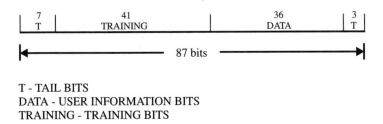

T - TAIL BITS
DATA - USER INFORMATION BITS
TRAINING - TRAINING BITS

Figure 7.18, Access Burst Slot Structure

The access burst has a larger number of training bits to allow rapid synchronization of the BTS. Once the BTS successfully receives the access burst, it can command the Subscriber Unit to adjust frequency and transmit time. The Subscriber Unit then uses the normal burst slot structure to send user information on the newly-assigned radio channel. Access bursts are also used during the handover process. Figure 7.18 illustrates the bit structure of the access (shortened) burst.

S Burst

Synchronization bursts are sent on the Synchronization CHannel (SCH), which is part of the common control channel. All Subscriber Units use this channel to initially synchronize to the common channel after they detect the FCCH. S bursts contain a long training sequence to assist the Subscriber Unit in synchronization. Figure 7.19 shows the bit structure of the S burst.

T - TAIL BITS
DATA - USER INFORMATION BITS
TRAINING - TRAINING BITS

Figure 7.19, S Burst Slot Structure

F Burst

F bursts (slots) are sent continuously on a control channel to allow synchronization. All of the data bits are replaced with zeros. When the BTS receives this information, it produces a pilot tone which Subscriber Units easily recognize. To simplify finding the F burst, a radio channel (carrier) which is transmitting F bursts is not allowed to frequency hop.

7.2.11 Short Message Service (SMS)

The ability for a GSM Subscriber Unit to receive short messages allows it to operate more like a pager. A GSM phone can receive short data messages of up to 160 characters from a short message service center (SC). Most of the SMS features are described in phase 2 of the specification.

Short messages can be sent while a phone is operating on a call or while it is idle. Short messages can be sent by a GSM device such as a Subscriber Unit, or by a short message entity (SME). An example of an outside message would be from a paging service company connected to the SC via a data link such as an X.25 network.

The SC's purpose is to store short messages, attempt to deliver them, and confirm their receipt. Short messages sent to a Subscriber Unit are typically stored in the SIM card, allowing the user to receive and keep messages, and display them on any GSM Subscriber Unit with message display capability. A typical SIM card can store five short messages [9]. Some Subscriber Units may also allow short messages to be stored in a designated memory area that is not part of the SIM card. Short messages may be accompanied by other information, including the originating data address, time, and date received.

Short messages can be broadcast (unacknowledged) or point-to-point (acknowledged). The Subscriber Unit may obtain status information from the SC about messages that it has previously sent. When the Subscriber Unit's short message memory is full, messages are typically kept in the SC until the memory is cleared. The SC can delete messages already sent to a Subscriber Unit. The SC can indicate that the message is urgent, allowing the message to be displayed immediately on the Subscriber Unit display rather than stored in memory to be viewed later. The SC can request a reply from the user, and indicate a response to be automatically directed back to the sender. If the Subscriber Unit could not receive a message (e.g. poor signal conditions or memory full), it can autonomously contact the SC and request its messages.

7.2.12 Subscriber Identity Module (SIM)

The subscriber identity module (SIM) is a removable card containing the subscriber's identity and feature information. This includes subscriber identity codes, personal features such as short-code (speed) dialing and short messages, and a Personal Identity Number (PIN). The PIN is used by the subscriber to restrict access to the SIM card to only people who know the code. Because SIM cards store the subscriber's unique information, the SIM card can be used in any phone that accepts a SIM card to place and receive calls. For example, a subscriber could use a GSM SIM card in a taxicab or a rented car.

SIM cards come in two sizes: credit card size and semi-permanent (small card) size. The full- sized card typically slides into a slot on the bottom of the phone, and the small card is usually located in the back cover of portable Subscriber Units. Because SIM cards are not radio technology specific, standards setting organizations are studying the possibility of using a SIM card with cellular phones of different radio access technologies (e.g. TDMA and CDMA).

7.3 Signaling

Signaling is the physical process of transferring control information to and from the Subscriber Unit. The radio channel is divided into several different logical control channels and traffic channels. Control channels and traffic channels carry different types of signaling.

7.3.1 Control Channel Signaling

When operating in idle mode, the Subscriber Unit listens to one (or more) control channel(s). Signaling on the control channel includes a Frequency Correction CHannel (FCCH), Synchronization CHannel (SCH), Broadcast Control CHannel (BCCH), and Paging and Access Grant CHannel (PAGCH). Optionally, a Cell Broadcast CHannel (CBCH) can be sent on the control channel. To coordinate

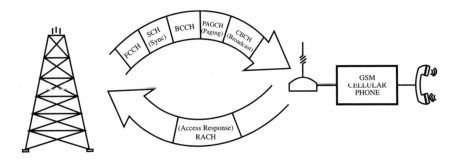

Figure 7.20, GSM Control Channels

access to the system, a Random Access Control CHannel (RACH) is used. Figure 7.20 shows the different types of control channels.

Frequency Correction CHannel (FCCH)

The Frequency Correction CHannel (FCCH) is the initial channel a Subscriber Unit finds when tuning to control channels. The FCCH channel uses F bursts of the same duration as normal bursts, but all bits are zero. Under the GSM system's GMSK modulation, demodulation of all zero bits produces an easily detected RF tone, similar to a pilot tone. The FCCH channel occurs on time slot number (TN) 0 and occurs one frame (8 burst periods) before the synchronization channel burst.

Synchronization CHannel (SCH)

The Synchronization CHannel (SCH) provides the Subscriber Unit with more information on how to access the cellular system. To allow the Subscriber Unit to rapidly time align to a messages on the traffic channel, the synchronization channel uses the S burst which contains a long training sequence. The SCH also provides the Subscriber Unit with an abbreviated frame count to help it identify all useful traffic channel cycles.

Broadcast Control CHannel (BCCH)

The Broadcast Control CHannel (BCCH) send system identification and access control information to all Subscriber Units. This information includes control channel location and configuration, initial Subscriber Unit access power level, the neighboring cell site radio channel frequency list, and cell site identity. The BCCH can send only one 23 octet message every 235 msec.

Paging and Access Grant CHannel (PAGCH)

The Paging and Access Grant CHannel (PAGCH) is composed of two channels; the Paging CHannel (PCH) and Access Grant CHannel (AGCH). A Subscriber Unit determines if a call is to be received by monitoring the PCH for its page messages. The AGCH sends messages to the Subscriber Unit that assign it to a traffic channel or to a Shared Dedicated Control CHannel (SDCCH).

Random Access Channel (RACH)

Requesting access (service) is a random event on the Random Access Channel (RACH). The RACH is the only control channel on the reverse (uplink) channel, so all slots on the RACH channel are usually dedicated for access.

Because one or more Subscriber Units may request access at the same time, a collision may occur. If the signal from one of the two contending Subscriber

Units is much stronger, it gets the cellular system's attention and receives a channel assignment. The other Subscriber Unit will wait a random amount of time prior to attempting to access the system again. If the signals of both Subscriber Units interfere with each other and neither could be decoded, no channel will be assigned, and both will wait a random amount of time before their next access attempt.

When accessing the system for the first time, the radio propagation time delay is not known. A normal burst received from the subscriber could overlap an adjacent burst period. To prevent this on initial access to the system, the Subscriber Unit transmits a shortened access burst with a longer training sequence. Access bursts contain only 87 bits so extra guard time is available to protect against overlapping the adjacent burst period. The larger number of training bits in the access burst helps the BTS to initially decode the first burst received from the Subscriber Unit.

TACH/8 and Stand-alone Dedicated Control CHannel (SDCCH)

Occasionally, a Subscriber Unit and the system need to establish a connection to only send signaling messages. The Traffic and Access Channel, eighth rate (TACH/8) is dedicated for signaling. Its signaling rate is very low (one eighth of a TACH/F), and no user data is sent on the TACH/8 except for short messages. Each TACH/8 channel has SACCH and FACCH messaging capability. The TACH/8 is composed of a TCH/8 and its SACCH channel.

The characteristics of the TACH/8 are similar to the TACH/F or TACH/H. Several different kinds of TACH/8 time organizations exist. The TCH/8 channels may be grouped and combined with other common channels to form the equivalent of a TACH/F. When grouped by 8, they form a Stand-alone Dedicated Control Channel SDCCH/8. When grouped by 4 and combined with other common control channels, they form a SDCCH/4.

The TACH/F can be used to manage Off Air Call Set Up (OACSU). OACSU is a process of assigning a Subscriber Unit to a low bit rate channel which maintains a connection (via TACH/8) while the call is being set up (ringing). When the call is answered, the Subscriber Unit is assigned to a TACH/F.

Cell Broadcast CHannel (CBCH)

The Cell Broadcast CHannel (CBCH) is an optional control channel that transfers short messages to Subscriber Units. A single CBCH channel can transfer about one 80 octet message every 2 seconds [10]. Because the CBCH shares the same control channel multiframe as the BCCH, the CBCH messages can be received without missing any BCCH messages.

7.3.2 Traffic Channel Signaling

As discussed in chapter 2, signaling on the traffic channel is divided into two channels; the Slow Associated Control Channel (SACCH) and the Fast Associated Control Channel (FACCH). The FACCH channel replaces speech with signal data. The SACCH channel uses dedicated (scheduled) frames within each burst.

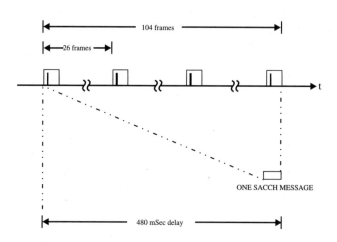

Figure 7.21, SACCH Full Rate Signaling

Slow Associated Control Channel (SACCH)

SACCH is a continuous stream of signaling data sent with speech data, called "out of band" signaling. SACCH messages are sent by dedicated slots in each traffic channel's multiframes, so that sending SACCH messages does not affect the quality of speech. The transmission rate for SACCH messages is very limited. For messages that require rapid delivery (e.g., handover), they are sent via the FACCH channel. A balance was maintained to allow the maximum number of bits to be devoted to speech, and to allocate a minimum number of bits to continuous signaling. Figure 7.21 illustrates SACCH signaling.

Fast Associated Control Channel (FACCH)

FACCH replaces speech data with signaling messages when required, a process called "in band" signaling. Only up to one out of six speech frames may be stolen for FACCH messages, because replacing speech frames with signaling information degrade speech quality. When a frame of coded speech is lost due to replacement by FACCH data, or when temporary fading or interference is severe, the

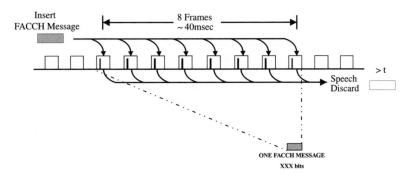

Figure 7.22, FACCH Signaling

GSM RELP speech coder can extrapolate the last good received speech coder data. Under these circumstances, listeners in GSM systems hear a brief prolongation of a speech sound rather than a silence, "click" or other gross disturbance of speech. The replacement of speech frames with FACCH messages results in non-linear degradation of speech quality as the number of stolen frames increases. Because data is bit-interleaved over 8 consecutive channel bursts, the data bits for a FACCH message are transmitted piece by piece over 8 sequential channels bursts (57 bits per channel burst). Figure 7.22 illustrates FACCH signaling. Figures 7.5 and 7.17 show that C bits (flag bits known as steeling bits) distinguish FACCH data from subscriber speech or data.

DTMF Signaling

Dual Tone Multi-Frequency (DTMF) audio frequency "touch tones," can be transmitted via the codec, but transmitting DTMF audio tones through the codec adds significant audio distortion. This distortion may cause improper operation of remote control DTMF decoding devices such as voice mail and answering machines. To overcome the DTMF distortion created by the speech coder, when the subscriber desires to send DTMF tones (typically by pressing a key on the keypad), the Subscriber Unit sends a DTMF control messages via the FACCH channel to the cellular system instead of the DTMF audio tone. Either the BTS or MSC receives these DTMF control message commands and converts them to the DTMF audio tones for transmission through the PSTN network. Figure 7.23

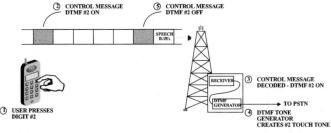

Figure 7.23, DTMF Signaling

shows an example of sending DTMF tones via the Subscriber Unit. Initially, a user presses the key number 2 (step 1). This creates a "DTMF ON" control message (step 2) that indicates digit #2 has been pressed. The receiver in the BTS decodes this control message (step 3) and commands a DTMF generator in the BTS to create a number 2 DTMF touch tone (step 4). When the user releases the #2 key, a "DTMF OFF" control message is created, which indicates the #2 key has been released, and the DTMF touch tone generator in the BTS is stopped.

7.4 System Parameters

GSM system parameters include a frequency band that partially overlaps the TACS and NMT 900 systems, frequency duplex channel separation, narrow spectrum bandwidth GMSK modulation, and variable RF power levels for both the BTS and Subscriber Units. European nations using these older systems have abandoned plans to clear the entire GSM band.

7.4.1 Frequency Allocation

Two 25 MHz bands at 890-915 MHz (Subscriber Unit transmit) and 935-960 MHz (Base Transceiver Station transmit) are separated by 45 MHz. Each frequency band is divided into 125 radio channel carriers. The bandwidth of each modulated carrier frequency is 200 kHz. One or more control channels can be located on any of the radio channels. In some systems, the entire frequency band may not be available, and in other systems, radio channels may be divided among

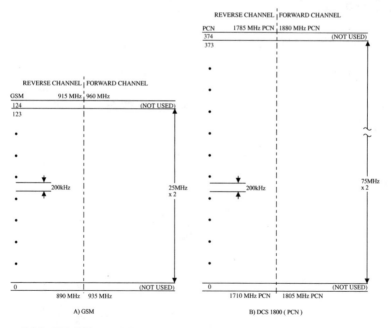

Figure 7.24, GSM Channel Frequency Allocation

multiple cellular service providers.

The GSM specification was modified to support the PCN frequency bands. The PCN frequency allocation has two bands: 1710-1785 MHz (Subscriber Unit transmit) and 1785-1880 MHz (BTS transmit) separated by 75 MHz. Each PCN frequency band is divided into 375 radio channels of 200 kHz each. Figure 7.24 diagrams GSM and PCN frequency allocation.

7.4.2 Duplex Channels

Although the digital channel is frequency duplex (transmit on one frequency and receive on a different frequency), the Subscriber Unit receives and transmits at the different times. With this time separation, a simple radio switch can separate

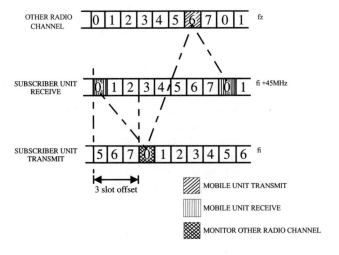

Figure 7.25, GSM Duplex Radio Channels

the transmitter and receiver sections in hand portables. Figure 7.25 shows the 3 slot time offset between the forward and reverse channel. The three slot offset allows for dynamic time alignment and for synthesizer tuning between transmit and receive bursts.

The duplex radio channel structure allows the Subscriber Unit to tune to the radio channels of a neighboring BTS, to synchronize, and to obtain radio channel signal strength and timing information. Figure 7.25 shows that the Subscriber Unit can tune to another radio channel during idle mode (non transmit or receive mode) to measure signal strength.

7.4.3 Modulation Type

The GSM digital traffic channel uses GMSK phase modulation to transfer voice and control signaling information. GMSK modulation was chosen to maintain the very narrow spectral bandwidth needed to limit interference with other GSM channels and neighboring radio systems such as TACS and NMT. Other forms of

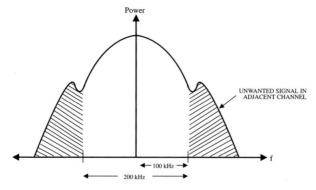

Figure 7.26, GMSK Modulation

modulation, such as Quadrature Phase Shift Keying (QPSK), would have increased the channel data rate, hence spectral efficiency, but would have interfered with adjacent radio channels [11]. QPSK also requires a more complex and less power efficient RF power amplifier. Figure 7.26 illustrates the relatively narrow GMSK modulation bandwidth.

7.4.4 RF Power Classification

Subscriber Units and BTSs are have power output classifications. The RF power class specifications set the maximum allowable power transmitted during a burst period. Both the Subscriber Unit and the BTS can decrease power from the maximum stepwise in 2 dB increments.

GSM Subscriber Units have 5 different power classes. The class 1 Subscriber Unit has a maximum output power of 20 Watts (+13 dBW or +43 dBm) and a minimum output power level of -4 dBm. The 20 Watt class 1 mobile classification was suppressed for the use of GSM [12]. Transportable Subscriber Units are class 2, with a maximum output power of 8 Watts. Hand portables are classified 3 through 5. Class 3 and class 4 hand portables' output power are 5 Watts and 2 Watts respectively. Class 5 hand portables are designated for microcellular networks, with a maximum output power of 0.8 Watts. Because Subscriber Units transmit in bursts, their output power is measured during the burst period. During full rate transmission, the average output power of the Subscriber Unit is approximately 1/8th the maximum burst transmit power. For a portable Subscriber Unit transmitting at 2 Watts, the average power is only 250 mWatts.

Power Class	Subscriber Unit Power (Watts)	Base Transceiver Power (Watts)
1	20	320
2	8	160
3	5	80
4	2	40
5	0.8	20
6		10
7		5
8		2.5

Table 7.1
GSM Power Classification

While the Subscriber Unit transmits in bursts, the BTS typically transmits continuously. However, the BTS can reduce power between bursts. The Subscriber Unit output power can be adjusted by commands from the BTS. Each RF power adjustment step is 2 dB with a total range of approximately 20-30 dB. The BSC is responsible for controlling power level settings of both the BTS and Subscriber Unit. Table 7.1 lists the GSM Subscriber Unit RF Power classifications.

7.5 System Operation

The GSM cellular system uses one type of radio channel which is divided into many different types of control channels and traffic channels (voice and data). The control channels allow the Subscriber Unit to retrieve system control information and compete for access.

When a Subscriber Unit is first powered on, it initializes by scanning for a control channel and tuning to the strongest one it finds. During initialization, it acquires all of the system information needed to monitor for paging messages and information about how to access the system. After initialization, the Subscriber Unit enters idle (sleep and wake cycle) mode and waits either to be paged for an incoming call or for the user to place a call (access). When a call is to be received or placed, the Subscriber Unit enters system access mode to try to access the system via a control channel. When access is granted, the control channel commands the Subscriber Unit to tune to a digital traffic channel. The Subscriber Unit tunes to the designated channel, and enters conversation mode. As the Subscriber Unit moves out of range of one cell site radio coverage area, it is handed over to a radio traffic channel at another nearby cell site.

7.5.1 Access

A Subscriber Unit attempts to gain service from the cellular system by transmitting a request on the Random Access CHannel (RACH). If the system is not busy, it attempts access by transmitting an access burst on the RACH channel. The access burst contains a 5 bit random number which temporarily identifies the Subscriber Unit attempting the access. The access burst also contains a 3 bit code

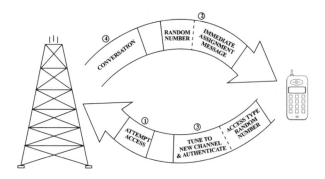

Figure 7.27, System Access

which identifies the type of access requested, such as page response, call origination, or reconnection of an accidentally disconnected call (due to poor quality radio signals). If the system successfully receives the access request message, it sends back the random number in the immediate assignment message, directing the Subscriber Unit to tune to a specific radio channel and time slot. After the Subscriber Unit tunes to its assigned channel, the system typically requests authentication. If the system authorizes service, conversation can begin. Figure 7.27 illustrates the system access process.

7.5.2 Paging

Paging is the process of sending a page message to the Subscriber Unit to indicate that a call is to be received. Page messages are sent on the Paging and Access Grant CHannel (PAGCH). To increase the number of paging messages that a control channel can deliver, a Subscriber Unit is assigned a Temporary Mobile Subscriber Identity (TMSI) when it registers in a system. The TMSI is shorter than the International Mobile Subscriber Identity (IMSI), which uniquely identi-

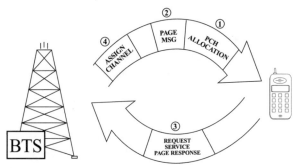

Figure 7.28, System Paging

fies the subscriber. If a Subscriber Unit has not been assigned a TMSI, the IMSI can be sent on the paging channel.

Figure 7.28 illustrates the paging process. Initially (step 1), the Broadcast CHannel (BCCH) locates the paging channel slots. Following the BCCH signals, paging messages will be sent on the PAGCH at a scheduled time. The Subscriber Unit then monitors the PAGCH for its page (step 2). After the Subscriber Unit is paged, it requests service from the cellular system (step 3), indicating that it is responding to a page message. The cellular system then assigns a traffic channel or a SDCCH channel (step 4) where subscriber will be authenticated and conversation may begin.

7.5.3 Handover

The transfer of a radio communication from one radio channel to another is called handover. A radio communication is handed over when the BSC determines that channel quality has fallen below a desired level, and another better radio channel is available. The BSC continuously receives radio channel quality information from the BTS and the Subscriber Unit.

Figure 7.29 illustrates the handover process. First, the Subscriber Unit monitors nearby cell sites' radio channels. A list of these channels is provided (step 1) via the Broadcast Control CHannel (BCCH) or via the SACCH if the Subscriber Unit is on a traffic channel in cell site #1. During idle periods between bursts, the Subscriber Unit measures the signal strength and channel quality of the radio channel serving it, and returns one of these measurements per second (approximately) to the BTS via the SACCH channel (time 2). When the BSC determines that the Subscriber Unit can be better served by another cell site, it sends a han-

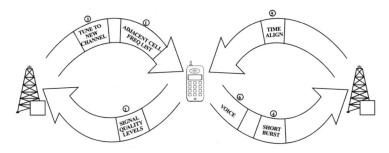

Figure 7.29, GSM System Channel Handover

dover message commanding the unit to tune to a new radio channel (time 3). When the Subscriber Unit receives the message, it mutes the audio and tunes to the new radio channel. It then begins sending Access Bursts (time 4) to avoid potential collisions with bursts sent from other Subscriber Units. If the

Subscriber Unit has pre-synchronized with the other cell site by decoding the FCCH and SCH channels, it begins to transmit normal bursts. If it has not pre-synchronized, the BTS determines the necessary time adjustment and commands the Subscriber Unit to time align (time 5). The Subscriber Unit can then unmute the audio and begin voice communications again (time 6). If adjacent cells in the system are similar in size, differing by less than about 1 km, no shortened access burst is required, and the Subscriber Unit sends normal bursts immediately (see section 7.2.3).

7.6 Call Processing

Call processing is the set of tasks performed to complete five different kinds of operations: 1) initialization of information when the Subscriber Unit is turned on, 2) monitoring the control channels, 3) attempting access, and 4) coordinating transmissions during conversation or while sending data.

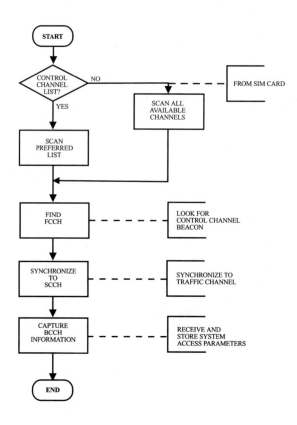

Figure 7.30, GSM Initialization Mode

7.6.1 Initialization

Initialization is the process of synchronizing with the system and obtaining system parameters to determine the information requirements for access and communication. As shown in figure 7.30, when the Subscriber Unit first powers on, it looks to the SIM card for a preferred control channel list. If there is no list, the Subscriber Unit scans all of the available radio channels to find a control channel. During scanning, the Subscriber Unit listens for the tone of a Frequency Correction CHannel (FCCH). When it finds the FCCH tone, it has found a control channel. Eight time slot periods later, it acquires the Synchronization CHannel (SCH) and uses it to time align decoding of the digital channel with the burst periods of the traffic channel. Next, the Subscriber Unit retrieves information the BCCH where it obtains system control and access information. The Subscriber Unit then enters the idle mode to monitor the paging channel.

7.6.2   Idle

During idle mode, the Subscriber Unit continuously monitors several different control channels to acquire system access parameters, to determine if it has been paged or received an order, or to initiate a call (if the operator is placing a call). Figure 7.31 illustrates idle mode call processing.

After obtaining the system parameters, the Subscriber Unit monitors the broadcast channel for changes in system parameters, including system identification and access information. If the Subscriber Unit has discontinuous reception (sleep mode) capability, and if the system supports it, the Subscriber Unit turns off its receiver and other non-essential circuitry for a fixed number of burst periods. The

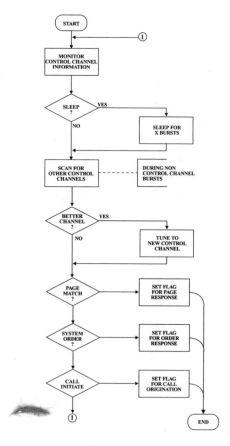

Figure 7.31, GSM Idle Mode

system knows when the Subscriber Unit will sleep, so it does not send pages designated for that Subscriber Unit during the sleep period. Control channels are on only one of the eight bursts in a frame, so during the other seven burst periods, the Subscriber Unit scans neighboring control channels. If a better control channel (higher signal strength or better bit error rate) is available, the Subscriber Unit tunes to it.

The Subscriber Unit monitors the paging control channel to determine if it has received a page. If a call is to be received, an internal flag is set indicating that the Subscriber Unit is entering access mode in response to a page. If the system sends an order such as a registration message, an internal flag is set indicating that the Subscriber Unit is attempting access in response to an order. When a user initiates a call, an internal flag is set indicating that the access attempt is a call origination, and dialed digits will follow the access request.

7.6.3 Initial Assignment/Access

When Subscriber Units respond to a page, set up a call, or attempt any other type of access to the cellular network, the attempt is at random (at any time), so multiple Subscriber Units could attempt access simultaneously. To avoid such access "collisions," a seizure collision avoidance procedure has been developed. The contention resolution process consists of four components: 1) a request sent on the RACH (slotted) channel to validate a response, 2) access class groups, 3) random time delays, and 4) maximum number of automatic access attempts

When initially requesting service, the Subscriber Unit indicates the type of service it is requesting (e.g. page response or call initiation) with a 5-bit random number and a 3- bit code word in the access burst. If the system receives the burst and has a radio channel available to assign, the BTS sends an immediate assignment message echoing the 5-bit random number. If, within a specified period of time, the Subscriber Unit does not receive the immediate assignment message with its random number identifier, it delays a random amount of time, and then

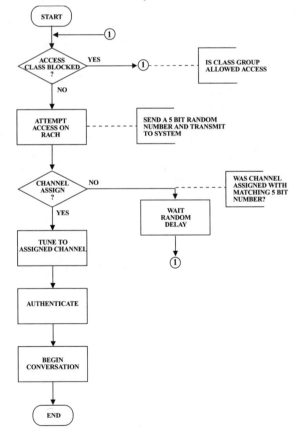

Figure 7.32, GSM Access Mode

attempts access again. The random delay prevents simultaneous access attempts from two or more competing Subscriber Units from being repeated. To prevent two Subscriber Units from using the same 5-bit random number to request service at the same time, Subscriber Units' unique identity information is also a part of the message.

To prevent a situation in which many Subscriber Units constantly attempt access and continually overload the system, the GSM system limits the number of automatic retries. To further limit simultaneous access attempts, GSM systems can also restrict access to customers with specific access classification. Every Subscriber Unit is assigned one of 16 possible access classifications. The first ten classifications are random, and the remaining six are for emergency or high priority customers. Before accessing the cellular system, the Subscriber Unit monitors the control channel to determine if its access class group has been restricted. If the access class is not blocked, the Subscriber Unit transmits an access burst on the Random Access CHannel (RACH).

7.6.4 Dedicated Mode

The BTS continues to control the Subscriber Unit during dedicated mode (e.g., during voice conversation). Control tasks include power level control, handover, alerting, etc. The BTS exercises control during conversation either through the FACCH, where the voice information is briefly replaced with signaling commands, or through out-of-band signaling (SACCH).

The Subscriber Unit continually receives data bursts from the BTS. If the burst data has a control flag set, the data will be decoded as a FACCH message. If the received frame has traffic channel multiframe sequence number 13 (or 26 for half-rate traffic channels), it is decoded as a SACCH message. After 4 SACCH bursts have been received, the SACCH message will be processed. All other received data is decoded as speech or other user data. Three burst periods after the Subscriber Unit receives data, it transmits data. Between the receive and transmit bursts, the Subscriber Unit tunes to other radio channels to measure channel quality, then sends this information in the next SACCH message back to the BTS to assist handover.

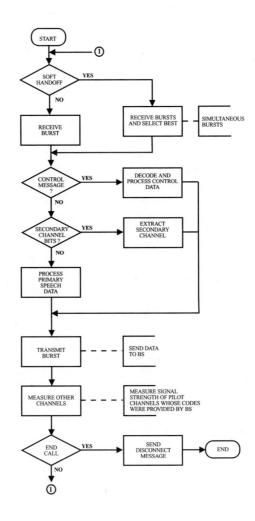

Figure 7.31, GSM Dedicated Mode

References

1. D.M. Balston, R.V. Macario, "Cellular Radio Systems," Artech House, 1993, pg. 188.
2. D.M. Balston, R.V. Macario, "Cellular Radio Systems," Artech House, 1993, pg. 154.
3. D.M. Balston, R.V. Macario, "Cellular Radio Systems," Artech House, 1993, pg. 126.

4. Michel Mouly, Marie-Bernadette Pautet, "The GSM System for Mobile Communications", M. Mouly et Marie-B Pautet, Palaiseau, France, pp. 218-221.
5. D.M. Balston, R.V. Macario, "Cellular Radio Systems," Artech House, 1993, pg. 181.
6. Michel Mouly, Marie-Bernadette Pautet, "The GSM System for Mobile Communications," M. Mouly et Marie-B Pautet, Palaiseau, France, pp. 257-258.
7. D.M. Balston, R.V. Macario, "Cellular Radio Systems," Artech House, 1993, pg. 183
8. D.M. Balston, R.V. Macario, "Cellular Radio Systems," Artech House, 1993, pg. 312.
9. Michel Mouly, Marie-Bernadette Pautet, "The GSM System for Mobile Communications." M. Mouly et Marie-B Pautet, Palaiseau, France, pg. 193
10. Michel Mouly, Marie-Bernadette Pautet, "The GSM System for Mobile Communications," M. Mouly et Marie-B Pautet, Palaiseau, France, pg. 254
11. Fax Correspondence with M.B. Pautet, 1 March, 1996, pg.3.

Chapter 8
Digital Cellular Telephones

8. Digital Cellular Telephones

Cellular telephones may be mobile radios mounted in motor vehicles, transportable radios (mobile radios configured with batteries for out-of-the-car use), or self-contained portable units. Whether mobiles, transportables, or portables, their functions are almost identical. Because all cellular telephone equipment provides benefits to the subscriber of the service, we refer to all types of cellular telephones as Subscriber Units. Dual mode Subscriber Units are capable of operating in more than one mode (usually an analog mode and a digital mode). In describing a dual mode Subscriber Unit, we divide its assemblies into the following components:

- User interface (previously called the control head)
- Radio frequency assembly
- Signal processing
- Power supply/battery

The user interface, sometimes called a man machine interface (MMI), allows the user to originate and respond to calls and messages. The Radio Frequency (RF) assembly converts the baseband signal (analog or digital voice) into RF signals for transfer between the Base Station and the Subscriber Unit. The signal processor section conditions the voice (audio and digital compression) and controls the internal operations of the Subscriber Unit (logic). The power supply provides the energy to operate the Subscriber Unit.

In addition to the key assemblies contained in a Subscriber Unit, accessories are often connected to adapt the Subscriber Unit to perform an optional feature such as hands free operation. The user interface, radio section, signal processing, power supply, and any attached accessories must work together as a system. For example, when a portable Subscriber Unit is connected to a hands free accessory, the Subscriber Unit must sense the accessory is connected, disable its microphone and speaker, and route the hands free accessory microphone and speaker to the signal processing section.

Subscriber Unit technology nuances exist for each digital technology. Time Division Multiple Access (TDMA) IS-54 Subscriber Units can operate on both EIA-553 analog systems and IS-54 digital systems. TDMA IS-136 Subscriber Units are multi-mode phones which can operate on EIA-553, IS-54 and IS-136 systems. Code Division Multiple Access (CDMA) IS-95 Subscriber Units can operate on EIA-553 and IS-95 systems. Narrowband Advanced Mobile Phone Service (NAMPS) Subscriber Units operate on EIA-553 and IS-88 systems. Global System for Mobile Communications (GSM) Subscriber Units are single mode technology which only operate on GSM systems. It is possible to produce Subscriber Units in any of the digital technologies that have multi-mode capability (e.g. GSM and TACS). Figure 8.1 illustrates the basic components of a dual mode digital and analog cellular radio.

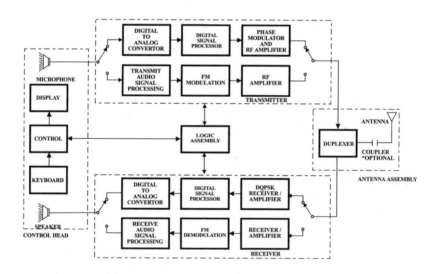

Figure 8.1, Dual Mode Subscriber Unit System Block Diagram

8.1 User Interface

The user interface consists of audio interface, display, keypad, accessory connection assemblies, and software to coordinate their operation.

8.1.1 Audio Interface

The audio interface assembly consists of a speaker and microphone which allow users to talk and listen to the Subscriber Unit. These assemblies are located in a handset although they can be replaced by units in the hands free assembly. For portable units, the entire unit acts as the handset with the speaker and microphone being incorporated into the single unit. Because of their small size, portable units have microphone systems with a substantial amount of gain. This allows the microphone to still pick up normal conversation even when it is not placed directly in front of the speakers mouth. This is especially important for the smaller portable units that have a length smaller that the distance between a person's ear and mouth. As in the land line system, a sidetone is generated to allow users the capability to hear what they are saying into the microphone.

8.1.2 Display

The display assembly allows the user to see dialed digits and call status information. Dialed digits can be displayed and altered before the call is initiated which is known as pre-origination dialing. An IN USE symbol is displayed when the call is initiated, indicating that RF power is being transmitted. The display may also indicate other available features such as a Received Signal Strength Indicator (RSSI), call timer, or other services. In recent years the display has been used to implement advanced service features which include calling number identification (caller ID), name and number storage, and selecting preferences for the Subscriber Unit operation.

8.1.3 Keypad

The keypad allows the user to enter information to control the phone. This includes dialed numbers, commands to receive and originate calls, along with selecting feature options. The keypad may sometimes be replaced by a voice activation unit.

Keypad layout and design vary greatly from manufacturer to manufacturer. A typical keypad will contain keys for the numbers 0 to 9, the * and # keys (used to activate many subscriber services in the network), volume keys and a few keys to control the user functions.

8.1.4 Accessory Interface

Accessories for Subscriber Units can be attached via an optional electrical interface connector (plug). The accessory interface connector typically provides control lines (for dialing and display information), audio lines (in and out), antenna connection, and power lines (in and out) to connect to and from accessory devices. No standard accessory interface connection exists for Subscriber Units. Each manufacturer, and often each model, will have a unique accessory interface. The accessory connector is normally on the "bottom" or end of the Subscriber Unit.

The types of accessories vary from active devices such as computer modems to passive devices like external antennas. A hands free kit includes an external microphone and speaker to allow the subscriber to talk to the phone without using the handset. An external power supply (such as a car battery) may be used to charge the battery. The Subscriber Unit power line may provide power from the phone's battery to external devices such as computer modems. An antenna connection allows the use of high gain external antennas which may be mounted on a car. When a data device such as a modem is used, this typically requires an audio connection (for the modem data) and a control connection (to dial and automatically set the Subscriber Unit's features). Other smart accessories (such as a voice dialer) may require audio and control line connection. Figure 8.2 shows a sample accessory connection diagram.

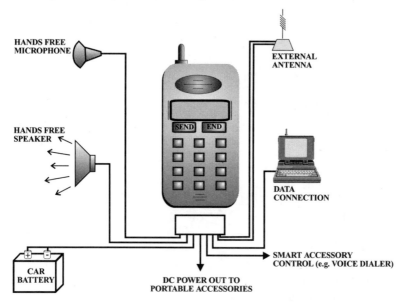

Figure 8.2, Typical Accessory Connection

8.2 Radio Frequency Section

The radio frequency section consists of transmitter, receiver, and antenna sections. The transmitter converts low level audio signals to proportional shifts in the RF carrier frequency. The receiver amplifies and demodulates low level RF signals into their original audio form. The antenna section converts RF energy to and from electromagnetic signals.

Most Subscriber Unit designs today use a microprocessor and Digital Signal Processors (DSPs) to initialize and control the RF section of the Subscriber Unit. RF amplifiers vary in their type and conversion efficiency. Analog Subscriber Units and some digital units use non-linear (Class C) amplifiers and most digital Subscriber Units use linear (class A or AB) amplifiers. The efficiency of the RF amplifier is the rating of energy conversion (typically from a battery) to RF energy. Because the RF amplifier typically is the largest power consuming section during transmit, the higher the conversion efficiency, the longer the battery life during conversation (transmission). While Class C amplifiers add some distortion to the radio signal and may be above 50% efficient, linear amplifiers add very little distortion and are only 30-40% efficient.

8.2.1 Transmitter

The transmitter section contains a modulator, a frequency synthesizer, and an RF amplifier. The modulator converts audio signals to low-level radio frequency modulated radio signals on the assigned channel. A frequency synthesizer creates the specific RF frequency the cellular phone will use to transmit the RF signal. The RF amplifier boosts the signal to a level necessary to be received by the Base Station.

The transmitter is capable of adjusting its power levels to transmit only the necessary power to be received by the Base Station. To conserve battery life in portables and transportables, the RF amplifier may turn off its power during periods when the mobile operator is not talking which is called Discontinuous Transmission (DTX).

8.2.2 Receiver

The receiver section contains a receiver amplifier, down-converting RF mixer, and a demodulator. Low-level radio signal received by the antenna assembly are amplified and routed to the demodulator. The demodulator converts the proportional frequency or phase changes into low level analog or digital signals.

8.2.3 Antenna Section

An antenna section focuses the transmission into a desired direction [1] and matches the impedance of the transceiver to that of free space. The antenna section consists of an antenna, cabling, duplexer, and possibly a coupling device for antenna connection through glass. An antenna system can enhance or seriously reduce the performance of the Subscriber Unit. The antenna may be an integral part of the transceiver section (such as a portable handset) or externally mounted (on the top of a car). Antennas can have a gain where energy is focused into a beamwidth area. This focused energy gives the ability to communicate over greater distances, but as the angle of the antenna changes, the direction of the beam also changes, reducing performance. For example, car-mounted antennas that have been tilted to match the style lines of the automobile often result in extremely poor performance.

Cabling that connects the transmitter to the antenna adds losses which reduce the performance of the antenna assembly. This loss ranges from approximately .01 to .1 dB per foot of cable. High gain antennas may be used to overcome these losses and possibly allow for a lower power output [2].

In early systems, separate antennas were used for transmitters and receivers to prevent the high power transmitter from overpowering the receiver. A duplexer or a Transmit-Receive (TR) switch allows a single antenna to serve both the transmitter and receiver. A duplexer consists of two RF filters; one for transmission and one for reception. A TR switch connects either the transmitter or the receiver to the antenna, but never at the same time.

8.3 Signal Processing

Digital Subscriber Units typically use DSPs or Application Specific Integrated Circuits (ASICs) to process all types of signals. If the Subscriber Unit has multi-mode capability (such as TDMA and AMPS), the same DSP can be used to process both the analog and digital signals. DSPs are high performance computing devices that are specifically designed to allow rapid signal processing. Advances in DSP technology allow increased signal processing ability of 40 to 60 Million Instructions Per Second (MIPS), lower operating battery voltage (3 volts -vs- 5 volts), and reduced cost. ASICs are custom designed integrated circuits and are typically created to combine several functional assemblies. For example, a single ASIC could contain all the control circuits necessary to connect the display and keypad to the DSP. By replacing several components with one ASIC, cost and size are reduced.

Digital cellular phones require approximately 40 to 60 MIPs of processing power compared to 0.5 MIPs required by AMPS phones [3]. The digital signal processing requirements come from speech coding (1.5 MIPS to 8 MIPS), modulation and demodulation, radio channel coding and decoding, and radio signal equalization or rake reception. High performance DSPs have only become available in the past few years which are required in digital Subscriber Units.

The power consumption of DSPs are roughly proportional to their operating voltage and processing speed. Efficient DSPs that operate at 5 volts consume approximately 14 mW per MIP. If a digital Subscriber Unit is using 40 MIPs of processing, it will consume approximately 0.5 Watts of energy. Using a lower battery voltage to supply a DSP reduces its power consumption significantly. A reduction in voltage of 40% will reduce the power consumption by approximately 64%. This has a significant impact on the standby time (receiver battery power consumption).

The cost of DSPs has dropped from over $50 each in the early 1990s to $15-28 in large manufacturing quantities [4] in 1995. Much of the cost of DSPs come from the amount of silicon wafer area required for a DSP chip. The more area required for a single DSP chip, the less chips that can be produced per wafer. As companies innovate ways to place more components on a single wafer, the costs will continue to decrease.

When DSPs are used in dual mode Subscriber Units, it is likely that both analog and digital signal processing will be performed by the same DSP. The DSP will use a different processing program to demodulate analog and digital signals.

8.3.1 Speech Coding

Digital Subscriber Units have a speech coder which compresses the digital signal information. On digital systems, the speech must be sent as digital data. When a subscriber speaks into the microphone it generates an analog signal which is digitized and input to a speech coding algorithm. This digitized voice is processed by to a speech coding algorithm to create a digital representation of the analog voice signal. When this compressed speech information is received, it is recreated to the original analog signal by decoding the speech data. Speech coders for digital systems are usually implemented in firmware which is executed by a DSP.

8.3.2 Channel Coding

Channel coding involves adding error protection and detection bits and multiplexing control signals with the transmitted information. Error protection and detection bits (they may be the same bits) are used to detect and correct errors which occur on the radio channel during transmission. The output of the speech coder is encoded with additional error protection and detection bits according to the channel coding rules for its particular specification. This extra information allows the receiver to determine if distortion from the radio transmission has caused errors in the received signal. Control signals such as power control, timing advances, frequency handoff must also be merged into the digital information to be transmitted. The control information may have a more reliable type of error protection and detection process which is different than the speech data. This is because control messages are more important to the operation of the Subscriber Unit than voice signals. The tradeoff for added error protection and detection bits is the reduced amount of data that is available for voice signals or control messages. The ability to detect and correct errors is a big advantage of digital coding formats over analog formats but it does come at the cost of the additional data required.

8.3.3 Audio Processing

Besides simply using the speech coder to translate between analog and digital formats, a digital Subscriber Unit can do other audio processing to enhance its overall quality. Some audio processing is very simple, such as filling in blank frames with data from the previous frame. This simple algorithm works remarkably well for short bursts of missing voice data. This is because the spoken voice for a single small frame (around 20 milliseconds) is very similar to the spoken voice for the previous 20 milliseconds.

More complex audio processing algorithms are used to implement other types of audio signal processing. The DSP may be involved in detecting speech, noise cancellation and emphasis of specific frequency bands of the audio. Echo is a particular problem in digital systems which can be introduced by the delay involved in the speech compression algorithm or through speaker phone operation. Cellular technologies that use speech compression (all except NAMPS) add delay to the signal.

The audio processing section can remove the echo by sampling the audio signal in brief time periods and looking for previous audio signal patterns. If the echo canceller finds a matched signal, it is subtracted, thus removing the echo. While this sounds simple, there may be several sources and levels of echo and they may change over the duration of the call.

8.3.4 Logic Section

In early applications, the logic section contained discrete logic components (such as AND and OR gates) for each processing section [5]. Today, logic sections usually contain a microprocessor operating from stored program memory. The logic section coordinates the overall operation of the transmitter and receiver section by allowing the insertion and extraction of control messages. Control signals can be analog (such as Signaling Tones) or digital (a FACCH control message). Control messages must be continually inserted and removed from the transmitter and receiver sections. The logic section encodes and decodes control signals and performs the call processing procedures.

The logic control section inserts and extracts special control messages that coordinate the transceiver and control head. The receiver extracts Base Station commands from the received radio signal and routes them to the logic section. In accord with the Base Station commands, the logic section then controls the transmitter and updates the control head display information if necessary. When the user initiates commands via the keypad (e.g. dials digits), the commands are transferred to the logic section. The logic section then controls the transmitter assembly.

8.3.5 Subscriber Identity

Each Subscriber Unit and end customer must have a unique identification code to allow cellular systems to deliver and process calls along with collecting billing information. Unique customer information is stored in a subscriber unit or a removable device. For early analog cellular phones, the subscriber identity was stored in a Programmable Read Only Memory (PROM) chip which required a separate chip programmer. The PROM chip would be installed by the programmer and was not convenient to remove. There are two basic types of identification use by the digital cellular technologies. North American cellular systems use a Number Assignment Module (NAM) and GSM systems use a Subscriber Identity Module (SIM) card. In addition to the unique information contained in the NAM and SIM, additional equipment identification codes such as Electronic Serial Numbers (ESNs) are included with the physical equipment.

8.4 Battery Technology

As the use of portable phones has increased, the importance of battery technology has increased. There are several types of batteries used in cellular Subscriber Units; Alkaline, Nickel Cadmium (NiCd), Nickel Metal Hybrid (NiMH) and Lithium (Li). A new type of battery technology, Zinc Air, is being explored that has increased energy storage capacity.

With the introduction of portable cellular phones in the mid 1980s, battery technology became one of the key technologies for users of cellular phones. In the mid 1990s, over 80% of all cellular phones sold were portable or transportable models rather than fixed installation car phones. Battery technology is a key factor in determining portable phones' size, talk time and standby time.

Batteries are categorized as primary or secondary. Primary batteries must be disposed of once they have been discharged while secondary batteries can be discharged and recharged for several cycles. Primary cells (disposable batteries) which include Carbon, Alkaline, and Lithium see limited use in cellular

Subscriber Units. Although they are readily available, disposable battery packs must be replaced after several hours of use and are more expensive than the cost of a rechargeable a battery. Disposable batteries have the advantages of a long shelf life and no need for a charging system.

One of the most common batteries used in portable cellular phones is the rechargeable NiCd cell which consists of two metal plates made of nickel and cadmium placed in a chemical solution. The package is vented to prevent explosions due to improper charging or discharging. A NiCd cell can typically be cycled (charged and discharged) 500 to 1000 times and is capable of providing high power (current) demands required by the radio transmitter sections of portable Subscriber Units. While NiCd cells are available in many standard cell sizes such as AAA and AA, the battery packs used in cellular phones are typically uniquely designed for particular models of Subscriber Units. The internal batteries of the battery pack typically cannot be replaced by the user. Some NiCd batteries can develop a memory of their charging and discharging cycles and their useful life can be considerably shortened if they are not correctly discharged. This is known as the "memory effect," where the battery remembers a certain charge level and won't provide more energy even if completely charged. Newer NiCd batteries use new designs that reduce the "memory effect."

NiMH batteries use a hydrogen adsorbing metal electrode instead of the cadmium plate. NiMH batteries can provide up to 30% more capacity than a similarly sized NiCd battery. However, for the same energy and weight performance, NiMH batteries cost about twice as much as NiCd batteries [6].

Li-Ion batteries are the newest technology that is being used in portable cellular phones. They provide increased capacity versus weight and size. A typical Li-Ion cell provides 3.6 volts versus 1.2 volts for NiCd and NiMH cells. This means that one third the number of cells are need to provide the same voltage.

Zinc-air batteries have shown they can provide more storage capacity than lithium-ion batteries. They use oxygen from the air to enable reactions that generate electricity. They also have no memory effect and a low self discharge, which allows for a longer shelf life after charging. Zinc-air batteries are being produced for portable computers and will require more testing and study before they are available for portable cellular phones. Figure 8.3 shows the relative capacity of different battery types.

The charge and discharge characteristics of batteries are important because every Subscriber Unit contains some type of charging system and battery status indicator. Customers typically desire to charge and discharge at any time. The charg-

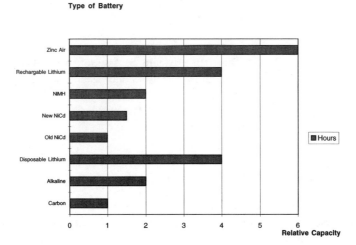

Figure 8.3, Battery Storage Capacity

ing system must determine when a battery is fully charged. The charging system stops charging the battery (or may change to a very low trickle charge rate) at a cutoff point. The cutoff point can be determined by temperature, maximum voltage level, a change in voltage (delta voltage), or a combination of these factors. The battery status indicator is typically a form of a voltage meter. Some portable phones actually indicate the voltage level of the battery but most choose to display a level (similar to a gas gauge). NiCd batteries have the most difficulty to accurately displaying battery levels because of their relatively flat discharge curves for the mid range discharge.

Some of the future trends in batteries are new packaging and higher mAH ratings for the same size and weight. Flat cells will allow denser packaging of batteries. Today between four and six round cells are placed into a single battery pack resulting in a lot of wasted "dead space" between the cells. As consumers use their phones more and more they are demanding longer standby and talk times. Because a .6 watt portable phone has to transmit at fixed power levels, the battery technology and more efficient power amplifiers are the only ways to increase talk time.

8.5 Accessories

Accessories are optional devices that may be connected to Subscriber Units to increase the functionality of the Subscriber Unit. Accessory devices include hands free speakerphone, data transfer adapter (modems), voice activation, battery chargers, antennas, and many others.

8.5.1 Hands Free Speakerphone

For safety reasons, the phone may have an option allowing the subscriber to use hands free operation [7]. A hands free system consists of: a speaker, usually located in the cradle assembly; a remote microphone, usually located near the visor; and interface circuitry which connects the audio paths and allows for sensing when the user requests hands free mode. Because digital cellular systems take time to process and convert audio signals, echoes caused by hands free operation can by very annoying to the user.

8.5.2 Data Transfer Adapters

Subscribers sometimes want to send digital information via their Subscriber Units. The Subscriber Unit can offer optional connections for a facsimile, modem, or a standard Plain Old Telephone Service (POTS) to the cellular phone. On the PSTN network, voice or data information can be sent reliably if it is within the 300-3000 Hz frequency range. While the audio frequency range on the cellular radio channel is the same as the PSTN, the varying nature of the RF channel is not well suited for efficient standard data transfer. Special error correction modems exist to increase the reliability and efficiency of data transfer via the cellular system [8]. A standard telephone interface may also allow a cellular phone to operate with a touch tone or rotary phone [9]. This interface simulates a dial tone and a call is initiated without the requirement of pre-origination dialing.

Digital cellular transmission allows new possibilities for data transmission. Instead of converting digital bits (pulses) into audio tones that are sent on the radio channel, the digital bits are sent directly into the phase modulator of the Subscriber Units digital transmitter. This allows much higher data transmission rates. Unfortunately, when the digital bits are received on the other end at the Base Station, they must be converted to a format that is able to be transmitted through the Public Telephone Switching Network (PSTN).

For the analog cellular system (e.g., AMPS and TACS), digital information is converted by a modem to audio signals. For the digital cellular system, digital information is only buffered and shifted in time for direct transmission. When the modem signal on the analog cellular system is received by the Base Station, it is ready to be sent to the PSTN. When the digital information is received on the digital cellular system by the Base Station, it must be converted to a signal that can be sent on the PSTN. This is typically performed by a modem. Modems are not necessary if the PSTN has the capability to directly send digital information such as Integrated Services Digital Network (ISDN).

8.5.3 Voice Activation

Another optional feature includes voice activation which allows calls to be dialed and controlled by voice commands. It is recommended a call should not be dialed by a handset while driving [10], but a call can be initiated via voice activation without significant distraction.

Two types of speech recognition exist—speaker dependent and speaker independent. Speaker dependent requires the user to store his voice command to be associated with a particular command. These recorded commands are used to match words spoken during operation. Speaker independent allows multiple users to control the phone without the recording of a particular voice. To prevent accidental operation of the Subscriber Unit by words in normal conversation, key words such as "phone start" are used to indicate a voice command [11].

8.5.4 Battery Chargers

There are two types of battery chargers, trickle and rapid charge. A trickle charger will slowly charge up a battery by only allowing a small amount of current to be sent to the Subscriber Unit. The battery charger may also be used to keep a charged battery at full capacity if the Subscriber Unit is regularly connected to an external power source (such as a car's cigarette lighter socket). Rapid chargers allow a large amount of current to be sent to the battery to fully charge it as soon as possible. The limitation on the rate of charging is often the amount of heat generated, the larger the amount of current sent to the battery the larger the amount of heat.

The charging algorithm can be controlled by either the phone itself or by circuits in the charging device. For some batteries, rapid charging reduces the amount of charge and discharge cycles. A charger will charge for a period of time, until a voltage transient occurs (called a knee voltage) and checks for temperature of the battery. The full charge is indicated by a couple of different conditions. Either the temperature of the battery can reach a level where the charging must be turned off, the voltage level will reach its peak value for that battery type, or the voltage level will stop increasing. Most chargers will then enter a trickle charge mode to keep the battery fully charged. Some chargers for NiCd batteries discharge the battery before charging to reduce the memory effect. This is called battery reconditioning.

8.5.5 Software Download Transfer Equipment

Some Subscriber Units store their operating software in re-programmable memory (flash memory). Having this type of re-programmable type of memory allows new or upgraded software to be downloaded to the phone to add feature enhancements or to correct errors. The new operating software sometimes can be downloaded with a service accessory that normally contains an adapter box connected to a portable computer. The new software is transferred from the computer through the adapter box to the Subscriber Unit. Optionally, an adapter box can contain a memory chip with the new software which eliminates the need for the portable computer. These types of devices have been very useful for the newer technologies where the systems and specifications are not stable. Changes can be made easily in the field without opening up a Subscriber Unit. Once a system is mature the Subscriber Units will probably contain a non-re-programmable type memory device to help reduce costs.

8.5.6 Antennas

Antennas convert radio signal energy to and from electromagnetic energy for transmission between the Subscriber Unit and Base Station. While there are several factors that will affect the performance of an antenna, only a fixed amount of energy is available for conversion. Antennas can improve their performance by focusing energy in a particular direction which reduces energy transmitted and received in other directions. The amount of gain is specified relative to a unity (omnidirectional) gain antenna. Car mounted antennas typically use 3 dB or 5 dB gain antennas. Portable antennas commonly use 0 to 1 dB gain antennas because people may turn the phone to many angles or leave the phone laying flat on the table.

8.6 IS-54 TDMA Dual Mode Subscriber Units

The complexity of an IS-54 TDMA Subscriber Unit is much greater than an AMPS Subscriber Unit. IS-54 TDMA Subscriber Units use the same 30 kHz wide radio channel however the baseband (digital audio) and broadband (RF) signals are very different. Some of the unique features of IS-54 TDMA Subscriber Units include digital signal processing and phase modulation.

Figure 8.4 shows the functional diagram of the TDMA portion of an IS-54 Subscriber Unit. The transmit section converts audio from the Subscriber Unit's microphone into a 64 kbps PCM signal which is divided into 20 msec slots. A speech coder compresses the data rate of the slots to 7950 bps. The channel coder then performs 1/2 rate convolutional coding to important bits (increasing the speech coding bit rate to 13 kbps), adds control information (SACCH, FACCH, and DVCC), and interleaves the blocks. This increases the gross data rate to 48.6 kbps (16.2 kbps for each of 3 subscribers). The burst signal is supplied to a pi/4 Differential Quadrature Phase Shift Keying (pi/4 DQPSK) modulator. Mixing the modulator with the output of the RF synthesizer, this produces a 30 kHz wide radio channel at the desired frequency (824-849 MHz). The RF amplifier boosts the signal for transmission. The RF amplifier gain is adjusted by the microprocessor control section which receives power level control signals from the Base Station.

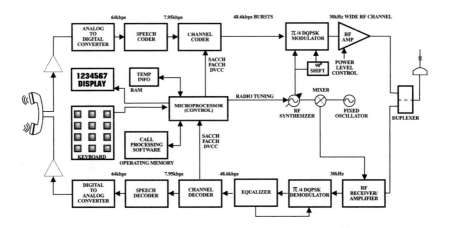

Figure 8.4, IS-54 TDMA Dual Mode Subscriber Unit Block Diagram

A received TDMA signal is down-converted and amplified by the RF receiver and amplifier section. Because the incoming radio signal is related to the transmitted signal, a fixed 45 MHz RF oscillator is mixed with the transmitter frequency synthesizer signal to produce the reference frequency for down-conversion. The down-conversion mixer produces an Intermediate Frequency (IF) signal which is digitized and supplied to a pi/4 DQPSK demodulator. The demodulator may be adjusted by an RF equalizer which helps adjust for distortions that may have occurred during transmission of the RF signal. The channel decoder extracts the data and control information and supplies the control information to the microprocessor and the speech data to the speech decoder. The channel decoder also provides Bit Error Rate (BER) information back to the Base Station which is used with the RSSI information to determine channel quality conditions received by the Subscriber Unit. The speech decoder converts the data slots into 64 kbps PCM signal which is then converted back to its original analog (audio) form.

The microprocessor section controls the overall operation of the Subscriber Unit. It receives commands from a keypad (or other control device), provides status indication to the display (or other alert device), receives, processes, and transmits control commands to various functional assemblies in the Subscriber Unit.

235

8.7   IS-136 TDMA Dual Mode Subscriber Units

IS-136 Subscriber Units (often called DCC units) are very similar to IS-54 Subscriber Units. Because the channel modulation and encoding schemes are so similar to IS-54 systems there are only a few areas that must be changed to allow conversion from an IS-54 Subscriber Unit to an IS-136 Subscriber Unit.

Figure 8.5 shows a functional block diagram of an IS-54 Subscriber Unit which has been converted to offer IS-136 capability. The call processing memory has been expanded to include the software necessary to operate on a Digital Control Channel (DCC). Memory has been added to allow short message storage. The microprocessor section can use the new sleep mode paging classes to selectively remove power from various electronic sections to increase standby time (sleep mode). And, during digital only operation, the duplexer can be replaced by a transmit/receive (T/R) switch.

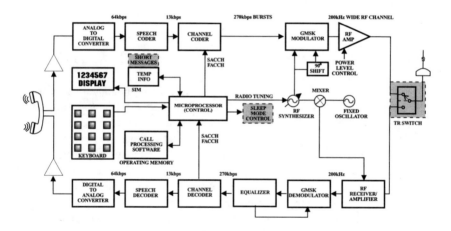

Figure 8.5, IS-136 TDMA Dual Mode Subscriber Unit Block Diagram

8.8   IS-95 CDMA Dual Mode Subscriber Unit

The complexity of an IS-95 CDMA Subscriber Unit is similar to an IS-136 Subscriber Unit. IS-95 CDMA Subscriber Units use wideband spread spectrum modulation (1.23 MHz) that allows simultaneous channel decoding. Some of the unique features include dual frequency bandwidths (30 kHz and 1.23 MHz)

which require different RF and IF filters and the addition of a rake receiver which requires additional DSP processing capability.

Figure 8.6 shows the functional diagram of the CDMA portion of an IS-95 Subscriber Unit. The transmit section converts audio from the subscriber into a 64 kbps PCM signal which is divided into 20 msec blocks. Based on the activity of the voice audio, the variable rate speech coder compresses the data rate of the blocks to 1.2 - 9.6 kbps. The channel coder then performs 1/3 rate convolutional coding, adds control information, and interleaves the blocks. This increases the data rate to 28.8 kbps. The baseband data signal is then XORed with a long code (unique subscriber ID) to produce a private data signal. The data signal is then multiplexed (called spreading) by a PseudoNoise (PN) code and a Walsh code. The PN code spreads the signal so that is unique to other radio channels in the CDMA system . The walsh code provides the unique code which identifies this Subscriber Units code from others operating on the same radio channel. The resultant spread signal is 1.2288 Mbps which is supplied to an Offset Quadrature Phase Shift Keying (O-QPSK) modulator. Mixing the modulator with the output of the RF synthesizer produces a 1.23 MHz wide radio channel

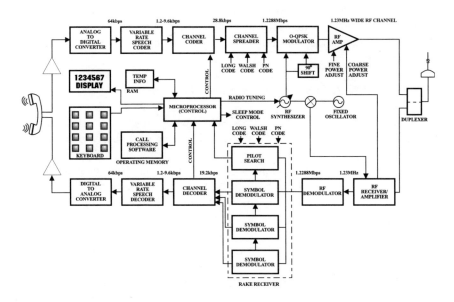

Figure 8.6, IS-95 CDMA Dual Mode Subscriber Unit Block Diagram

at the desired frequency (824-849 MHz). The RF amplifier boosts the signal for transmission. The RF amplifier gain is course adjusted by the received signal level and fine adjusted by the control signal received from the Base Station.

A received CDMA signal is down-converted and amplified by the RF receiver and amplifier section. A fixed 45 MHz RF oscillator is mixed with the transmitter frequency synthesizer signal to produce the reference frequency for frequency down-conversion. The resultant 1.23 MHz signal is digitized by the RF demodulator and supplied to the rake receiver. The rake receiver is composed of several correlators which digitally decode the primary and delayed CDMA signals. The channel decoder uses one or more (coherent combination) of these received signals for channel decoding. One of the correlators is dedicated to decoding a pilot channel. The pilot channel is used as a reference to assist in the demodulation (coherent phase reference) of the desired data signal. The gross data rate of the decoded channel is only 19.2 kbps. This is because only 1/2 rate convolutional coding is required on the forward channel due to the benefit of the pilot channel. The channel decoder extracts the data and control information. It supplies the control information to the microprocessor and the data blocks to the variable rate speech decoder. The speech decoder converts the data blocks into 64 kbps PCM signal. The PCM signal is then converted to its original analog (audio) form.

The microprocessor section controls the overall operation of the CDMA Subscriber Unit. It receives commands from a keypad, provides status indication to one or more displays, receives, processes, and transmits control commands to various functional assemblies in the Subscriber Unit. During the sleep mode cycles, the microprocessor section will selectively remove power from various electronic sections.

8.9 IS-88 NAMPS Dual Mode Subscriber Unit

The complexity of a IS-88 NAMPS Subscriber Unit is similar to an AMPS Subscriber Unit. IS-88 NAMPS Subscriber Units have the ability to use narrowband FM modulation (10 kHz) and have the capability of 100 bps out of band signaling (DSAT -vs- SAT). Some of the unique characteristics include dual frequency bandwidths (10 kHz and 30 kHz) which require different IF filters and the addition of a sub band audio channel which requires combining and extracting low frequency audio signals.

Figure 8.7 shows a functional block diagram of an NAMPS Subscriber Unit narrowband section. The transmit audio processes section prepares (enhances) the audio signal for FM transmission. The compressor section adjusts the dynamic range of the audio signal so speakers with different voice intensities (audio level) have approximately the same level of audio signal. By limiting the amount of audio level variance, the average amount of RF signal deviation is increased for low audio volume which enhances transmission performance. The pre-emphasis section provides additional gain for high frequencies. Because speech audio signals contain much of their energy at low frequencies, the pre-emphasis section increases the signal to noise ratio for the high frequency components. The limiter section ensures high levels of audio do not over-modulate the radio signal

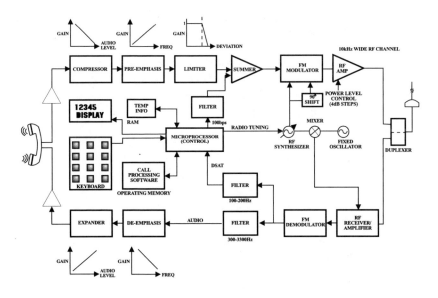

Figure 8.7, IS-88 NAMPS Dual Mode Subscriber Unit Block Diagram

(splatter signal outside regulated bandwidth). The limiter has a unity gain until the audio level exceeds a level that would cause the deviation to exceed 5 kHz. The processed audio signal is then combined with the low frequency sub band digital supervisory audio tone (DSAT). The modulator uses the RF synthesizer output with the complex audio signal (voice and data) to produce a FM modulated 10 kHz wide RF radio channel. The RF amplifier then boosts the signal the signal for transmission. The RF amplifier output power is adjusted by the microprocessor control section in 4 dB steps.

A received NAMPS radio channel signal is down-converted and amplified by the RF receiver and amplifier section. A fixed 45 MHz RF oscillator is mixed with the synthesizer to produce the reference frequency for frequency down-conversion. The resultant 10 kHz signal is FM demodulated. The demodulated audio signal is supplied to filters in the sub band digital signal so they can be separated from the audio signal. The audio signal is then converted to its original analog (audio) form by processing it through the de-emphasis and expander sections which reverses the effects of the pre-emphasis and compressor sections.

The microprocessor section of the NAMPS Subscriber Unit controls the overall operation. It receives commands from a keypad (or other control device), provides status indication to the display (or other alert device), receives, processes, and transmits control commands to various functional assemblies in the Subscriber Unit.

8.10 GSM Subscriber Unit

GSM Subscriber Units are similar to IS-136 digital only Subscriber Units. Figure 8.8 shows a functional block diagram of a GSM Subscriber Unit. The transmit section converts audio from the subscriber into a 64 kbps PCM signal which is divided into 20 msec frames. A speech coder compresses the data rate of the slots to 13 kbps. The channel coder then performs 1/2 rate convolutional coding to important bits (increasing the speech coding bit rate to 22.8 kbps), adds control information (SACCH slots and FACCH messages), and interleaves the blocks. This increases the gross data rate to 270 kbps (33.75 kbps for each of 8 subscribers). The burst signal is supplied to a Gaussian Minimum Shift Keying (GMSK) modulator. Mixing the modulator with the output of the RF synthesizer produces a 200 kHz wide radio channel at the desired frequency 890-915 MHz (1710-1785 MHz for PCN). The RF amplifier boosts the signal for transmission. The RF amplifier gain is adjusted by the microprocessor control section which receives power level control signals from the Base Station.

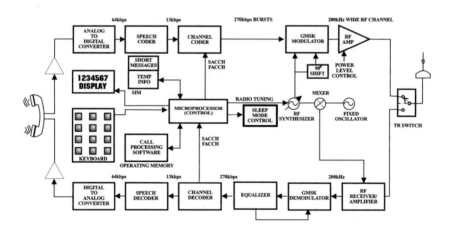

Figure 8.8, GSM Subscriber Unit Block Diagram

A received TDMA signal is down-converted and amplified by the RF receiver and amplifier section. A fixed 45 MHz RF oscillator (95 MHz for PCN) is mixed with the synthesizer to produce the reference frequency for frequency down-conversion. The resultant intermediate frequency signal is digitized and supplied to a GMSK demodulator. The demodulator may be adjusted by an RF equalizer which helps adjust for distortions that may have occurred during transmission of the RF signal. The channel decoder extracts the data and control information. It supplies the control information to the microprocessor and the speech slots to the speech decoder. The channel decoder also provides BER information back to the Base Station which is used with the RSSI information to determine channel quality conditions received by the Subscriber Unit which may better assist a handover (handoff) decision. The speech decoder converts the data slots into 64 kbps PCM signal. The PCM signal is then converted to its original analog (audio) form.

The microprocessor section controls the overall operation of the Subscriber Unit. It receives commands from a keypad (or other control device), provides status indication to the display (or other alert device), receives, processes, and transmits control commands to various functional assemblies in the Subscriber Unit. A removable memory card (SIM card) has been added to store subscriber information and provide short message storage. The microprocessor section can use the new sleep mode paging classes to selectively remove power from various electronic sections to increase standby time (sleep mode). Because the GSM system does not require simultaneous transmission and reception, the duplexer is typically replaced by a transmit/receive (T/R) switch.

References

1. William Sinnema, "Electronic Transmission Technology", pp.201-244, Prentice Hall, 1979.
2. MRT editorial staff, "Why Cellular Mobiles Use `High-Gain' Antennas", Mobile Radio Technology, pp.44-46, Volume 5, Issue 5, May 1987.
3. CTIA Narrow AMPS Forum, Chicago, IL, 9 December 1990.
4. Lawrence Harte, "Techniques", Cellular Integration Magazine, Argus Publishing, January 1995.
5. The Bell System Technical Journal, January 1979, Vol. 58, No. 1, American Telephone and Telegraph Company, Murray Hill, New Jersey.
6. Purchasing agent, Ericsson, personal interview, industry expert.
7. CTIA Winter Exposition, "Safety", San Diego, February 17, 1991.
8. U.S. Patent 4,697,281, Cellular Telephone Data Communication System and Method, Harry M. O'Sullivan, 1987.
9. U.S. Patent 4,658,096, System for Interfacing a Standard Telephone Set with a Radio Transceiver, William L. West Jr. and James E. Shafer, 1987.
10. CTIA, winter exposition, "Safety", San Diego, 1991.
11. U.S. Patent 4,827,520, Voice Actuated Control System for Use in a Vehicle, Mark Zeinstra, 1989.

Chapter 9
Cellular System Networks

9. Cellular System Networks

A key to the success of cellular systems is their ability to integrate radio technology and network intelligence. This chapter describes the basic network elements, system operation, inter-systems connection, cellular system planning, and network options. The chapter also describes how the cellular system network connects and integrates subscriber units, private networks, and the public switched telephone network (PSTN).

A cellular system is composed of cell sites, a mobile switching center (MSC), subscriber databases, and a public switched telephone network (PSTN) interconnection. In a typical cellular system, a proprietary interface provides the intra-system connections that link cell sites to the MSC. The MSC coordinates the overall allocation and routing of calls throughout the cellular system. Inter-system connections link different cellular systems to allow subscriber units to move from system to system. Inter-system connections can be proprietary, or they may use standard ITU GSM (European) or the EIA/TIA interim standard IS-41 (American) interface. The design and layout of a cellular system is a continually changing and complex process. New technologies allow for variations in the type and use of cellular system equipment.

Figure 9.1 illustrates a the fundamental interconnections in a cellular system network. Cell sites convert radio signals from subscriber units to a form suitable for transfer to the MSC. The radio signals may be analog (e.g. AMPS or TACS) radio technology or digital radio transmissions (e.g. TDMA or CDMA). If the cellular system is dual mode (e.g. digital and AMPS), each cell site's base station radio equipment can process more than one type of radio technology. Calls can be transferred between cells within the system (intra-system hand-off) or to cell sites in an adjacent system (inter-system hand-off). Currently, the MSC coordinates these processes, although this may change with the increased use of distributed switching. However, regardless of the type of radio transmission, the MSC routes calls to and from cell sites and the PSTN.

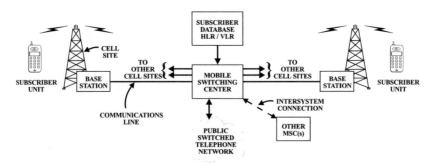

Figure 9.1, Cellular System Network Block Diagram

9.1 Cell Sites

Cell sites are composed of an antenna system (typically a radio tower), building, and base station radio equipment. Base station radio equipment consists of RF equipment (transceivers and antenna interface equipment), controllers, and power supplies. Base Station transceivers have many of the same functional elements as a subscriber unit.

The radio transceiver section is divided into transmitter and receiver assemblies. The transmitter section converts a voice signal to RF for transmission to the subscriber unit. The receiver section converts RF from the subscriber unit to voice signals routed to the MSC.

The controller section commands insertion and extraction of signaling information. Unlike the subscriber unit, the transmit, receive, and control sections of a base station are grouped into equipment racks. For example, a single equipment rack may contain all of the RF amplifiers or voice channel cards. For some analog or early-version digital cellular systems, one transceiver in each base station is dedicated for a control channel. In most digital cellular systems, control channels and voice channels are mixed on a single radio channel.

Figure 9.2 illustrates the components of a base station. Note that each assembly (equipment rack) contains multiple modules, one for each RF channel. Base station components include the following: voice cards (sometimes called line cards), radio transmitters and receivers, power supplies, and antenna assemblies. Base stations have several radio channels that subscribers use for communications. The analog (AMPS or TACS) radio channels are called voice channels, and the digital ones are called traffic channels. When cellular systems convert from analog to digital service, they add digital radio channels or use digital channels to replace analog channels.

Each base station has at least one radio channel or a portion of a digital radio channel dedicated as a control channel. The control channel provides an access channel for subscriber units attempting to use the system, and it sends system information and paging messages.

Base stations are equipped with a radio channel scanning receiver (sometimes called a locating receiver) to measure subscriber units' signal strength and channel quality during hand-off. The digital hand-off process has the advantage of using subscriber units to provide radio signal strength and channel quality information back to the switching system. This information greatly improves the MSC's hand-off decisions.

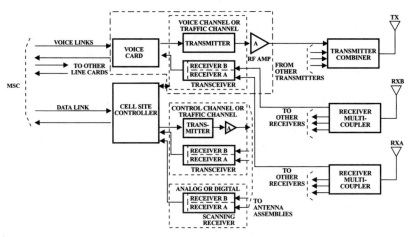

Figure 9.2, Base Station Block Diagram

9.1.1 Radio Antenna Towers

Cell site antennas can vary in height from about 20 feet to more than 300 feet. Radio towers raise the height of antennas to provide greater area coverage. Antenna systems mounted on radio towers can include equipment such as a paging system antenna or a microwave antenna to link cell sites.

The basic antenna options are monopole mount, guided wire, free standing, or man made structures such as water towers, office buildings, and church steeples. Monopole heights range from 30-70 feet; free standing towers range from 20-100 feet; and guided wire towers can exceed 300 feet. Cell site radio antennas can also be disguised to fit in with the surroundings. Figure 9.3 illustrates several antenna systems. Figure A is a free standing single pole called a monopole.

Figure B is a tower supported by guide wires. Free standing antennas (not shown) are self-supporting structures with three or four legs. Cellular system antennas can also be located on building tops (figure C) to focus their radio energy to specific areas, or disguised inside a building (figure D).

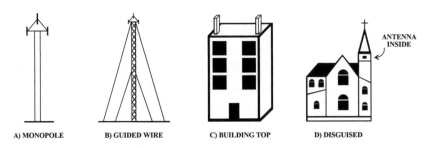

A) MONOPOLE   B) GUIDED WIRE   C) BUILDING TOP   D) DISGUISED

Figure 9.3, Cell Site Radio Towers

9.1.2 Transmitter

A Base Station transmitter contains audio processing, modulation, and RF power amplifier assemblies. An audio processing section converts audio signals from the communications link to frequency and audio levels (see subscriber unit audio processing) optimized for FM or phase shift modulation (digital). The transmitter audio section also inserts control information. A modulation section converts the audio signals into proportional phase shifts at the carrier frequency. The RF power amplifier boosts the signal to much higher power levels (typically 40 to 100 watts) than the subscriber unit (typically less than 1 Watt). Once initially set, the transmitter power level is normally fixed unless it is changed to vary cell boundaries either manually or automatically (by control section software).

9.1.3 Receiver

The Base Station receiver sections consist of an RF amplifier, demodulator, and audio processor. The RF amplifier boosts low level signals received from subscriber units to a level appropriate for input to the demodulator. The demodulator section converts the RF to audio or digital voice signals. Audio processing converts the optimized audio to its original frequency and amplitude levels. Receiver audio processing also extracts control information and converts the output audio level for transmission on the voice channel communication links to the MSC. In addition to all of these functions, most receivers are also able to select or combine the strongest radio signals that are received on several antennas at the base station, a process called diversity reception.

9.1.4 Controller

The controller sections consist of control signal routing and message processing. Controllers insert control channel signaling messages, set up voice channels, and operate the radio location/scanning receiver. In addition, controllers monitor equipment status and report operational and failure status to the MSC. Typically, there are three types of controllers: base station controller, base station communications controller, and transceiver communications controller.

The base station controller coordinates the operation of all base station equipment based on commands received from the MSC. The base station communications controller buffers and rate-adapts voice and data communications from the MSC. The transceiver communications controller converts digital voice information (PCM voice channels) from the communications line to RF for radio transmission and routes signals to the subscriber units. The transceiver controller section also commands insertion and extraction of voice information and digital signaling messages to and from the radio channel.

9.1.5 RF Combiner

The RF combiner is like the duplexer in a subscriber unit (see chapter 8). Each of a base station's many radio channels is served by a dedicated RF amplifier. The RF combiner allows multiple RF amplifiers to share one antenna without their signals interfering with each other. RF combiners are narrow bandpass filters with directional couplers that allow only one specified frequency to pass through. The filtering and directional coupling prohibits signals from one amplifier from leaking into another. A later section in this book discusses an application of a broadband linear amplifier that may eliminate the need for multiple amplifiers and RF combiners.

9.1.6 Receiver Multi-Coupler

To allow one antenna to serve several receivers, a receiver multi-coupler must be attached to each receiving antenna. Figure 9.5 illustrates a receiver multi-coupler assembly. Because a receiver multi-coupler output is provided for each receiver antenna input, the splitting of received signal reduces its total available power. By increasing the number of receivers, the signal to noise ratio to each receiver section is reduced. Often, low noise RF preamplifiers are included to boost the low level received signals prior to the RF multi-coupler splitter.

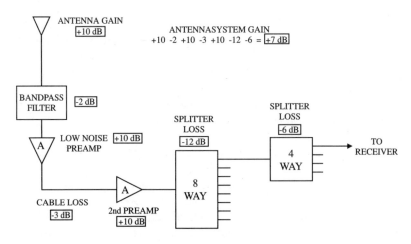

Figure 9.5, Receiver Multi-Coupler

9.1.6 Communication Links

Communication links carry both data and voice information between the MSC and the cell sites. Options for the physical connections include wire, microwave, or fiber optic links. Alternate communication links are sometimes provided to prevent a single communication link failure from disabling communication [1]. Some terrain conditions may prohibit the use of one type of communication link. For example, microwave systems are not usually used in extremely hilly or earthquake areas because they require precise line-of-sight connection. Hills or other obstructions can block microwave signals, and small shifts in the earth can misalign microwave transceivers to break communications.

Regardless of the physical type of communication link, the channel format is usually the same. Communication links are typically digital time-multiplexed to increase the efficiency of the communication line. The standard format for time-multiplexing communication channels between cell sites in North America is the 24 channel T1 line, or multiple T1 channels. The standard format outside of North America is the 32 channel (30 useable channels) E1 line.

Figure 9.6 illustrates T1 (North American) and E1 (European) standard communication links. A T1 communication link is divided into time frames which contain 24 time slots plus a framing bit. To allow for control signaling, some bits are stolen from various slots. An E1 communication link is divided into time frames which contain 32 time slots. Two of the E1 time slots are dedicated as synchronization and signaling control slots.

Each slot contains eight bits of information. For standard land line voice transmission, each analog voice channel is sampled 8000 times per second and converted to an 8 bit PCM digital word. The 8000 samples x 8 bits per sample results in a data rate of 64 kbps, and it is called a DS0 or PCM (one channel).

For T1 communication lines, 24 DS0's plus a framing bit are time multiplexed onto the high speed T1 channel frame. Therefore, with a frame length of 193 bits, the data rate is 8000 x 193 = 1.544 Mbps. For E1 communication lines, 32 PCM channels are time multiplexed, resulting in a gross data rate of 2.048 Mbps.

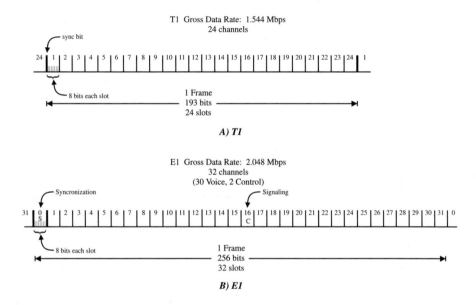

Figure 9.6, T1 and E1 Communication Links

9.1.8 Antenna Assembly

When a cellular system is first established, base station antenna assemblies usually employ omni-directional antennas [2]. As the system matures, directional (sectored) antennas replace the original antennas to reduce interference. An antenna assembly in each sector usually consists of one transmitting antenna and two receiving antennas.

Separate transmit and receive antennas are used to keep excessive amounts of the transmitter's RF energy from being coupled into receive antennas. The few feet of separation between the antennas provides more than 40 dB of isolation. In some installations, where antenna tower platform space is limited, and three antennas cannot be used, one antenna must also be used for transmitting. In this case, a very deep notch isolation filter is used to prevent the transmit signal from leaking into receivers on the shared antenna.

Two receive antennas are used for diversity reception to minimize the effects of Rayleigh signal fading (discussed later). Using two receiving antennas enables a technique that employs two receivers per channel to select the antenna which is receiving the stronger signal or allows the combination of the RF energy of both antennas. The technique improves power reception up to 6 dB, improves the signal to noise ratio (S/N) up to 3 dB, and reduces the effects of fading signals.

9.1.9 Scanning or Locating Receiver

A scanning or locating receiver measures subscriber unit signal strength for hand-off decisions. It can tune to any channel and measure the received signal strength, from which it determines a subscriber unit's approximate distance from the base station.

When the signal strength falls below a level determined by the serving base station transceiver, the base station signals the MSC that a hand-off will be necessary soon. The MSC then commands one or more adjacent cell sites to tune their scanning receiver(s) to monitor the subscriber unit's radio channel and continually measure and report the signal strength. The MSC (or other base station controller) compares the reported signal strength with other signal levels to decide hand-off.

All of the new cellular technologies provide for hand-off decisions assisted by the subscriber unit. Using signal and interference levels from subscriber units, the cellular system can better decide when a hand-off may be necessary. The information from subscriber units also reduces the burden of communications between adjacent base stations while coordinating (passing signal quality information) the hand-off process. The digital subscriber unit's hand-off information can possibly eliminate the need for a base station scanning or locating receiver.

9.1.10 Power Supplies and Backup Energy Sources

Power supplies convert the base station power source to regulated and filtered AC and DC voltage levels required by the base station electronics assemblies. Batteries and generators are used to power a base station when primary power is interrupted. Backup power is also needed for radio equipment and cooling systems. In 1989, a hurricane destroyed almost all land line communications in parts of Puerto Rico. Due to good planning, cellular communications were unaffected, and became the primary communication link [3].

9.1.11 Maintenance and Diagnostics

Base station radio equipment must be maintained and repaired as equipment assemblies fail, and cell sites are often remote and scattered throughout a cellular system. In response to maintenance needs, advanced maintenance and diagnostic tools have been created.

Base stations require software to operate. Base station operating software can be installed at the factory or loaded by the maintenance technicians after the cell site's radio equipment has been installed. As base station software improves, new software must be installed into the base station controllers. Some systems do not require technicians to visit the cell site, but instead download new software via the communication links. During the software download process, one or more communication link voice channels (some time slots) are dedicated to transfer software programs.

The base station continuously monitors its status of operation and performance and sends it to the MSC. If spare equipment is available when base station equipment fails, the controller can reconfigure to continue service. Other maintenance tasks include routine testing to detect faults before they affect service. Routine test functions operate in a background mode and are suspended when a faults are detected. Diagnostics begin and status reports may be continuously printed to inform system operators that maintenance may be required.

If a system fails, the suspect equipment assemblies must be tested to isolate faults. To verify equipment performance and monitor operational status, test signals are inserted at various points. Loop-back testing inserts test signals on one path (such as the forward direction) of the system and monitors the response of the signals on a return path.

Figure 9.8 illustrates two loop-back test paths used to test a base station from the MSC. To test the communication link between the MSC and base station, the MSC sends an audio test signal to the base station on a voice channel (path 1). The line card then returns the test signal to the MSC via another voice path. If the return is unsuccessful, the fault is in one of the two voice channels or the voice channel card interface. To determine if the radio transmitter and receiver are working correctly, a second test path (path 2) routes a test audio signal through transmission and reception equipment. The test samples a portion of the output signal before the 45 MHz frequency shift and transmission via the antenna section, then directs the signal to the receiver section. If the MSC receives the test signal, the RF transmission and reception equipment are operational. Other loop-back paths can isolate other network equipment assemblies.

An alternative method of determining if the base station transmission equipment is operating correctly is to install a subscriber unit in each base station. If the subscriber unit can be operated remotely (possibly via a standard telephone voice channel), test signals which are transmitted by the base station can be sampled and verified (path 3) [4].

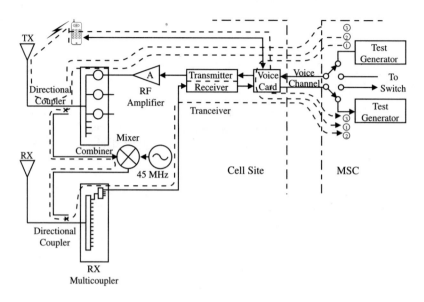

Figure 9.8, Cellular System Loop-Back Testing

9.2 Mobile Switching Center

Formerly called Mobile Telephone Switching Office (MTSO), the Mobile Switching Center (MSC) coordinates all communications channels and processes. The MSC processes requests for service from subscriber units and land line callers, and routes calls between the base stations and the public switched telephone network (PSTN). The MSC receives the dialed digits, creates and interprets call processing tones, and routes the call paths.

Figure 9.9 illustrates an MSC's basic components: system and communication controllers, switching assembly, operator terminals, primary and backup power supplies, subscriber unit database registers, and, in some cases, an authentication (subscriber validation) center.

A system controller coordinates the MSC's operations. A communications controller adapts voice signals and controls the communication links. The switching assembly connects the links between the base station and PSTN. Operator terminals are used to enter commands and display system information. Power supplies and backup energy sources power the equipment. Subscriber databases include a Home Location Register (HLR), used to track home subscriber units, and a Visitor Location Register (VLR) for subscriber units temporarily visiting the system. The authentication center (AC) stores and processes secret keys required to authenticate subscriber units.

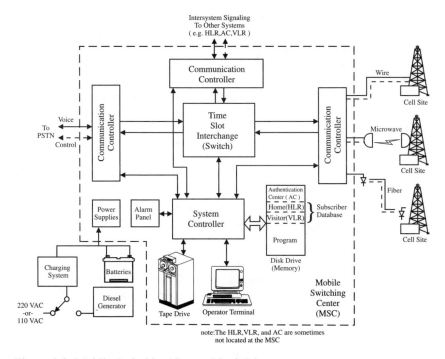

Figure 9.9, Mobile Switching Center Block Diagram

9.2.1 Controllers

Controllers coordinate Base Stations, MSC switching functions, and PSTN connections. A system controller (or a subsection) creates and interprets commands between the MSC and the base stations, controls the MSC switch, validates customers requesting access, maintains air time and PSTN billing records, and monitors for equipment failures. Communication controllers process and buffer voice and data information between the MSC, Base Stations, and PSTN [5].

The communication controllers combine the channels from multiple voice channels on T1 or E1 lines in one high speed data channel. They also reverse the process by separating the voice paths from the high speed data channel and routing them to T1 or E1 lines. The high speed data channel contains time slots representing each voice path. The time slot for each voice path is switched in or out of a memory location in the switching matrix. The communication controllers also route call control commands (e.g. hang up, dialed digits) to the control assembly.

9.2.2 Switching Assembly

The switching assembly connects base stations and the PSTN with either a physical connection (analog) or a logical path (digital). Early analog switches required a physical connection between switch paths. Today's digital cellular switches use digital communication links.

A switching assembly is a high speed matrix memory storage and retrieval system that provides virtual connections between the base station voice channels and the PSTN voice channels. Figure 9.10 illustrates a simplified switching matrix system. Time slots of voice channel information are input through switch S1 to be sequentially stored in the PCM data memory. Time slots that are stored in the PCM memory are retrieved and output through S2 to the slots which are routed to another network connection (such as a particular PSTN voice channel). Switch 1 (S1) is linked to switch 3 (S3) and switch 4 (S4) so that each moves to predetermined memory locations together (e.g. all at position 1). The address in the control memory determines the position of switch 2 (S2). The switch matrix controller determines which addresses to store in control memory slots. This address system matches input and output time slots.

9.2.3   Communication Links

Communication links are dedicated lines which transfer several channels (usually 64 kbps voice channels) of information between the base station and switch-

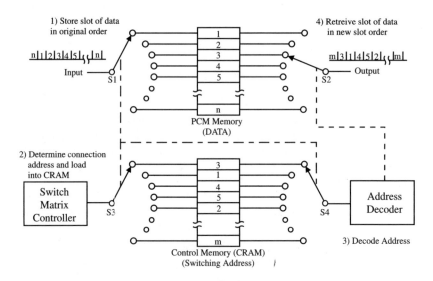

Note: Switches (S1-S4) are electronic and do not use mechanical parts.

Figure 9.10, Time Slot Interchange (TSI) Block Diagram

ing system. In addition to the primary purpose of carrying voice traffic, one channel is typically reserved for control information (data) which allows the MSC to command the base station. To ensure reliability, alternate communication links can route channels through different network points if one or more communication links fails or is unavailable. Most communication links between cell sites and MSC's use a T1 or E1 Time Division Multiplex (TDM) Pulse Coded Modulation (PCM) digital transmission system.

When many communication channels are required, high speed links can be used. High speed communication links are combined lower speed links. Even though MSCs may connect to a PSTN with hundreds of voice channels, MSC's use several T1 or E1 lines to provide the many voice channels needed. Table 9.1 illustrates the commonly used communication links.

9.2.4 Operator Terminals

Operator terminals control most maintenance and administrative functions. The operator terminal is usually a computer monitor and keyboard dedicated to con-

No of Channels	Digital Signal Number	System Input	Bit Rate (Mbps)
1	DS0	1 PCM Voice	.064
24	DS1 (T1)	24 DS0	1.544
48	DS1C	2 DS1	3.152
96	DS2	4 DS1	6.312
672	DS3	28 DS1	44.736
4032	DS4	6 DS3	274.176
32	E1	30 PCM + 2 control	2.048
128	E2	4 E1	8.448
512	E3	4 E2	34.368
2048	E4	4 E3	137.462

Table 9.1, Communication Line Types

trolling equipment (e.g. turn on a radio channel) and modifying the subscriber database. Operator terminal(s) might or might not be at the MSC, and there may be more than one.

9.2.5 Backup Energy Sources

Backup energy sources are required to operate the cellular network system when primary power is interrupted. Backup energy to power switching equipment, subscriber databases, and cooling systems is usually a combination of batteries and diesel generators. During normal operations, batteries are charged with a charger using primary power. The batteries are directly connected to the cellular system, and when outside power is interrupted, they immediately and continuously power the system. After a short period of power loss, a diesel generator automatically begins to power the battery charger.

9.2.6 Home Location Register

The Home Location Register (HLR) is a subscriber database containing each customer's Mobile Identification Number (MIN) and Electronic Serial Number (ESN) to uniquely identify each customer. Each customer's user profile includes the selected long distance carrier, calling restrictions, service fee charge rates, and other selected network options. The subscriber can change and store the changes for some feature options in the HLR (such as call forwarding). The MSC system controller uses this information to authorize system access and process individual call billing.

The HLR is a magnetic storage device for a computer (commonly called a hard disk). Subscriber databases are critical, so they are usually regularly backed up, typically on tape, to restore the information if the HLR system fails.

9.2.7 Visitor Location Register

The Visitor Location Register (VLR) contains a subset of a subscriber's HLR information for use while roaming. The VLR eliminates the need for the visited MSC to continually check with the visitor's HLR each time access is attempted. The visitor's information is temporarily stored in the VLR memory, and then erased either when the subscriber unit registers in another system or after a specified period of inactivity.

9.2.8 Billing Center

A separate database, called the billing center, keeps records on billing. The billing center receives individual call records from the HLR. The billing records are converted into automatic message accounting (AMA) format to collect and process the information. The billing records are then transferred via tape or data link to a separate computer (typically "off-line") to generate bills and maintain a billing history data base.

9.2.9 Authentication Center

The Authentication Center (AC) stores and processes information required to authenticate a subscriber unit. During authentication, the AC processes information from the subscriber unit and compares it to previously stored information. If the processed information matches, the subscriber unit passes.

9.3 Public Switched Telephone Network

The Public Switched Telephone Network (PSTN) is the land line telephone system connects a subscriber unit to any telephone connected to the PSTN. Cellular subscriber units, land line plain old telephone service (POTS), and other networks such as private automatic branch exchanges (PABX), all have different capabilities. Unfortunately, some control messages (such as calling line indicator) cannot be sent between the cellular subscriber unit and different telephone networks, prohibiting some advanced features that digital systems could offer.

Figure 9.11 is an overview of the PSTN system. Two types of connections are shown: voice and signaling. End office (EO) and tandem office (TO) switches route voice connections. The end office (EO) switch is nearest to the customer terminal (telephone) equipment. Tandem office (TO) switching systems connect end office (EO) switches when direct connection to an end office is not economically justified. Tandem office switches can be connected to other tandem office switches. Signaling connections are routed through a separate signaling network called Signaling System number 7 (SS7). The SS7 network is composed of Signaling Transfer Points (STPs) and Signaling Control Points (SCPs). A Signaling Transfer Point (STP) is a telephone network switching point that routes control messages to other switching points. Signaling Control Points (SCP) are databases that allow messages to be processed as they pass through the network (such as calling card information).

Two types of land line networks are shown: a local exchange carrier (LEC) and inter-exchange carrier (IXC) network. LEC providers furnish local telephone service to end users. An Inter-Exchange Carrier (IXC) is the long distance service provider. In some countries, local and long distance providers are operated by the same company or government agency. In the US, government regulations prohibit directly connecting an IXC to some end-user equipment, requiring a Point of Presence (POP) connection. The POP connection is a location within a Local Access and Transport Area (LATA) designated to connect a Local Exchange Carrier (LEC) and an Inter-exchange Carrier (IXC). A LATA is typically a geographic service area boundary for the Local Exchange Carriers (LECs).

The MSC is the gateway from the cellular system to the public switched telephone network (PSTN). Various types of connections can connect the MSC to the PSTN: Type 1, Type 2A or Type 2B. Type 1 connects the MSC with an End Office (EO). A Type 2A ties the MSC into a tandem office. Type 2B occurs in conjunction with a Type 2A, and connects to an end office to allow alternate routing for high usage.

9.4 Cellular Network System Interconnection

Subscribers can only visit different cellular systems (ROAM) if the systems communicate with each other to hand-off between systems, verify roamers, automatically deliver calls, and operate features uniformly. Fortunately, cellular systems

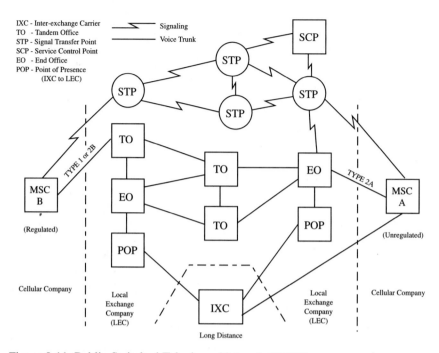

Figure 9.11, Public Switched Telephone Network (PSTN)

can use standard protocols to directly communicate with each other. These inter-system communications use brief packets of data sent via the X.25 packet data network (PDN) or the SS7 PSTN signaling network. SS7 and X.25 are essentially private data communication networks. SS7, which is used by the telephone companies, is available only to telephone companies for direct routing using telephone numbers. The X.25 network does not route directly using telephone numbers. Some MSCs also use other proprietary data connections. No voice information is sent on the SS7 or X.25 networks. Only inter-system signaling (such as IS-41) is sent via these networks to establish communication paths. Voice communications are routed through the PSTN.

Ideally, inter-system signaling is independent of cellular network radio technology, but this can be difficult between systems where radio technologies differ. Consider inter-system hand-off between a CDMA-capable and a TDMA-capable cell site (assuming the subscriber unit were capable of both). The CDMA system uses soft hand-off while TDMA does not. As new features in cellular systems

change, inter-system signaling messages must change to support them.
Communication between MSCs is performed either by a proprietary or standard protocol. Standard protocols such as IS-41 allow MSCs of different makes to communicate with few or no changes to the MSC. Regardless of whether a standard (e.g. IS-41) protocol or a manufacturers private (proprietary) protocol is used, data transfer via inter-system signaling is the same. If changes are required to communicate with a different protocol, an interface (protocol converter) changes the proprietary protocol to standard protocol. The interface has a buffer which temporarily stores data elements being sent by the MSC and reformats it to the IS-41 protocol. Another buffer stores data until it can be sent via the control signaling network.

9.4.1 Inter-System Hand-off

Inter-system hand-off links the MSCs of two adjacent cellular systems during the hand-off process. During inter-system hand-off, the MSCs involved continuously communicate their radio channel parameters. Figure 9.12 illustrates inter-system hand-off between two different manufacturers' MSCs. The process begins when the serving base station (#1) informs the MSC (system A) that a hand-off is required. The MSC determines that a base station in an adjacent system is a potential candidate for hand-off. The MSC requests the adjacent MSC (system B) to measure the subscriber unit's signal quality. Both base stations (#1 and #2) measure the subscriber unit's signal quality until hand-off. In many cases, hand-off may be immediate. The serving MSC (system A) compares its measured signal strength with the signal strength that the MSC in system B measures. When the system B MSC measures a sufficient signal, the system A MSC requests the hand-off. Base station #1 issues the hand-off command, informing the subscriber unit to tune to another frequency, and base station #2 begins communicating to the subscriber unit on the new frequency. The voice path is then connected from the anchor (original) MSC to the system B MSC, and the call continues. After the anchor MSC receives a confirmation message that the subscriber unit is successfully operating in system B, the radio resources in the original cell site become available.

During inter-system hand-off, adjacent MSCs are typically connected by a T1 or E1 link, providing both inter-system messaging and voice communications. An external land line connection between the MSCs is technically possible, but the setup time between them is prohibitive.

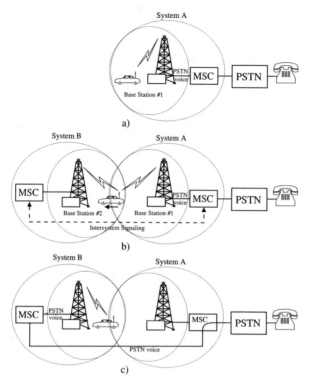

Figure 9.12, IS-41 Inter-System Hand-off

9.4.2 Roamer Validation

Roamer validation is the verification of a subscriber unit's identity using registered subscriber information. Validation is necessary to limit fraudulent use of cellular service. The two types of roamer validation are "post-call" and "pre-call". Post-call validation occurs after a call is complete, and pre-call validation occurs before granting access to the system.

During the early deployment of cellular systems, the limited connections between systems resulted in delays of minutes or even hours before roaming cellular subscribers could be validated. To allow customers to use the phone immediately, early systems used post-call validation. Pre-call validation became possible when improved inter-system interconnection greatly reduced validation time.

Figure 9.13 illustrates roamer validation. When a subscriber unit initiates a call in a visited system (step 1), the cellular system attempts to find the subscriber unit's ID in its visitor location register (VLR). The visited system determines that the subscriber unit is not registered in its system (step 2). Using the subscriber units ID (phone number), the visited cellular system sends a message to the subscriber unit's home system requesting validation (step 3). The HLR compares the ESN and MIN to determine if it is valid (step 4). If the subscriber proves valid, the HLR responds to MSC-A to indicate that validation was successful (step 5). After MSC-A receives confirmation that the visiting subscriber unit is valid, the call is processed (step 6). MSC-A's VLR may then temporarily store the subscriber unit's registration information to validate the subscriber's identity rather than requesting validation from the home system again for the next call. After a pre-determined period of subscriber unit inactivity, the information stored in the VLR will be erased. If the subscriber unit was recently operating in another cellular system, the home system informs the old visited system that the subscriber unit has left. This allows the old visited system to erase the subscriber unit's identification information.

9.4.3 Authentication

Authentication is the exchange and processing of stored information to confirm a subscriber unit's identity. Authentication is significant because roamer valida-

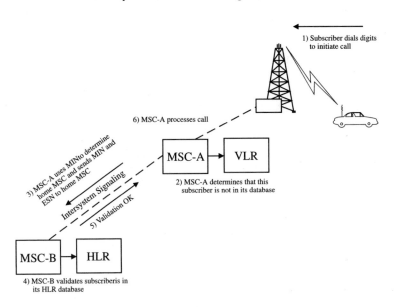

Figure 9.13, Roamer Validation

tion cannot detect cloned (duplicated) subscriber units.

New technologies offer a new authentication process to verify a subscriber's identity. The new process transfers stored information between the subscriber unit and an authentication center (AC). The two primary options for inter-system authentication are: 1) the visited MSC can use a temporary key, or 2) the MSC can request the authentication center (AC) to validate subscriber units each time. If the AC validates the SU each time, the visited MSC must send all the authentication parameters. If the AC provides a temporary key, the visited MSC can use the key while the subscriber is in the system without validation from the home system for each call.

Figure 9.14 illustrates subscriber unit authentication. When a subscriber unit detects a new cellular system (new system identifier), it attempts to register with the system (step 1). The visited system searches for the subscriber unit's ID in it's visitor location register (VLR) and determines that the subscriber unit is not yet registered. The visited cellular system uses the subscriber unit's ID (phone number) to request authentication of identity (step 2) from the subscriber's home cellular system. If the home system information processes correctly, the authentication center (AC) validates the registration request (step 4). The AC either confirms validation or creates a key for future authentication using information received from the home system (step 5). MSC-A's VLR then temporarily stores the subscriber's registration information (step 6) for future authentication without contacting the subscriber's home HLR for the next call. After a predetermined period of inactivity, the temporary authentication information stored in the VLR will be erased.

9.4.4 Automatic Call Delivery

Ideally, call delivery is completely automatic whether the subscriber unit is in its

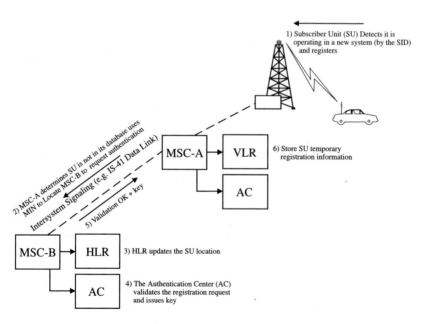

Figure 9.14, Authentication

home system or visiting another system. Such automatic delivery requires the home system to continuously track the subscriber unit's location. Roamer validation is the means for providing this information back to the home system.

To enable voice connection between the home system and visited system, a Temporary Location Directory Number (TLDN) is assigned for each automatic call delivery request.

Figure 9.15 illustrates basic inter-system call delivery. When a home system (MSC-A) receives a call for its subscriber, the MSC checks its home location register (HLR) to determine if the subscriber unit is operating in another cellular system (step 1). The home MSC then sends a request to the visited MSC for a TLDN (step 2). The TLDN is cross-referenced with the MIN in the VLR (step 3). The home system (MSC-H) then initiates a call to the TLDN (step 4). The visited system (MSC-A) receives the call and finds the TLDN number is listed in it's VLR (step 5). It replaces the TLDN with the subscriber unit's MIN which was previously stored in the VLR, and pages the subscriber unit (step 6). When the subscriber answers, the call is connected (step 7). The TLDN can be placed back into a pool of TLDNs to be used for other calls.

9.4.5 Subscriber Profile

Features such as call forwarding, three-way calling, and call waiting activation options may operate differently in different cellular systems. Although EIA/TIA

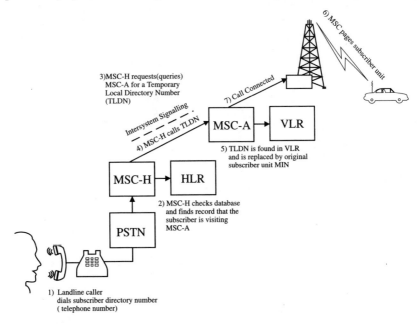

Figure 9.15, Automatic Call Delivery

IS-53 and other standards have standard system feature specifications to resolve many of these issues, it may be undesirable to change established feature controls (such as dialing *69). Also, some features are linked to special services that go beyond functional operation (such as speed dialing). For subscribers to experience the same feature operation while roaming, system feature operations can be transferred to the visited system using inter-system messaging.

Figure 9.16 illustrates how subscriber profile features operate in a visited system. During initial registration in the visited system, the subscriber unit's feature profile can be transferred to the visited system's VLR. When the roaming subscriber activates a feature by dialing a digit pattern (step 1), the VLR determines that the subscriber has selected a feature that its home system must process (step 2). MSC-A sends the dialed digit feature request to the subscribers home system (HLR) for processing (step 3). The HLR reviews the subscriber's feature profile, processes the request (e.g. call forwarding), and determines what tasks the visited MSC must perform (step 4). MSC-B then sends any additional operation instructions to MSC-A (step 5). Finally, MSC-A completes processing the feature request (step 6).

Inter-system specifications continue to evolve as new features are added. If the inter-system signaling protocol does not include the profile feature request (such as IS-41 revisions 0 through B), a failure tone is given and the advanced service are not provided.

9.5   Cellular System Planning

Cellular system design is tedious, ongoing, and unique for each cellular system. A cellular system is built upon radio coverage areas created by RF transmission

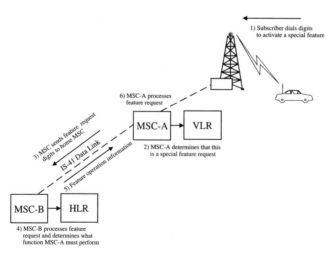

Figure 9.16, Roaming Subscriber Feature Operation

from interconnected base stations. Each separate radio coverage area must be planned to work together for the good of the whole system. The selection of frequencies and power levels at each base station determines the cell site and system serving capacity limits. Transmission power levels are influenced by antenna directivity, gain, and type, and by surrounding terrain.

Radio propagation factors, strategic planning (e.g. key locations), system testing and validation, and system expansion (capacity) requirements all affect cellular system design. Radio propagation characteristics, which change with the seasons, affect radio signal quality. Strategic planning requirements include data acquisition, MSC and cell site location selection, equipment procurement, and validation. System capacity expansion involves optimizing the existing system and adding/dividing radio service coverage areas.

9.5.1 Radio Propagation

Cellular system radio propagation has unique attenuation, signal fading characteristics, and signal quality requirements. Attenuation varies as a function of distance and terrain. Fading characteristics result in variations in received signal levels over short distances. Signal quality is limited by both received level and interfering signal strength.

Attenuation

For RF energy radiating into free space, power density decreases at more than 20 dB/decade of distance. At 10 times the distance from the transmitter (e.g. 10 meters to 100 meters), received power decreases by a factor of 100. RF energy its partially transmitted through obstructions such as buildings, but attenuation through such obstructions is about 40 dB per decade [6], and attenuation in heavy foliage can be more than 60 dB per decade [7]. Therefore, as seasons change and leaves fall, decreased attenuation can increase cell site areas.

Fading Characteristics

Radio signal fading can cause weak or dead spots in radio coverage areas so that subscribers hear a noticeable drop in audio quality and volume. In a deep signal fade, transmission can become distorted or interrupted, and subscribers must move to an area with better signal strength. A particular type of radio signal fading called Rayleigh fading causes dead spots every few inches.

Rayleigh fading results from multipath propagation when the same signal is received at slightly different times. Combining in- and out-of-phase signals varies signal strength. When two radio signals are added in phase, their combined maximum power is higher than either individual signal. Adding two equally strong signals that are out of phase by 180 degrees can nearly eliminate the radio signal. The statistical variation of such fluctuations is Rayleigh fading. Figure 9.17 illustrates a simplified Rayleigh fading variation where the fades occur approximately every 1/2 wavelength. At 840 MHz (US cellular frequency band), deep fades (weak spots) are about 7 inches (18 cm) apart. At 1900 MHz (PCS/PCN frequency band), deep fades are about 3 inches (8 cm) apart.

Radio signal fades are frequency and time dependent. It is unlikely that a signal fade on one frequency will be occurring on another frequency at the same time. The new digital technologies overcome some of the effects of Rayleigh signal fading either by using a very wide radio channel, by slow frequency hopping, or by handing the call off to another radio channel that is not experiencing the fad-

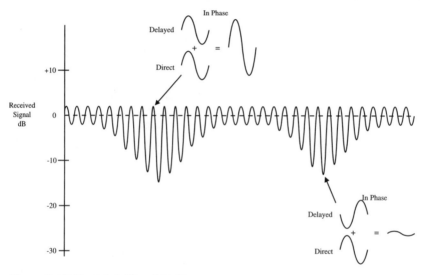

Figure 9.17, Rayleigh Signal Fading

ing characteristics.

Signal Quality

Signal quality varies throughout a cellular system coverage area, and may be degraded by a low carrier-to-noise (C/N) ratio or a low carrier-to-interference (C/I) ratio. In a C/N limited environment, communications quality decreases as the subscriber unit signal strength declines and thermal noise (natural background noise) becomes more audible. In a C/I limited environment, communications quality decreases as the subscriber unit moves nearer to a strong interfering signal. In a single cell system where the subscriber unit can wander out of cellular service completely, signal strength available to overcome the background noise (C/N ratio) is the limiting factor. In mature cellular systems, as most in the US are rapidly becoming, the limiting factor is signal strength available to overcome interference from other cell sites (C/I ratio).

266

To reuse frequencies, cellular systems rely on attenuation to prevent interference. The minimum allowable distance between cell sites is determined by the maximum desired carrier to interference ratio (C/I). Once the distance to cell radius (D/R) ratio is calculated, frequencies are selected to avoid interference among cell sites. For AMPS cellular systems, the reuse distance is typically 4.6. This means that if a cell site with a one-mile radius is assigned channel 424, then channel 424 could not be reused at any cell site nearer than 4.6 miles.

9.5.2 Strategic Planning

Strategic planning for a cellular service provider involves setting company goals, such as subscriber growth, quality of service, and cost objectives. It also involves making plans to realize those goals. Building and expanding a cellular system requires collecting demographic information, targeting key locations, selecting potential cell and MSC sites, conforming to government regulations, purchasing equipment, construction, and testing validation.

Most providers gather physical and demographic information first. For example, transportation thoroughfares, industrial parks, convention centers, railway centers, and airports may be identified as possible high usage areas. Estimates of traffic patterns help to target coverage areas for major roadway corridors. Terrain maps, marketing data, and demographic data are all used to divide the cellular system into RF coverage areas. The object is to target gross areas where cell site towers may be located. The raw data needed might include system specifications, road maps, population density distribution maps, significant urban center locations, marketing demographic data, elevation data, and PSTN and switch center locations.

For the US market, government regulations include quality of service and time intervals for service offerings [8]. While business considerations may indicate that radio coverage is not necessary (e.g. an unpopulated rural area), government regulations may require that area to be covered within a specified period.

After systems are planned, equipment manufacturers and their systems are reviewed and purchase contracts are signed. During various stages of equipment installation, validation testing is performed to ensure that all of the planning goals are being realized.

After the system is planned and cell site locations are selected, RF simulation begins. Data gathered from elevation and terrain data are used to estimate expected signal strengths and quality levels.

Typically, different colors indicate signal strength levels overlaid on maps of the service area. System simulations may predict estimated signal coverage and performance levels, but to be certain, temporary cell sites are often tested using a crane to lift a temporary antenna to the planned tower height.

9.5.3 Frequency Planning

When deploying narrowband radio channels, nearby cell sites cannot use the same frequency. Therefore, the frequencies to be used in each cell site must be planned to account for differing cell site boundaries and terrain conditions that enhance or reduce interference from nearby cell sites.

Radio channels in the same cell site can be separated by as little as 90 kHz for AMPS and US TDMA, only 30 kHz for NAMPS channels, 200 kHz for GSM. No separation is required for IS-95 CDMA. However, frequency planning requirements typically specify that narrowband radio channels attached to the same antenna be separated by several radio channel frequency bands. For analog (AMPS, TACS, and NAMPS) and US TDMA (IS-54 and IS-136), the separation is typically 21 channels. For GSM, separation is typically 4 channels. For IS-95 CDMA, the same radio channel can be used in each cell.

Sectorization

Sectorization adds radio channel capacity by dividing a cell site's radio coverage area into sectors. The number of radio channels used in the cell site is multiplied by the number of sectors. For example, when three 120-degree sectors are used, antennas focused in one direction interfere less with antennas in the opposite direction. Therefore, frequencies in other cell sites can be reused more often.

9.5.4 System Testing and Verification

System testing and verification determine if RF coverage quality is good and if the system is operating correctly. Signal quality is tested for individual cells first, then adjacent cells are measured to determine system performance. System operation is verified by measuring the signal quality level at hand-off, blockage performance, and the number of dropped calls.

RF coverage area verification confirms that a minimum percentage of coverage area is being served. Received signal strength and interference levels are measured to locate holes where terrain and obstructions cause the signal levels to fall below an acceptable level. Such coverage holes may require another cell site or repeaters to amplify the existing signal and refocus the energy into the dead spot (e.g. a parking garage).

System operation can be verified by recording the signal levels received by a test subscriber unit during hand-off and access. Blocking probability can be estimated from the number of access attempts rejected by the system. The test equipment shown in figure 9.19 monitors and records the signal quality levels as the test subscriber unit moves throughout the system area. The resultant data can then be plotted to determine where signal levels and hand-off thresholds need to

Figure 9.19, Cellular System Monitoring Equipment
Source: LCC

be increased and decreased.

9.5.5 System Expansion

Cellular systems are expanded to allow more customers to obtain service in a given area. This can be performed either by adding cell sites (adding radio coverage areas) or by increasing radio spectrum efficiency. All new technologies increase radio spectrum efficiency by allowing multiple users to share a single radio channel or by allowing more radio channels into the same frequency spectrum.

Cellular systems can expand by adding cell sites (called cell division), but at a high cost. For example, replacing a 15-mile radius cell site with 1/2-mile radius cell sites would require more than 700 new cell sites. All of the new cell sites would need to be interconnected, and hand-off switching would increase dramatically. Figure 9.20 illustrates how cellular systems can expand through dividing the coverage area.

Cellular system capacity can be expanded with new radio technologies. By adding or replacing low capacity radio channels (such as AMPS) with high efficiency radio channels (such as TDMA or CDMA), each cell site can serve more customers. Unfortunately, such expansion requires new hardware and network software for new technologies.

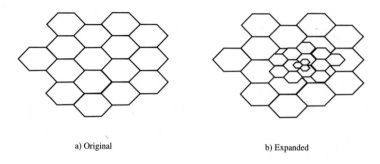

Figure 9.20, Cellular System Expansion Via Cell Division

9.6 Network Options

While industry specifications may define how radio channels and messages are coordinated, manufacturers and cellular system operators have many options. New technologies can be integrated or set up as separate overlay systems. Integrated systems use a single switching system to control different types of radio channels while overlay systems use two or more switches to control the different types of radio channels. Communication channels can be sub rate multiplexed to decrease communication link costs. Broadband linear amplifiers allow the use of different types of radio channel cards without changing RF amplifier equipment. Distributed switching can reduce the burden on an MSC and provide for advanced features. Repeaters can extend the range of any type of cellular radio system.

9.6.1 Integrated and Overlay Systems

Dual mode digital cellular systems support two implementation systems: integrated and overlay. The integrated approach makes one system responsible for assigning both analog and digital radio channels. The overlay approach uses two separate systems to assign radio channels. For example, in an overlay approach, the analog system may coordinate AMPS radio channels and the overlay system could coordinate digital radio channels. Overlay systems permit continued use of older cellular equipment that may not be replaced or upgraded to digital service. Figure 9.21 compares an integrated and an overlay cellular system.

9.6.2 Sub Rate Multiplexing

Whatever transmission medium is used between cell sites and the MSC, maintaining the network's voice and data communication links is costly. Various methods, such as sub rate multiplexing, can improve the efficiency of these links. Sub rate multiplexing combines multiple compressed voice channels on a single communication channel (such as a 64 kbps DS0).

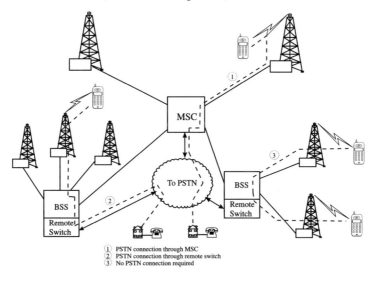

Figure 9.21, Integrated and Overlay Systems

Several digital cellular equipment configuration options can optimize sub rate multiplexing. One option is where to locate the speech coder and channel coder: in the cell site or MSC. Data transfer on a radio channel includes error protection bits along with the compressed voice data bits. The error protection bits protect against the radio signal distortions common in cellular radio transmissions. PSTN communication links between the base station and MSC are either wire, fiber, or fixed microwave, so distortion is low and few error protection bits are needed. The low requirement for error protection allows for the removal of cellular radio channel coding bits (error protection bits) at the base station prior to transmission to the MSC. The resulting low bits rates (1.2 - 9.6 kbps) allow several compressed channels to be shared on a single DS0 communications channel. Figure 9.22 illustrates how several sub rate multiplexed channels can share a single DS0 channel.

Since the cost of the communication links between the MSC and cell sites can be a significant part of the system cost, using each DS0/PCM channel in a T1 or E1 communication link for multiple conversations is an important potential system cost reduction. For analog cellular systems, the using 32 kbps ADPCM digital voice coding on each voice communications channel combines 2 conversations on each DS0/PCM channel. This increases the number of voice channels per communication link to 48 for T1 links and 60 for E1 links.

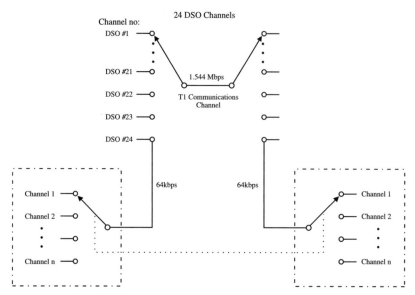

Figure 9.22, Sub Rate Multiplexing

9.6.3 Broadband Linear Amplified Base Stations

With a single linear RF amplifier, it is technically possible to mix any combination of cellular radio signal types (AMPS, NAMPS, TDMA, CDMA). This linear amplifier allows any of the cellular technologies to by installed by adding transceiver cards into the equipment frames. This mix of radio channels can then be controlled by the cellular system.

9.6.4 Distributed Switching

Distributed switching is the transfer of some coordination of network switching to locations distributed throughout a cellular system network. Distributed switching may also allow advanced features such as 4 digit dialing.

Small, easily installed cell sites called microcellular systems have increased the use of distributed switching. Microcells can be installed in high usage areas such as airports or office parks to expand coverage. Many of these systems relieve the burden on the MSC through distributed switching. Since many calls on a private system are within the system and the MSC does not need to be involved, private cellular systems often distribute switching to reduce the work of the MSC.

Figure 9.23 illustrates how distributed switching allows calls to be placed between subscriber units without the central MSC. In communication path 1, the subscriber unit communicates to a land line customer through the MSC. Using communications path 2, the subscriber unit communicates to a land line telephone customer through a distributed switch. The distributed switch is connected directly to the PSTN. Using communication path 3, two subscriber units are connected with each other via a distributed switch. No connection to the PSTN is required.

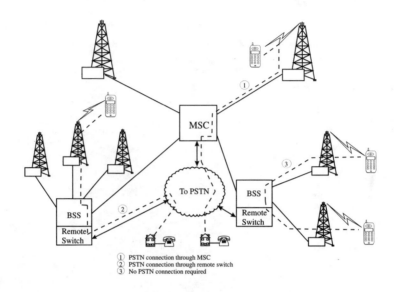

Figure 9.23, Distributed Switching

9.6.5 Cell Site Repeaters

Repeaters are vital in providing cost-effective cellular service in rural areas by making it possible to extend the range of analog and digital channels. Repeaters receive a radio signal from a nearby cell site, amplify that signal, and re-transmit it in a new direction. Some repeaters receive, amplify, and re-transmit the same frequency while others receive, decode, and re-transmit on a new frequency.

It is possible to extend the range over a hundred miles, but the increased range may introduce propagation delays and phase distortion. The propagation delays from repeaters may exceed the maximum delay offset that digital subscriber units can accommodate. In addition, repeaters that use class C amplifiers introduce phase distortion and high or unusable bit error rates (BERs) for TDMA and CDMA digital radio channels. Figure 9.24 illustrates a cell site repeater that can repeat analog and digital radio channels

Figure 9.24, Cellular System Repeater
Source: Allen Telecom

9.6.6 Network Communication Links

Cellular systems are interconnected via communication links. Figure 9.25 illustrates a sample cellular system communication link structure. The links between cell sites typically have alternate routing (e.g. via another cell site) for redundancy.

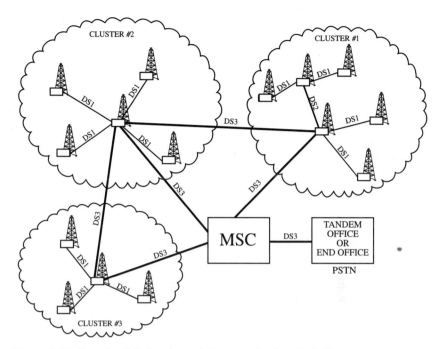

Figure 9.25, Sample Cellular System Communication Link Structure

References

[1]. CTIA Winter Exposition, "Disaster Experiences", Reno Nevada, February 6, 1990.
[2]. William Lee, "Mobile Cellular Telecommunications Systems", p.160, McGraw Hill, 1989.
[3]. CTIA Winter Exposition, "Disaster Experiences", Reno Nevada, February 6, 1990.
[4]. Interview, Richard Levine, Beta Laboratories, 9 January 1996.
[5]. U.S. Patent 4,887,265, Packet-Switched Cellular Telephone System, Kenneth Felix, December, 1989
[6]. William Lee, "Mobile Cellular Telecommunications Systems", McGraw Hill, 1989.
[7]. ibid.
[8]. FCC Regulations, Part 22, Subpart K, "Domestic Public Cellular Radio Telecommunications Service," 22.903, (June 1981).

Chapter 10
Cellular Economics

10. Cellular Economics

This chapter describes cost and revenue factors related to digital subscriber units (mobile telephones) and system equipment (radio towers and switching equipment), the costs of building and operating cellular systems, and marketing considerations that may affect consumer demand for digital cellular equipment.

Subscriber unit wholesale costs have dropped by approximately 20% per year over the past 7 to 10 years [1]. While the technology and mass production cost reductions for analog subscriber units are mature, new digital subscriber units are more complex and typically do not have the large sales volume that promotes cost savings through mass production. System equipment costs for digital cellular equipment must also compete against a mature, competitive Advanced Mobile Phone Service (AMPS) equipment market, which already has the advantage of cost reductions due to large production runs.

The economic goal of a cellular system is to effectively serve many customers at the lowest possible cost. The ability to serve customers is determined by the capacity of the cellular system. The capacity is determined by two key factors; the size of the cell sites and the spectral efficiency of the radio channels. When using any radio access technology, system capacity is increased by the addition of smaller cell site coverage areas, which allows more radio channels to be reused in a geographic area. If the number of cell sites remains constant, the efficiency of the radio access technology (e.g., the number of users that can share a single radio channel) determines the system capacity.

Wireless service providers usually strive to balance the system capacity with the needs of the customers. Running under capacity results in blocked calls, while running over capacity results in the purchase of unused equipment, which increases cost. Any cellular system (including AMPS) can be designed for very high capacity use through very small cell site coverage areas. The objective of the new digital cellular technologies is to achieve cost-effective service capacity, using techniques such as narrow radio channels, digital voice compression, or voice activity.

Purchasing and maintaining cellular system equipment is only a small portion of the cost of a cellular system. Administration, leased facilities, and tariffs may play significant roles in the financial success of cellular systems.

The cellular marketplace is undergoing a new change. New service providers, such as Specialized Mobile Radio (SMR) and Personal Communications Service (PCS) providers, are entering the market. This is likely to increase wireless services competition. Sales and distribution channels may become clogged with a variety of wireless product offerings. Advanced wireless digital technologies offer a variety of new features that may increase the total potential market and help service providers to compete. These new features may offer added revenue and provide a way to convert customers to more efficient digital service. The same digital radio channels that provide voice services may offer advanced messaging and telemetry applications.

10.1 Subscriber Unit Equipment Costs

The initial digital subscriber units introduced in 1992 were approximately two times the cost of their analog equivalents, while Narrowband AMPS (NAMPS) subscriber units were introduced at approximately the same cost as their analog only equivalents. Time Division Multiple Access (TDMA) and Code Division Multiple Access (CDMA) digital subscriber units reaching the market in 1996 were approximately two to three times the cost of comparable analog units. However, the average wholesale cost of analog subscriber units (AMPS) has dropped from $307 in 1992 to approximately $104 in 1996 [2]. The high cost of digital subscriber units is due to the following primary factors: development cost, production cost, patent royalty cost, marketing, post-sales support, and manufacturer profit.

10.1.1 Development Costs

Development costs are non-recurring costs that are required to research, design, test, and produce a new product. Unlike well-established FM technology, non-recurring engineering (NRE) development costs for digital cellular can be high, due to the added complexity of digital cellular phones. Several companies have spent millions of dollars developing digital cellular products [3]. Figure 10.1 shows how the non-recurring development cost varies from two to ten million dollars and quantity of production varies from 20,000 to 100,000 units. Even small development costs become a significant challenge if the volume of production of the digital subscriber units is low (below 20,000 units). At this small volume, NRE costs will represent a high percentage of the wholesale price.

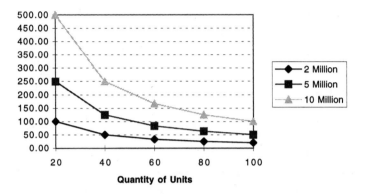

Figure 10.1, Subscriber Unit Development Cost

The market introduction of the first digital subscriber equipment was lower than expected due to the initial high cost of subscriber units. Because such a limited number of subscriber units were sold, it has been estimated that over $100 per unit had been dedicated to recover NRE costs [4].

The introduction of a new technology presents many risks in terms of development costs. Development costs include: market research; technical trials and evaluations; industrial, electrical, and software design; prototyping; product and FCC testing; creation of packaging, brochures, user and service manuals; marketing promotion; sales and customer service training; industry standards participation; unique test equipment development; plastics tooling; special production equipment fabrication; and overall project coordination.

When a completely new product is created (not a revision of an existing product), a cost-effective design is sometimes compromised by using readily available components. Cost effective design is achieved by integrating multiple assemblies into a custom chip or hybrid assembly. Custom integrated circuit chip development is used to integrate many components into one low-cost part. Excluding the technology development effort, custom Application Specific Integrated Circuit (ASIC) development typically requires a development setup cost that ranges from $250,000 to $500,000 [5]. More than one ASIC may be used in a digital subscriber unit.

10.1.2 Cost of Production

The cost to manufacture a subscriber unit includes the component parts (bill of materials), automated factory assembly equipment, and human labor. Digital subscriber units are more complex than the older analog units. A digital subscriber unit is composed of a radio transceiver and a digital signal processing section. The primary hardware assemblies that affect the component cost for digital subscriber units are Digital Signal Processors (DSP's) and radio frequency assemblies. A single DSP, and several may be used, costs $15 to $28 [6]. The Radio Frequency (RF) assemblies used in digital subscriber units include more precise linear amplifiers and fast switching frequency synthesizers. These RF components cost approximately $15 to $30. Other components that are included in the production of a subscriber unit include printed circuit boards, integrated circuits and electronic components, radio frequency filters, connectors, a plastic case, a display assembly, a keypad, a speaker and microphone, and an antenna assembly. In 1995, the bill of materials (parts) for a digital subscriber unit was approximately $100 [7].

The assembly of subscriber units requires a factory with automated assembly equipment. Each production line can cost between two and five million dollars [8]. Typically, one production line can produce a maximum of 500 to 2,000 units per day [9] (150,000 to 600,000 units per year). The number of units that can be produced per day depend on the speed of the automated component insertion machines and the number of components to be inserted. Production lines are often shut down one day per week for routine maintenance and two weeks per year for major maintenance overhauls. This leaves about 300 days per year for the manufacturing line to produce products. Between interest cost (10 to 15% per year) and depreciation (10 to 15% per year), the cost to own such equipment is approximately 25% per year. This results in a production facility overhead of $500 thousand to $1.25 million per year for each production line. Figure 10.2 shows how the cost per unit drops dramatically from approximately $10 to $25

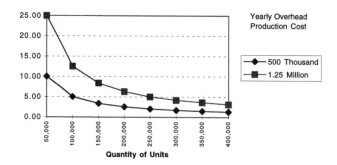

Figure 10.2, Factory Assembly Equipment Cost

per unit to $1 to $3 per unit as volume increases from 50,000 units per year to 400,000 units per year.

While automated assembly is used in factories for the production of subscriber units, there are some processes that require human assembly. Efficient assembly of a subscriber unit in a modern factory requires 1/2 to 1 hour of human labor [10]. The amount of human labor is a combination of all workers involved with the plant, including administrative workers and plant managers. The average loaded cost of labor (wages, vacation, insurance) varies from approximately $20 to $40 per hour [11], which is based on the location of the factory and the average skill set of human labor. This results in a labor cost per unit that varies from $10 to $40. Because digital subscriber units may have more parts to assemble due to the added complexity, the labor cost may increase.

10.1.3 Patent Royalty Cost

Another significant cost factor may be patent royalties. Cellular technology was originally developed and patented by AT&T [12]. To the author's knowledge, AT&T has never requested a single royalty payment for this fundamental technology, which is not the case for the new digital standards. Several companies have disclosed that they believe they have some proprietary technology that is required to implement the IS-54 and IS-95 standards [13]. Prior to the creation of the GSM standard, companies agreed to freely license the GSM technology to other companies participating in the standards development process.

Some large manufacturing companies exchange the right to use their patented technology with other companies that have patented technology the want to use. For manufacturers who do not exchange patent rights, in 1995, the combined royalties for IS-54 digital cellular phones was up to $70 per unit, and the combined royalties for IS-95 digital cellular phones was up to $100 per unit [14]. Patents from other companies that may be desirable or essential to implement the standard specifications may not have been discovered or disclosed.

10.1.4 Marketing Costs

The marketing costs, which are included in the wholesale cost of cellular subscriber units, include a direct sales staff, manufacturer's representatives, advertising, trade shows, and industry seminars.

Cellular subscriber unit manufacturers typically dedicate a highly paid representative for key clients. Much like the sales of other consumer electronics products, manufacturers employ several technical sales people to answer a variety of technical questions prior to the sale.

Some cellular subscriber unit manufacturers use independent distributors to sell their products. This practice is more prevalent for smaller, lesser known manufacturers who cannot afford to maintain a dedicated direct sales staff. These representatives typically receive up to four percent of the sales volume for their services.

Advertising programs used by the subscriber unit manufacturers involve broad promotion for brand-recognition and advertisements targeted for specific products. The budget for brand recognition advertising typically ranges from less then one percent to over four percent. Product-specific advertising is often performed through co-operative advertising. Co-operative advertising involves dedicating a percentage of the sales invoice (typically 2 to 4 percent) as an advertising allowance. When the advertising allowance meets the manufacturers requirements, the allowance is paid back to the customer. This approach allows customers to determine the best type of advertising for their specific markets. The typical advertising budget for subscriber unit manufacturers varies from approximately three to six percent.

Cellular system manufacturers exhibit at trade shows, typically three to four times per year. Trade show costs are high. Cellular subscriber unit manufacturers exhibiting at these shows typically have large trade show booths, gifts, and theme entertainment. Medium to large hospitality parties at the trade shows are also common. Cellular subscriber unit manufacturers often bring 15 to 40 sales and engineering experts to the trade shows to answer customer questions.

To help promote the industry and gain publicity, cellular subscriber unit manufacturers participate in a variety of industry seminars and associations. The manufacturers typically have a few select employees who write for magazines and speak at industry seminars.

All of these costs and others result in an estimated marketing cost for subscriber unit manufacturers of 10 to 15 percent of the wholesale selling price.

10.1.5 Post Sales Support

The sale of cellular subscriber units involves a variety of costs and services after the sale of the product, including warranty servicing, customer service, and training.

A customer service department is required for handling distributor and customer questions. Because the average customer for a cellular telephone is not technically trained in cellular technology, the amount of non-technical questions can be significant. Fortunately, customer questions can typically be answered during normal business hours.

Distributors and retailers require training for product feature operation and servicing. The post sales support cost for cellular subscriber units is typically between four to six percent.

10.1.6 Manufacturer's Profit

Manufacturers must make a profit as an incentive for producing products. The amount of profits a manufacturer can make typically depends on the risk involved with the manufacturing of products. As a general rule, the higher the risk, the higher the profit margin.

The cellular telephone market in the early 1990's became very competitive due to manufacturers' ability to reduce cost through mass production. To effectively compete, manufacturers had to invest in factories and technology, which increased the risk and the required profit margin. In 1995, the estimated gross profit in the subscriber unit manufacturing industry was 15 to 35 percent [15].

10.2 System Equipment Costs

The cost for digital system equipment is due to the following primary factors: development cost, production cost, patent royalty cost, marketing, post sales support, and manufacturer profit.

10.2.1 Development Costs

Cellular system equipment development costs are much higher than subscriber unit development costs. When a completely new technology is introduced, cellular system development costs can exceed $500 million [16] because the complexity of an entire cellular system is significantly greater than a subscriber unit and more testing and validation is required.

While the performance of the base station radio is similar to a subscriber unit, the coordination of all the subscriber units involves many systems. Additional assemblies include communication controllers in the base station and switching center, scanning locating receivers, communication adapters, switching assemblies, and large databases to hold subscriber features and billing information. All of these assemblies require hardware and very complex software.

Unlike subscriber units, when a cellular system develops a problem, the entire system can be affected. New hardware and features require extensive testing. Testing cellular systems can require thousands of hours of labor by highly skilled professionals [17]. Introducing a new technology is much more complex than adding a single new feature.

10.2.2 Cost of Production

The physical hardware cost for digital cellular system equipment should be more expensive than analog equipment due to the added technological complexity. However, the physical hardware cost for digital cellular system equipment may actually be less than older analog equipment. This is a result of a more competitive market and economies of scale.

The costs to manufacture a cellular system include the component parts, automated factory equipment, and human labor. The number of cellular system assemblies produced is much smaller than subscriber units. Setting up automated factory equipment is time consuming. For small production runs, much more human labor is used in the production of assemblies because setting up the automated assembly is not practical. The production of system equipment does require a factory with automated assembly equipment for specific assemblies. However, because the number of units produced for system equipment is typically much smaller than subscriber units, production lines used for cellular system equipment are often shared for the production of different assemblies, or remain idle for periods of time.

With over 150 countries developing and expanding their cellular systems (see Appendix 4), the demand for cellular system equipment is increasing exponentially. This increased demand allows for larger production runs, which reduce the average cost per unit. Large production runs also permit investment in cost-effective designs, such as using application-specific integrated circuits (ASICs) to replace several individual components.

The telecommunications industry has begun to standardize interfaces between system equipment. Different manufacturers can develop and sell parts of the cellular system without the cost burden of investing in the development of all the cellular network system components. It is possible in the Global System for Mobile Communications (GSM) system for one manufacturer to supply the cellular system base station control equipment and another manufacturer to supply the base station radios. The ability for a manufacturer to focus on the high demand components without a large development investment increases competition and reduces prices to the cellular system customer.

The maturity of digital technology is promoting cost reductions through the use of cost-effective equipment design and low-cost commercially available electronic components. In the early 1990s, many technical system equipment changes were required due to changes in radio specifications. Manufacturers had to modify their equipment based on field test results. For example, complex echo cancellors were required due to the long delay time associated with digital speech compression. Manufacturers typically did not invest in cost-effective custom designs because of the rapid changes. As the technology has matured, the investment in custom designs is possible with less risk. In the early 1990s, it was also unclear which digital technology would become commercially viable, which limited the availability of standard components. Today, the success of digital systems has created a market of low-cost digital signal processors and RF components for digital cellular systems.

Like the assembly of subscriber units, the assembly of system radio and switching equipment involves a factory with automated assembly equipment. The primary difference is the smaller production runs, multiple assemblies, and more complex assembly.

The number of equipment units that are produced is much smaller than the number of subscriber units produced because each radio channel produced can serve 20 to 32 subscribers. The result is much smaller production runs for cellular system equipment. While a single production line can produce a maximum of 500 to 2,000 assemblies per day [18], several different assemblies for radio base stations are required. A change in the production line from one assembly process to another can take several hours or several days. Cellular system radio equipment requires a variety of different connectors, bulky RF radio parts, and large equipment case assemblies. Due to the low-production volumes and many unique parts, it is not usually cost effective to use automatic assembly equipment. For unique parts, there are no standard automatic assembly units available. Because of this more complex assembly and the inability to automate many assembly steps, the amount of human labor is much higher than for subscriber units.

Each automated production line can cost two to five million dollars. The number of units that can be produced per day depend on the speed of the automated component insertion machines, the number of components to be inserted, the number of different electronic assemblies per equipment, and the amount of time it takes to change/setup the production line for different assemblies. If we

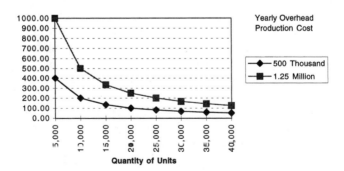

Figure 10.3, Factory Assembly Equipment Cost

assume there are four electronic assemblies per base station radio equipment (e.g., controller, RF section, baseband/diagnostic processing section, and power supply), the automated production cost for base station equipment should be over four times that of subscriber units.

Figure 10.3 shows how the production cost per unit drops dramatically from approximately $400 to $1,000 per unit to $50 to $125 per unit as the volume of production increases from 5,000 units per year to 40,000 units per year. This chart assumes production cost is four times that of subscriber units due to the added complexity and the use of multiple assemblies.

While automated assembly is used in factories for the production of subscriber units, there are some processes that require human assembly. Efficient assembly of base station units in a modern factory requires between five and ten hours of human labor [19]. The amount of human labor includes all types of workers from administrative workers to plant managers. The average loaded cost of labor (wages, vacation, insurance) varies from approximately $20 to $40 per hour [20], depending upon the location of the factory and average workers skill set. The resultant labor cost per unit varies from $100 to $400.

10.2.3 Patent Royalty Cost

There are only a few large manufacturers that produce cellular system equipment. Cellular system equipment requires the use of many different technologies. Large manufacturers have a portfolio of patents that they commonly trade. Cross-licensing is common and tends to reduce the cost of patent rights. When patent licensing is required, the patent costs are sometimes based on the wholesale price of the assemblies in which the licensed technology is used.

10.2.4 Marketing Costs

The marketing costs that are included in the wholesale cost of cellular system equipment includes a direct sales staff, sales engineers, advertising, trade shows, and industry seminars. Cellular system manufacturers often dedicate several highly paid representatives for key clients. Cellular system sales are much more technical than the sale of subscriber units. Manufacturers employ several people to answer a variety of technical questions prior to the sale.

Advertising used by the cellular system equipment manufacturers involves broad promotion for brand recognition and advertisements targeted for specific products. The budget for brand recognition advertising is typically small and is targeted to specific communication channels because the sale of cellular system equipment involves only a small group of people who typically work for a cellular service provider. Product-specific advertising is also limited to industry specific trade journals. Much of the advertising promotion of cellular system equipment occurs at trade shows, industry associations, and direct client presentations. The advertising budget for cellular system equipment manufacturers is typically less than two percent.

Cellular system manufacturers exhibit at trade shows typically three to four times per year. The trade show costs are sometimes much higher than the trade show costs for subscriber unit manufacturers. Cellular manufacturers exhibiting at trade shows often have large hospitality parties that sometimes entertain thousands of people. Cellular system manufacturers often bring 60-100 sales and engineering experts to the trade shows to answer customer questions.

To help promote the industry and gain publicity, cellular system manufacturers participate in many industry seminars and associations. Manufacturers use trained experts to present at industry seminars.

All of these costs and others result in an estimated marketing cost for system equipment manufacturers of approximately eight to ten percent of the wholesale selling price.

10.2.5 Post-Sales Support

The sale of cellular systems involves a variety of costs and services after the sale of the product. This includes warranty servicing, customer service, and training. A 24-hour customer service department is required for handling customer questions. Customers require a significant amount of training for product operation and maintenance after a system is sold and installed. The post-sales support costs for cellular system equipment is typically three to five percent.

10.2.6 Manufacturer's Profit

Standardization of systems and components, particularly GSM, has led to a rapid drop in the wholesale price of system equipment. While the increased product volume of cellular system equipment has resulted in decreased manufacturing costs, the gross profit margin for cellular system equipment has decreased. The estimated gross profit in the subscriber unit manufacturing industry is ten to fifteen percent [21].

10.3 Network Capital Costs

The cellular carrier's investment in network equipment includes cell sites, base station radio equipment, switching centers, and network databases. One of the primary objectives of the new technologies was to decrease the network cost per customer which was made possible because the new technologies can serve more customers with less physical equipment.

In theory, existing analog cellular technology can serve almost an unlimited number of subscribers in a designated area by replacing large cell site areas with many Microcells (small cell coverage areas). However, expanding the current analog systems in this way increases the average capital cost per subscriber due to the added cost of increasing the number of small cells and interconnection lines to replace a single large cell. For example, when a cell site with a 15 km radius is replaced by cell sites with a 1/2 km radius, it will take over 700 small cells to cover the same area.

One of the reasons that digital cellular technologies were developed was to allow cost-effective capacity expansion. Cost-effective capacity expansion results when existing cell sites can offer more communication channels, which allows more customers to be served by the same cell site. As systems based on such new technologies expand, the average cost per subscriber decreases.

10.3.1 Cell Site

The cell site is composed of a radio tower, antennas, a building, radio channels, system controllers, and a backup power supply. The cell site radio tower is typically 100 to 300 feet tall. The cost ranges between 30 thousand and 300 thousand dollars [22]. While some of the largest towers can cost $300 thousand, an average cost of $70 thousand is typical because as systems expand, smaller towers can be used.

Many cell sites can be located on a very small area of land. Land is either purchased or leased. In some cases, existing tower space can be leased for $500 to $1,000 per month [23]. If the land is purchased, the estimated cost of the land is approximately $100 thousand.

A building on the cell site property is required to cover the cell site radio equipment. This building must be bullet proof, have climate control, and various other non-standard options. The estimated building cost is $40 thousand [24].

Cell sites are not usually located where high-speed telephone communication lines are available. Typically, it is necessary to install a T1 or E1 communications line to the cell-site which is leased from the local phone company. If a microwave link is used in place of a leased communication line, the communications line installation cost will be applied to the installation of the microwave antenna. The estimated cost of installing a T1 or E1 communications line is $5 thousand [25].

The land where the cell site is to be located must be cleared, foundations poured, fencing installed, and building and tower installed. A construction cost of $50 thousand is estimated. Table 10.1 shows the estimated cost for a typical cell site without the radio equipment.

Item	Cost x $1,000's
Radio Tower	$70
Building	$40
Land	$100
Install Comm Line	$5
Construction	$50
Antennas	$10
Backup Power Supply	$10
Total	$285

Table 10.1, Estimated Cell Site Capital Cost Without Radio Equipment

In addition to the tower and building cost, radio equipment must be purchased. The cost of the radio equipment usually varies based on technology and the number of customers that use the system.

After the total investment of each cell site is determined, the cell site capital cost per customer can be determined by dividing the total cell site cost by the number of subscribers that will share the resource (cell site). Because not everyone uses every radio channel at the same time, cellular systems typically add 20-32 subscribers per voice channel. For the different cellular technologies, each RF channel can supply one or more voice paths. The number of voice paths per radio channel is multiplied by the number of radio channels per cell site. If 20 subscribers are added to the system for each voice path, the average number of subscribers per cell site varies from 1,000 to 19,200.

Since one RF channel can serve many users, either the number of base station RF channel equipment assemblies is decreased, or the system capacity is increased. For example, a cell site that has 24 analog RF channels can convert 12 channels to full-rate TDMA (three users to one 30 kHz RF channel) and support 48 voice channels. With the introduction of the half rate TDMA channels (six users to one 30 kHz RF channel), the same cell site with 24 analog RF channels could convert 12 analog RF channels to half rate TDMA and support 84 voice channels. A future consideration is the acceptance of Digital Speech Interpolation (DSI) to allow the transmission of speech data only when a person is actively speaking. DSI doubles the capacity (assuming 50% voice activity) by allowing the same 24-channel cell site to convert 12 channels to digital and have up to 156 voice channels.

Table 10.2 shows a sample of system equipment costs as digital technology evolves. In column 1, we see analog FM technology that supports one voice channel per carrier. Because each subscriber will only access the cellular system for a few minutes each day, approximately 20 subscribers (customers) can share the service of a single radio channel. NAMPS radio channels also provide one voice channel per carrier. The advantage of NAMPS is that because the radio channels are one-third the width of AMPS channels, more channels can be placed in each cell site. This increases the total number of subscribers per cell site to approximately 3,060. For the IS-54/IS-136 system, each RF channel supports up to three users (full rate), so each RF channel cost is shared by 60 subscribers. The evolution of IS-136 will include a half rate system where six users can share each radio channel. For IS-95 CDMA where each RF channel supports approximately 20 users, each RF channel cost is shared by up to 400 subscribers. For GSM, each RF channel can provide service to 8 simultaneous users which allows up to 160 subscribers to share the radio channel cost. The GSM phase 2 specifi-

	AMPS/ ETACS	NAMPS	IS-54/IS-136 TDMA	IS-95 CDMA	GSM
Cost per RF Radio Channel	10,000	10,000	15,000	45,000	6,000
Number of Radio Channels per Cell Site (3 sector)	51	153	51	24	30
Total Radio Channel Cost	510,000	1,530,000	765,000	1,080,000	180,000
Tower and Building Cost	285,000	285,000	285,000	285,000	285,000
Total Cell Site Cost	795,000	1,815,000	1,050,000	1,365,000	465,000
Number of Voice Paths per Radio Channel	1	1	3	20(est)	8
Number of Voice Paths per Cell Site	51	153	153	480	240
Number of Subscribers per Voice Channel	20	20	20	20	20
Number of Subscribers per Cell Site	1020	3060	3060	9600	4800
Cell Site Capital Cost per Subscriber	$779	$593	$343	$142	$97

Table 10.2, Cell Site Capital Cost per Subscriber

cation (discussed in chapter 7) allows up to 16 users to share each radio channel (half rate). While it is not suggested that all of the available RF channels can be converted to digital in the near future, the following table shows target costs that project the reasonable costs of a cellular system equipment in the future.

The multiplexing of several radio channels through one RF equipment reduces the number of required RF equipment assemblies, power consumption, and system cooling requirements. Multiplexing in this way typically reduces cell site size and backup power supply (generator and battery) requirements, and ultimately, cost.

10.3.2 Mobile Switching Center

Cell sites must be connected to an intelligent switching system (called the "switch"). An estimate of $25 per subscriber is used for the cellular switch equipment and its accessories [26],based on one Mobile Switching Center (MSC) costing $2.5 million that can serve up to 100,000 customers.

The switching center must be located in a long-term location (10 to 20 years) near a Local Exchange Carrier (LEC) Public Switched Telephone Network (PSTN) central office connection. The building contains the switching and communication equipment. Commonly, a customer database called the Home Location Register (HLR) is located in the switching center. The switching center software and associated cellular system equipment typically contain basic software that allows normal subscriber unit operation (place and receive calls). Special software upgrades that allow advanced services are available at additional cost.

10.4 Operational Costs

The costs of operating a cellular system includes leasing and maintaining communication lines, local and long distance tariffs, billing, administration (staffing), maintenance, and cellular fraud. The operational cost benefits of installing digital equipment includes a reduction in the total number of leased communication lines, a reduction in the number of cell sites, a reduction in maintenance costs, and a reduction of fraud due to advanced authentication procedures.

10.4.1 Leasing and Maintaining Communications Lines

Cell sites must be connected by leased communication lines between radio towers, or by installing and maintaining microwave links between them. The typical cost for leasing a 24-channel line between cell sites in the US in 1992 was $750/month [27]. Microwave radio equipment can cost from $20 thousand to $100 thousand.

The number of subscribers that can share the cost of a communication line (loading of the line) varies with the type of service. For cellular-like subscribers who typically use the phone for two minutes per day, approximately 480 customers can share a T1 (20 subscriber per voice path x 24 voice paths per communication line) or 600 customers per E1 (20 subscribers per voice path x 30 voice paths per communication line). For residential-type service, where customers use the phone for approximately 30 minutes per day, approximately 120 customers can be loaded onto a T1 or 150 per E1. For office customers who use the phone for approximately 60 minutes per day, approximately 60 customers can be loaded onto a T1 or 75 for E1.

The monthly cost per subscriber is determined by dividing the monthly cost by the total number of subscribers. Table 10.3 shows the estimated monthly cost for interconnection charges. The estimated monthly cost is based on 100% use of the communication lines. If the communication lines are not used to full capacity (it is rare that communication lines are used at full capacity), the average cost per line increases.

Service	Line Cost per Month	No Chan	Load	Total Cost per Month
Cellular	750	24	20	1.56
LEC (residential)	750	24	5	6.25
Office	750	24	2.5	12.5

Table 10.3, Monthly Communications Line Cost

Digital signal processing for all the proposed technologies allows for a reduction in the number of required communications links through the use of sub-rate multiplexing. Sub-rate multiplexing allows several users to share each 64 thousand bit per second (kbps) communications (DS0/PCM) channel. This is possible because digital cellular voice information is compressed into a form much smaller than existing communication channels. If 8 kbps speech information is sub-rate multiplexed, up to 8 voice channels can be shared on a single 64 kbps channel, which can reduce the cost of leased lines significantly.

10.4.2 Local and Long Distance Tariffs

Telephone calls in cellular systems are often connected to other local and long distance telephone networks. When cellular systems are routed to existing landline telephone customers, they are typically connected through the wired telephone network (usually via the Local Exchange Company (LEC)). The local telephone company typically charges a small monthly fee and several cents per minute (approximately 3 cents per minute) for each line connected to the cellular carrier. Because each cellular subscriber uses their subscriber unit for only a few minutes per day, the cellular service provider can use a single connection (telephone line) to the PSTN to service hundreds of subscribers.

In the United States and other countries that have separate long-distance service providers, when long distance service is provided through a local telephone company (LEC), a tariff is paid from its cellular service provider to the local exchange company (LEC). These tariffs can be up to 45% of the per minute charges. In 1993, MCI paid $5.3 billion in the United States for local exchange tariffs out of the $11 billion in revenues [28]. Due to government regulations limiting the bundling of local and long-distance service, it is necessary for some cellular service providers to separate their local and long-distance service. In the future, regulations may permit cellular carriers to bypass the LEC and save these tariffs.

10.4.3 Billing Services

Cellular systems exist to provide services and collect revenue for those services. Billing involves gathering and distributing billing information, organizing the information, and invoicing the customer.

As customers initiate calls or use services, records are created. These records may be provided in the customer's home system or a visited system. Each billing record contains details of each billable call, including who initiated the call, where the call was initiated, the time and length of the call, and how the call was terminated. Each call record contains approximately 100-200 bytes of information [29]. If the calls and services are provided in the home system, the billing records can be stored in the company's own database. If they are provided in a visited system, the billing information must be transferred back to the home system. In the mid 1980s, cellular systems were not interconnected, which required the use of a clearing house. The clearing house was used to accumulate and balance charges between different cellular service providers. Billing records from cellular systems were typically transferred via tape in a standard Automatic Message Accounting (AMA) format.

In the 1990s, standard intersystem connection allowed billing records to be transferred automatically, which made advanced billing services such as advice of charging or debit account billing possible. Advice of charging provides the customer with an indication of the billing costs. Debit account billing allows a cellular service provider to accept a prepaid amount from a customer (perhaps a customer that has a poor credit rating) and decrease his or her account balance as calls are processed.

With the introduction of advanced services, billing issues continue to become more complicated. The service cost may vary between different systems. To overcome this difficulty, some service providers have agreed to bill customers at the billing rate established in their home system. In the United States, Cellular Digital Packet Data (CDPD) services are billed at the home subscribers rate [30].

Each month, billing records must be totaled and printed for customer invoicing, invoices mailed, and payments received and posted. The estimated cost for billing services is $1 per month [31]. This billing cost includes routing and summarizing billing information, printing the bill, and the cost of mailing. To help offset the cost of billing, some cellular service providers have started to bundle advertising literature from other companies along with the invoice. To expedite the collection, some cellular service providers offer direct billing to bank accounts or charge cards.

10.4.4 Operations, Administration, and Maintenance

Running a cellular service company requires people with many different skill sets. Staffing requirements include executives, managers, engineers, sales, customer service, technicians, marketing, legal, finance, administrative, and other personnel to support vital business functions. The present staffing for local telephone companies is approximately 35 employees for each 10,000 customers [32]. Cellular companies have approximately 20 to 25 employees per 10,000 customers [33], and paging companies employ approximately 10 employees per 10,000 customers [34]. If we assume a loaded cost (salary, expenses, benefits, and facility costs) of $40,000 per employee, this results in a cost of $3.33 to $11.66 per month per customer ($40,000 x (10-35 employees) /10,000 customers/12 months).

Maintaining a cellular system requires calibration, repair, and testing. System growth involves frequency planning, testing, and repair. When a frequency plan is changed, manual tuning of 20-30 radio channels per cell site is required. Urban systems may have over 400 cell sites. Because it is desirable to perform frequency re-tuning in the same evening, some cellular service providers borrow technicians from neighboring systems when their small staff cannot perform this task. All of the technologies offer potential automatic RF frequency planning (see MRI, MAHO, or CDMA frequency planning in chapters 3 thru 7).

During the year, the geographic characteristics of a cellular system change (e.g., leaves fall off trees). This changes the radio coverage areas. Typically, a cellular carrier tests the radio signal strength in its entire system at least four times per year. Testing involves having a team of technicians drive throughout the system and record the signal strength.

Maintenance and repair of cellular systems is critical to the revenue of a cellular system. In large cellular systems, a staff of qualified technicians are hired to perform routine testing. Smaller cellular systems often have an agreement with a cellular service provider or a cellular system manufacturer to provide these technicians when needed. Cellular systems have automatic diagnostic capabilities to detect when a piece of equipment fails. Most cellular systems have an automatic backup system, which can provide service until the defective assembly is replaced.

10.4.5 Land and Site Leasing

In rural areas, exact locations for cell site towers are not required. The result is that land leasing is not a significant problem. In urban areas, and as systems mature, more exact locations for cell sites are required. This results in increased land leasing costs. By using a more efficient RF technology, one cell site can be used to serve more channels, which limits the total number of required cell sites.

Land leasing is typically a long-term lease for a very small portion of land (40 to 200 square meters) for approximately 20 years or longer. The cost of leasing land is dependent on location. Premium site locations such as sites on key buildings or in tunnels can exceed the gross revenue potential of the cell site [35].

Another leasing option involves leasing space on an existing radio tower. Site leasing on an existing tower is approximately $500 per month. Site leasing eliminates the requirement of building and maintaining a radio tower.

10.4.6 Cellular Fraud

It is estimated that cellular fraud in the United States during 1994 was in excess of $460 million [36]. This was approximately three percent of the $14 billion yearly gross revenue generated [37]. Each of the new cellular standards has an advanced authentication capability, which limits the ability to gain fraudulent access to the cellular network.

The type of cellular fraud has changed over the years. Initially, cellular fraud was subscription fraud. With advances in technology available to distributors of modified equipment, cellular fraud has changed to various types of access fraud. Access fraud is the unauthorized use of cellular service by changing or manipulating the electronic identification information stored inside the subscriber unit.

Subscription fraud occurs when a cellular phone is registered by a person using false identification. After the required documentation is provided, unlimited service is often provided by the cellular service provider. When the bill is not paid, the fraudulent activity is determined and service is disconnected. Some cellular service providers now require valid identification and credit checks prior to service activation, which reduces subscription fraud.

In the mid 1980's, roamer fraud was possible. Roamer fraud occurs when a subscriber unit is programmed with an unauthorized telephone number and home system identifier so that it looks like a visiting customer. Because some of the cellular systems in the mid 1980's were not directly connected to each other, these systems could not immediately validate the visiting customer. In the 1990's, intersystem connection provides validation of the phone number, which limits space (or eliminates) roamer fraud.

To allow fraudulent access to valid cellular customer accounts, criminals began to modify the electronic serial number (ESN) of subscriber units to match a valid subscriber's ESN. This duplication of subscriber information is called "cloning."

To enable the cloning process, subscriber units have to be modified to accept a new ESN and a valid ESN must be acquired. ESNs are typically stored in a hard-secure memory area of a subscriber unit. It typically takes a very technical person to be able to override the security system in the subscriber unit to modify the ESN. Obtaining valid ESN's is possible by reading the ESN on the label or by using a commercially available test set that commands the subscriber unit to send its ESN. Cellular carriers are able to detect changing patterns of use when a cloned phone has been created. If the subscriber's billing account jumps dramatically, the cellular customer can be contacted to have their phone number changed. The original ESN is then marked invalid.

To overcome the barriers of ESN's becoming invalid, fraudulent users designed their systems to change the ESN during each call. This changing ESN process is called "tumbling." Tumbling uses valid ESN's that are pre-stored in the subscriber unit or captured from the radio channels during a valid subscriber's regular access. These valid ESN's are used by the modified subscriber units either during each call or randomly each time they send or receive calls.

Most of these methods can be detected and blocked by the use of Subscriber Unit authentication information. Authentication is a process of using previously stored information to process keys that are transferred via the radio channel. Because the secret information is processed to create a key, the security information is not transferred on the radio channel. The secret information stored in the subscriber unit can be changed at random either by manual entry, by the customer, or by a command received from the cellular system. Authentication is supported in all of the digital technology specifications.

10.5 Marketing Considerations

The cellular service providers commissioned the development of digital cellular technology to allow cost-effective system capacity expansion and to provide more revenue from advanced services. To obtain this cost savings and new feature service revenue, it is necessary to have a percentage of subscribers who have digital-capable equipment that can access the advanced technology. Digital cellular marketing programs have focused on converting existing subscribers to digital service, enticing new customers to purchase digital over analog, and targeting new customers for advanced services. The key marketing factors that may determine the success of digital cellular include the type of new services, system cost savings, pricing of voice and data service, subscriber unit cost, consumer confidence, new features, retrofitting existing customer equipment, availability of equipment, and distribution channels.

10.5.1 Service Revenue Potential

At the end of 1995, there were over 74 million cellular telephone customers in the world. In the United States, there were over 24 million customers with an average monthly service bill of $58.65 [38]. While the average cellular telephone bill is not much higher than the average wired residential telephone bill, the amount of usage for a cellular telephone is approximately one tenth of residential usage.

The average cellular telephone bill in the US has declined eight to nine percent each year over the last five years [39]. The average charge per minute has not decreased much; however, the amount of usage has declined because new customers entering into the market are consumers that do not use the cellular telephone much.

The number of subscribers on the US cellular systems have been increasing by over 45% per year over the past five years [40]. Some of this growth is due to new system service areas and the decreased price of cellular telephones. It is not reasonable to assume a continued 45% yearly growth period as the total number of cellular telephones would exceed the total world population in only 12 years. Figure 10.4 shows the subscriber growth in Europe, Asia, and the Americas over the past five years.

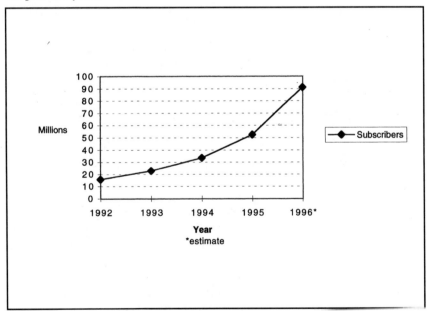

Figure 10.4, World Cellular Subscriber Growth
Source: US Dept of Commerce

The main revenue for cellular service providers is derived from providing telecommunications service. In 1995, a majority of the service revenue came from voice services. Digital cellular provides for increased service revenue to come from a variety of sources such as advanced services and system cost reduction.

10.5.2 System Cost to the Service Provider

One of the advantages of digital service is to allow more customers to share the same system equipment. Cellular system equipment costs account for approximately 10 to 15 % of the service provider's revenue. Digital cellular can offer a reduction of approximately 60% of system equipment cost per customer. Some of the advanced features of digital cellular (such as authentication to reduce fraud) also provide for reductions in Operations, Administration, and Maintenance (OA&M) costs. These system cost reductions offered by digital cellular technology may be necessary to allow existing cellular service providers to effectively compete against other wireless service providers. As more companies (such as PCS companies) begin to offer cellular-like services, the potential for a surplus of voice channel time exists.

10.5.3 Voice Service Cost to the Consumer

Over the past few years, the average cost of airtime usage to a cellular subscriber has not changed very much. To help attract subscribers to digital service, some cellular carriers have offered discounted airtime plans to high usage customers [41]. This discount provides a significant incentive to these customers. By shifting a small portion of them to digital service, the loading on the older analog radio channels is reduced.

10.5.4 Data Service Cost to the Consumer

There are two types of data services that are provided to customers; continuous (called "circuit switched data") or brief packets (called "packet switched data"). Typically, continuous data transmission is charged at the same rate as voice transmission. Packet data transmission is often charged by the packet or by the total amount of data that has been transferred.

Numeric paging systems can serve 40 to 100 thousand paging customers per radio channel [42], while cellular can serve only 20-32 customers per radio channel [43]. In the United States during 1994, the average revenue for a paging customer was approximately $10 per month and the average revenue for a cellular customer was $56.21 per month [44]. If paging is viewed as packet data service, the total revenue potential per radio channel is significantly higher than voice. D is a packet data service that is offered on cellular. Using the pricing of 7 to 23 cents per kilobyte of data [45], the average cost to the customer is $3 to $10 per minute.

10.5.5 Subscriber Unit (Mobile Phone) Cost to the Consumer

In 1984-85, cellular subscriber unit prices varied from $2000 to $2500 [46]. In 1991, you could get a free cellular phone with the purchase of a hamburger at selected Big Boy Restaurants in the United States [47]. One of the primary reasons for the continued penetration of the cellular market is the declining terminal equipment costs and stable airtime charges [48]. In 1995, the wholesale price of digital subscriber equipment was higher than analog subscriber equipment. The cellular service providers often subsidize the sale of end-user telephones, which reduces revenue. Cellular service providers do not usually anticipate revenues from the sale of telecommunications equipment.

Many of the service providers are not concerned with the profit on mobile equipment because their goal is to gain monthly service revenue. Cellular service providers determine the deployment of digital equipment in the field. In the Americas, it is common for the cellular service provider to subsidize the sale of cellular subscriber equipment. It is not common for such subsidies in Europe. To help introduce digital cellular into cellular markets, subsidies may be increased for digital subscriber unit equipment. Some manufacturers use groups of serial numbers to identify subscriber units that have digital capability [49].

The subscriber unit service activation subsidy and the type of distribution channels used usually affects the retail price paid by the consumer. Figure 10.5 shows the wholesale subscriber unit cost in the United States and Europe over the past seven years [50].

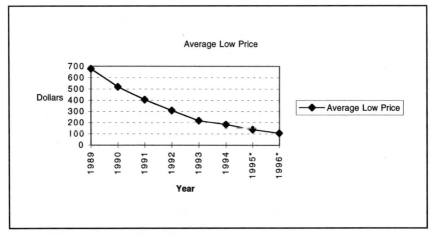

Figure 10.5, Wholesale Subscriber Unit Cost in the United States
Source: Herschel Shosteck Associates, Wheaton, Maryland, USA

10.5.6 Consumer Confidence

To effectively deploy a new technology, the consumer must have confidence that the technology will endure. The next generation of cellular technology was required to be compatible the old and new systems to maintain this consumer confidence [51]. In some parts of the world, there are multiple digital technologies. These technologies offer different features, services, and radio coverage areas. Consumers must be willing to choose a technology that may not exist in future years.

10.5.7 New Features

Customers purchase subscriber units and cellular service based on their own value system, which estimates the benefits they will receive. New features provide for new benefits to the consumer. These features can be used for product differentiation and to increase service revenue. New features available for digital cellular include longer battery life, caller ID, and message services. These new features are sometimes used to persuade customers to convert to digital or to pay extra for these new advanced services.

While the first FM cellular telephone weighed over 80 pounds and required almost all available trunk space, the first digital subscriber units were only slightly larger than their analog predecessors. The size of digital subscriber units continues to be reduced as production volumes allow for custom (application specific integrated circuit) ASIC development, which integrates the analog and digital processing sections. There is no reason that the digital subscriber units cannot approach the same size as the analog cellular phones [52].

New features allow for different types of customers. With advanced data capabilities, cellular service may focus its products, services, and applications on non-human applications (many non-human applications are discussed in chapter 12). Without a change in focus, wireless service providers could become limited to the voice services market which may eventually reach saturation.

10.5.8 Retrofitting

Retrofitting involves subscriber unit hardware conversion and equipment exchange programs. Due to the limited capability of analog subscriber units, they cannot be easily converted to digital capability. While it may be possible to replace the analog subscriber unit (mobile car phone) transceiver with a digital transceiver, the retrofit market is considered small.

One of the easiest ways to reduce system blockage (system busy signals) and quickly increase system efficiency is by retrofitting high-usage subscribers [53]. Retrofitting high-usage subscribers can increase the total demand for digital subscriber units over several years. There have been some retrofit incentive programs created that allow a customer to exchange their analog subscriber units for a digital subscriber unit. These programs have required the customer to exchange or designate another customer to receive the old analog unit prior to receiving a digital unit. In return for this exchange, the digital subscriber unit would be provided to the customer at a reduced cost or no cost.

Higher commissions have been paid for digital subscriber units to entice new customers to select digital subscriber units over their analog equivalents. Some manufacturers group ESN numbers to allow the cellular service provider to know if the subscriber unit has digital capability that allows the carrier to differentiate activation commissions between analog and digital subscriber units.

10.5.9 Churn

Churn is the percentage of customers that discontinue cellular service. Churn is usually expressed as a percentage of the existing customers that disconnect over a one-month period. Churn is often the result of natural migration (customers relocating) and switching to other service providers. Because some cellular service providers provide an activation commission incentive to help reduce the sale price of the phone, this can be a significant cost if the churn rate is high. The percentage of churn in North America over the last five years has remained relatively constant at approximately 2.8% per month [54].

Cellular carriers and their agents have gone to various lengths to reduce churn. This includes programming in lockout-codes to lengthy service agreements. Programming lockout-codes (typically four to eight digits) can be entered by the programmer into some subscriber units to keep the subscriber unit from being reprogrammed by another cellular service provider. Cellular service providers sometimes require the customer to sign a service agreement, which typically requires them to maintain service for a minimum of one year. These service agreements have a penalty fee in the event the customer disconnects service before the end of the one year period.

10.5.10 Availability of Equipment

The design and production of digital cellular equipment requires significant investment by a manufacturer. Digital phones are more complex, and portable digital cellular phones are typically larger and more expensive. There are fewer manufacturers that offer digital cellular equipment that has reduced competition and maintained higher prices than their analog equivalents.

The first cellular portable, introduced in 1984 by Motorola, weighed 30 ounces and had approximately 30 minutes of talk time. In early August 1991, Motorola released the Micro-Tac Light which weighed 7.7 ounces [55] and had 45 minutes of talk time. The additional DSP circuitry that is required for a TDMA dual mode portable weighs approximately 3 ounces using commercially available DSP's [56], which allowed the introduction of a 10 oz IS-54 TDMA dual mode portable phone in 1993. In 1995, a 6.9 ounces digital cellular phone was introduced for GSM and IS-54 TDMA. In 1996, the minimum size and weight of digital cellular handportable subscriber units was close to analog cellular handportables.

10.5.11 Distribution and Retail Channels

Products produced by manufacturers are distributed to consumers via several channels. The types of distribution channels include: wholesalers, specialty stores, retail stores, power retailers, discount stores, and direct sales.

Wholesalers purchase large shipments from manufacturers and typically ship small quantities to retailers. Wholesalers typically specialize in particular product groups, such as pagers and cellular phones.

Specialty retailers are stores that focus on a particular product category such as a cellular phone outlet. Specialty retailers know their products well and are able to educate the consumer on services and benefits. Specialty retailers usually get an added premium via a higher sales price for this service.

Retail stores typically provide a convenient place for the consumers to view products and make purchases. Retailers often sell a wide variety of products, but a salesperson may not have an expert's understanding of or be willing to dedicate the time to explain the features and cellular service options. In the early 1990s, mass retailers began selling cellular phones. Mass retailers sell a very wide variety of products at a low margin, and do not provide customer education on new features.

Power retailers specialize in a particular product group such as consumer electronics. Power retailers look for particular product features that match their target market and carry only a select group of products. Because there are only a few products for the consumer to select from, the demand for a single product is higher than if several different models were on display. This tends to increase sales for a particular product, which leads to larger quantity purchases and discounts for the power retailer.

Discount stores sell products at a lower cost than competitors. They achieve this by providing a lower level of customer service. Because there is limited customer service, many of the subscriber units sold in discount stores are pre-programmed or are debit (pre-paid) units.

Some cellular service providers employ a direct sales staff to service large customers. These direct sales experts can offer specialty service pricing programs. The sales staff may be well-trained and typically sell at the customer's location.

Distribution channels are typically involved in the activation process. The application for cellular service can take a few minutes to several hours, and programming of the subscriber unit for the consumer must be performed. With several new cellular service providers such as Personal Communications Service (PCS) and Personal Communications Network (PCN), the complexity of sale and activation can increase. To simplify this process, some cellular service providers have streamlined the application process and pre-programmed the subscriber units to make the process more like a typical retail sale.

Because there are several new technologies and different models of phones, access to particular distribution channels is limited. In 1995, each retailer stocked approximately three to four different manufacturer's brands. Retailers can only dedicate a limited amount of shelf space for each product or service, which may limit the introduction of new digital products and services into the marketplace.

References

[1]. Herschel Shosteck, "The Retail Market of Cellular Telephones", Herschel Shosteck Associates, Wheaton, Maryland, USA, 1st Quarter, 1996.

[2]. ibid.

[3]. Personal interviews, development experts at cellular handset manufacturing companies.

[4]. Telecommunications Industries Association Transition to Digital Symposium, "New Services and Capabilities", TIA, Orlando, FL, Sep 1991.

[5]. Personal interview, manufacturing purchasing agent, industry expert.

[6]. Cellular Integration Magazine, "Tech-niques", Argus Business , January 1996

[7]. Personal interview, cellular telephone manufacturing manager, industry expert.

[8]. Personal interview, manufacturing manager, industry expert.

[9]. Personal interview, manufacturing manager, industry expert.

[10]. Personal interview, manufacturing manager, industry expert.

[11]. Personal interview, manufacturing manager, industry expert.

[12]. United States Patent 3,663,762, "Mobile Communication System" Assigned to ATT, May 1992,

[13]. Letter to the TIA voting members from Eric J. Schimmel, 21 November, 1990.

[14]. Lawrence Harte Personal Interview, Industry Expert.

[15]. Personal interview, "Jeffrey Schlesinger", UBS Securities, New York, February 12, 1996.

[16]. Personal interview, cellular system development expert.

[17]. Personal interview, cellular system expert, system testing of thousands of hours.

[18]. Personal interview, Bob Glen, Sparton Electronics, Raleigh NC, January 1996.

[19]. Personal interview, manufacturing manager, industry expert.

[20]. Personal interview, manufacturing manager, industry expert.

[21]. Personal interview, "Jeffrey Schlesinger", UBS Securities, New York, February 12, 1996.

[22] Personal interview, cellular service provider, industry expert.

[23]. Personal interviews, tower leasing companies, industry experts.

[24]. Personal interview, cell site building manufacturer, industry expert.

[25]. Personal interview, cellular system executive, industry expert.

[26]. Personal interview, cellular system sales engineer, industry expert.

[27]. EMCI, "Digital Cellular, Economics and Comparative Analysis", Washington DC, 1993.

[28]. Paine Webber Conference, New York City, 1993.

[29]. D.M. Balston, R.V. Macario, "Cellular Radio Systems", Artech House, 1993, pg. 223.

[30]. Wireless Internet conference, Council for Entrepreneurial Development, Raleigh, NC USA, Sep 95.

[31]. Personal interview, cellular system manager, industry expert.

[32]. Personal interview, LEC manager, industry expert.

[33]. Personal Interviews, Cellular System Managers.

[34]. Personal Interview, Elliott Hamilton, EMCI Consulting, Washington DC, 25 February 1996.

[35]. Personal interview, Bell Atlantic manager, industry expert.

[36]. North Carolina Electronics Information Technologies Association, "Cellular Fraud", Raleigh, NC, November, 1995.

[37]. CTIA, "Wireless Factbook", Washington DC, Spring 1995.

[38]. CTIA, Mid-Year Results, Cellular Telecommunications Industry Association, Washington DC, , 1995.

[39]. CTIA, "Wireless Factbook", Washington DC, Spring 1995.

[40]. ibid.

[41]. Personal interview, McCaw manager, October, 1994.

[42]. Personal interview, Lee Horsman, Allen Telecom, February, 1996.

[43]. Personal interview, cellular service provider manager, industry expert.

[44]. CTIA, "Wireless Factbook", Washington DC, Spring 1995.

[45]. Sprint, Wireless Data Symposium, Raleigh, NC, December, 1995.

[46]. Dr. George Calhoun, "Digital Cellular Radio", p.69, Artech House, MA. 1988.

[47]. Stuart F. Crump, Cellular Sales and Marketing, p.2, Creative Communications Inc., Vol. 5, No. 8, Washington DC, August 1991.

[48]. Hilbert Chan, C. Vinodrai, "The Transition to Digital Cellular", p.191, IEEE 1990 Vehicular Technology Conference.

[49] Personal interview, cellular handset manufacturer engineer, June, 1994.

[50] Herschel Shosteck, "The Retail Market of Cellular Telephones", Herschel Shosteck Associates, Wheaton, Maryland, USA, 1st Quarter, 1996.

[51]. CTIA Winter Exposition, John Stupka, "Technology Update", Reno Nevada, 1990.

[52]. Telecommunications Industries Association Transition to Digital Symposium, "New Services and Capabilities", TIA, Orlando, FL, Sep 1991.

[53]. ibid.

[54]. Interview, Elliott Hamilton, EMCI, Washington DC, 6 March 1996.

[55]. Stuart F. Crump, Cellular Sales and Marketing, p.1, Creative Communications Inc., Vol. 5, No. 8, Washington DC, August 1991.

[56]. Telecommunications Industries Association Transition to Digital Symposium, "New Services and Capabilities", TIA, Orlando, FL, Sep 1991.

Chapter 11

Future Cellular and PCS Technology

11. Future Cellular and PCS Technology

There are several new wireless and network technologies that will impact future cellular telephone systems. New frequency allocations are being assigned to provide for these advanced services.

In 1995 the United States began licensing bands of frequencies for advanced wireless services. It is perceived that these technologies combine the services of paging, cellular, data, and cordless telephones into one portable device. Personal Communications Services (PCS) will occur on these bands of frequencies that have been recently auctioned by the federal government. While the technologies may offer some unique services, PCS companies will provide similar services as cellular carriers.

Integrated Dispatch Enhanced Network (iDEN) is a wireless technology which is being deployed in 800 MHz Specialized Mobile Radio (SMR) channels. Originally called Motorola's Integrated Radio System (MIRS), iDEN is a new digital TDMA technology that provides both voice and dispatch services. Nextel (originally called Fleetcall) has been acquiring and partnering with SMR operators to begin to offer nationwide voice and dispatch services using iDEN technology.

Extended Time Division Multiple Access (ETDMA) is a technology that was created to extend the capability of the IS-54 and IS-136 TDMA systems. ETDMA provides for an increase in the number of subscribers that can be served by an IS-54 TDMA system.

Voice paging is a new service that is being deployed in the cellular system to transfer voice messages to a cellular voice pager. Cellular system voice paging uses store and forward messaging to allow the queuing of messages and increased information compression.

Cellular Digital Packet Data (CDPD) is a packet data system that overlays on a cellular system to allow efficient delivery of short digital messages.

Spatial Division Multiple Access (SDMA) is a new technology that can be applied to any of the wireless technologies and frequencies to improve its performance and ability to serve more subscribers. SDMA allows an increase in the number of subscribers that can be served by any cellular system by using smart antennas to focus the radio energy into narrow radio coverage beam widths.

Satellites can provide voice and data services to wide regional geographic areas which terrestrial (land based) services do not reach. Satellite voice systems have been available for many years. However, these systems required large antennas and expensive subscriber communications equipment. New satellite cellular systems are planned for deployment in the late 1990's which will allow low cost handheld Subscriber Units.

11.1 Personal Communications Services

Personal Communications Services (PCS) are new wireless networks that use cellular technology (frequency reuse) to provide telecommunication services in North America to customers through the use of new radio frequency bands in the 1900 MHz range. The PCS radio spectrum has two types of frequency bands - licensed and unlicensed.

To use the PCS licensed radio spectrum as a resource, a PCS company must obtain a license from the Federal Communications Commission (FCC) and relocate existing users of the radio spectrum. A licensee is permitted to use a radio frequency band with certain restrictions such as serving public users compared to creating private wireless networks.

To use the unlicensed PCS frequency band, a PCS company must use equipment that will conform to the FCC unlicensed requirements and pay a fee to relocate existing users. Unlicensed requirements include low power transmission that does not typically interfere with other users in the same frequency band. The fee for relocating existing users is paid by the manufacturers of the unlicensed radio equipment and will be arbitrator who will assist in the relocation process.

Licenses for new PCS radio channel frequency bands were auctioned by the federal government. Because there were existing licensed users in the radio spectrum chosen for PCS service operators, the government will require the existing users to relocate over time. The existing users of the PCS radio spectrum are primarily microwave point-to-point communications. The existing users will be reassigned to other frequencies and the equipment will be changed at no cost to them. The new users must pay to convert the existing users to new frequencies. As of December 1995, the government had raised $7.7 billion from the sale of radio spectrum [1].

A PCS license allows a PCS carrier to provide service to a specific geographic area. The FCC divided the US license areas into 51 major trading areas (MTAs) and 493 basic trading areas (BTAs) [2]. MTAs and BTAs are defined by Rand McNally maps. BTA's are subdivisions of MTA's. Unlicensed users are allowed to transmit in any geographic area provided they conform to the FCC regulations which allow the coordination of unlicensed transmissions.

Figure 11.1 shows the broadband PCS frequency bands that were auctioned by the federal government in 1995. PCS frequency bands range from 1850 to 1990 MHz. The PCS spectrum has been divided into blocks A-F and unlicensed bands.

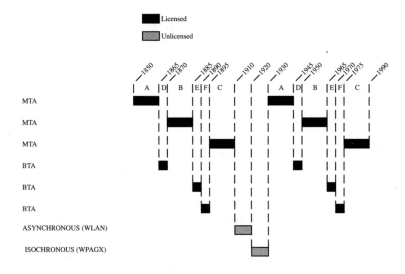

Figure 11.1, PCS Frequency Bands

Blocks A-F were divided into frequency pairs to allow simultaneous transmission and reception (frequency duplex). Unlicensed bands were not divided into frequency pairs because their type of use allowed time division duplex (TDD) operation.

PCS system Base Stations and Subscriber Units typically operate at lower maximum power levels than existing cellular telephones. Because of the higher radio signal attenuation at PCS frequencies, this results in smaller cell sizes in rural areas. In urban areas where cellular is mature, the power limitation is of lesser concern.

The FCC limited who could compete for bands of spectrum. Blocks A and B were available for any company which did not have a controlling interest in a cellular system in that region. Ownership of the cellular system and PCS system in the same region would minimize competition. Block C was set aside for designated entity (DE) companies who received special spectrum license financing terms. Initially, DE companies who had a majority of ownership by minorities received additional financing incentives. The bidding rules were modified to allow any small company (below $145 million in sales) to bid on the radio spectrum. Entrepreneurial companies (less than $40 million in sales) received additional special financing and license fee payment terms. Blocks D and E were available to any company provided the combination of PCS or cellular frequency bands they control is less than 40 MHz. Block F was also set aside for DE companies.

There are two unlicensed (anyone can use) 10 MHz bands that have been created to allow various consumer devices. While anyone can use this radio spectrum without buying a license, a company must obtain FCC type approval for these types of products. When the company produces or sells products operating in this frequency band, they will have to share the cost of moving existing frequency users.

The unlicensed radio spectrum (1.91 GHz to 1.93 GHz) was divided into 10 MHz asynchronous and isochronous bands. The asynchronous band is targeted to allow Wireless Local Area Network (WLAN) computing and similar data transmission devices. This type of transmission is intermittent bursts of information. The Isochronous band was created to allow for advanced wireless cordless telephones such as Wireless Private Automatic Branch eXchange (WPABX). This type of transmission is continuous or repetitive bursts.

While several frequency bands have been specified that allow wide area voice communications, no specific technology was required. A Joint Technology Committee (JTC) was established to create one or more standard radio specifications that can provide PCS services. This effort is expected to standardize the industry so only a few variations of PCS equipment will be required. These standards define how the licensee allows users to share this spectrum. This is called the access method. There were seven access methods proposed in 1995; PCS 1900 (Upbanded GSM), IS-95 CDMA, IS-136 TDMA, DCT, B-CDMA, IS-661 CDMA, and PACS.

PCS 1900 (GSM)

PCS 1900 is based on a European digital standard Global System for Mobile Communications (GSM). While GSM was initially offered in Europe, GSM is available in many countries outside of Europe (e.g. Australia). In 1995, there were are over 60 countries offering GSM service. Due to the large number of GSM customers and suppliers, the cost of system equipment and Subscriber Units has become very competitive.

The PCS 1900 system divides the radio bandwidth channels into 200 kHz bands which is the same as the GSM standard. It uses Time Division Multiple Access (TDMA) technology which divides the 200 kHz radio channels into short time slots and allows several users to share these time slots. Differences between PCS1900 and GSM start with different frequency bands for PCS 1900, GSM, and PCN (DCS 1800 version of GSM). The PCS 1900 system also has a new speech coder called EFR. Standard GSM and PCN Subscriber Units are not compatible with the US PCS system. To become compatible, redesign of the Subscriber Units and system equipment will be required. The GSM system infrastructure equipment is based on linking to the European telephone systems which are different than the US. These differences range from the speed of digital transmission in the communication lines to software control of the telephone network.

CDMA (IS-95)

IS-95 Code Division Multiple Access (CDMA) is a technology that began commercial deployment in 800 MHz systems in 1995. A detailed description of IS-95 CDMA technology is provided in chapter 5. The CDMA system which is offered for PCS includes a high bit rate speech coder (13 thousand bits per second (kbps)) that provides improved voice quality.

TDMA (Upbanded IS-136)

The only major difference between existing IS-136 equipment and its PCS version is the frequency band. This will only require a redesign of the Subscriber Units and Base Station radios in the cell sites. A detailed description of IS-136 technology is provided in chapter 4.

Digital Cordless Telephone (DCT)

Digital Cordless Telephone (DCT) is based on another European standard Digital European Cordless Telephone (DECT). DCT was designed for wireless office applications that have small cell coverage areas (typically less than 1 mile). While it is possible to cover an entire city with a DCT system by using many small cells, the number of cells could be prohibitive.

Differences between DCT and standard European systems include different frequency bands and power levels. The DCT system divides the radio bandwidth channels into wide 1.1 MHz bands where DECT has 1.728 MHz radio channels.. This may require a redesign of the Subscriber Units and Base Stations. Like the GSM system equipment, DCT is based on linking to the European telephone systems which are different than the US. However, because DCT has been deployed in Canada, some equipment has already been adapted for the North American market.

B-CDMA

B-CDMA is a wideband CDMA technology proposed for both 800 MHz cellular and PCS technology. B-CDMA technology has been optimized to overlay on top of existing radio channels without the need to replace them. Due to the lack of testing and commercial equipment, B-CDMA is receiving limited support from service providers and suppliers. The B-CDMA system divides the radio bandwidth channels into very wide 5, 10, or 15 MHz bands.

CDMA (IS-661)

IS-661 CDMA is a combined TDMA and CDMA standard which was created by Omnipoint Data Corporation. Due to advances Omnipoint has made in the development of this technology, the FCC has awarded a pioneers preference license for PCS to Omnipoint for the New York market. The Omnipoint CDMA system divides the radio spectrum into 5 MHz radio channels. The technology combines both CDMA and TDMA technologies to gain advantages not possible with a single access technology.

Personal Access Communications System (PACS)

Personal Access Communications System (PACS) is a combination of the US Wireless Access Communications System (WACS) and Japan's Personal Handy-Phone (PHP). WACS was developed by Bellcore for wireless local telephone service. PHP was developed in Japan for the digital cordless market which is functionally similar to DCT.

The PACS technology divides the radio bandwidth channels into 300 kHz wide frequency bands. It uses Time Division Multiple Access (TDMA) technology which divides radio channels into short time slots and allows several users to share these time slots. Similar to DCT, PACS technology is not well suited for large area coverage. PACS would most likely serve as a residential wireless telephone replacement.

11.2 Integrated Dispatch Enhanced Network (iDEN)

Integrated Dispatch Enhanced Network (iDEN) is a digital only TDMA radio system. Formerly called Motorola Integrated Radio System (MIRS), iDEN is primarily being used for 800 MHz Specialized Mobile Radio (SMR) channels which range from 806 MHz to 821 MHz and 851 MHz to 866 MHz. Originally, MIRS technology was developed by Motorola for a company called Fleet Call. Fleet Call changed its name to Nextel in 1994. Since it began purchasing or partnering with other SMR companies in 1988, Nextel and its partners had radio coverage in almost all of North America in 1995.

The MIRS technology is a TDMA system which divides the frequency band into 25 kHz radio channels. Each radio channel is divided into frames which contain 6 slots. When used for cellular like service, three users can share each radio channel. When used for dispatch (where a lower voice quality is acceptable), up to 6 users can share each radio channel. The iDEN radio channel uses 16 level Quadrature Amplitude Modulation (QAM) to allow up to 64 kbps of data on a single 25 kHz radio channel. iDEN technology combines voice and data capability and is sometimes called Digital Integrated Mobile Radio System (DIMRS).

11.3 Extended TDMA (ETDMA)

ETDMA was proposed by Hughes Network Systems in 1990 as an extension to the existing IS-54 TDMA standard. ETDMA uses the existing TDMA radio channel bandwidth and channel structure. ETDMA Subscriber Units are tri-mode as they can operate in AMPS, TDMA, or ETDMA modes. While a TDMA system assigns a Subscriber Unit fixed time slot numbers for each call, ETDMA dynamically assigned time slots on an as needed basis. The ETDMA system contains a half-rate speech coder (4 kbps) that reduces the number of information bits that must be transmitted and received each second. This makes use of voice silence periods to inhibit slot transmission so other users may share the transmit slot. The overall benefit is more users can share the same radio channel equipment and improved radio communications performance. The combination of a low bit rate speech coder, voice activity detection, and interference averaging increases the radio channel efficiency to beyond 10 times the existing AMPS capacity. Figure 11.2 shows a ETDMA Subscriber Unit. This phone is capable of AMPS, IS-54, and ETDMA operation.

Figure 11.2, ETDMA Subscriber Unit
Source: Hughes Network Systems

11.3.1 ETDMA Operation

ETDMA radio channels are structured into the same frames and slots structures as the standard IS-54 radio channels. Some or all of the time slots on all of the radio channels are shared for ETDMA communication. EDTMA slots can be shared on the same radio channel which is similar to IS-54 and IS-136 radio channels, or slots can be shared on different frequencies. When a Subscriber Unit is operating in extended mode, the ETDMA system must continually coordinate time slot and frequency channel assignments. The ETDMA system performs this by using a time slot control system. On an ETDMA capable radio channel, some of the time slots are dedicated as control slots. These control slots coordinate access to time slots on an as needed basis.

The existing 30 kHz AMPS control channels are used to assign analog voice and digital traffic channels. ETDMA systems can assign either an AMPS channel, a TDMA full-rate or half-rate channel, or an ETDMA channel.

Figure 11.3 shows an ETDMA radio system. Some of the radio channels include a control slot which coordinates time slot allocation. Control time slots account for an estimated 15% of available time slots in a system. The control time slots assign an ETDMA subscriber to voice time slots on multiple radio channels. Notice that the voice path in figure 11.3 moves between voice time slots (shaded) on multiple radio channels.

ETDMA uses the following process to allocate time slots from moment to moment as needed. The cellular radio maintains constant communications with the Base Station through the control time slot (marked "C" in figure 11.3). When a conversation begins, the cellular radio uses the control slot to request a voice time slot from the Base Station. Through the control slot, the Base Station assigns a voice time slot and sets the cellular radio to transmit in that assigned voice time slot. During each momentary lull in phone conversation, the transmitting cellular radio gives up its voice time slot, which is then placed back into the Base Station's pool of available time slots.

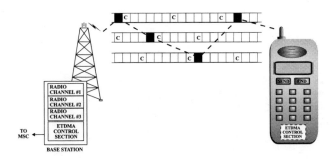

Figure 11.3, ETDMA System

When a cellular radio is ready to receive a voice conversation, the Base Station uses the control slot to tell it which voice time slot has the conversation being sent. The cellular radio receiver then tunes to the appropriate slot. Through the control slot, the Base Station constantly monitors the cellular radio to determine whether it has given up a slot or needs a slot. In turn, the cellular radio constantly monitors the control slot to learn which time slot contains voice conversation being sent to it.

11.4   Voice Paging

A new type of device for cellular systems is a dedicated cellular/PCS voice pager. The difference between a standard Subscriber Unit and a cellular voice pager (CVP) is the CVP stores voice messages in its internal memory. The operation of a voice paging system allows for unique non-real time message delivery.

Figure 11.4 shows a voice messaging system which is connected to a cellular system. In step 1, the caller dials a voice pager telephone number. After a recorded message, the caller leaves a message which is stored in a voice mailbox (step 2). The voice mail system then dials the CVP's phone number and attempts to deliv-

315

er the message (step 3). After the CVP has answered the call, the voice messaging system checks to see if enough memory is available in the CVP for the new message(s). If there is enough memory, the voice mailbox will transfer the message (step 4) and it will be stored in the CVP voice message memory (step 5). After the message is completely delivered, the CVP will acknowledge the successful receipt of the message (step 6) and it can be deleted from the voice mail system's memory. The subscriber will be alerted by visual and/or audio that a message is waiting. The message can then be played at any time from Cellular Voice Pager's memory (step 7).

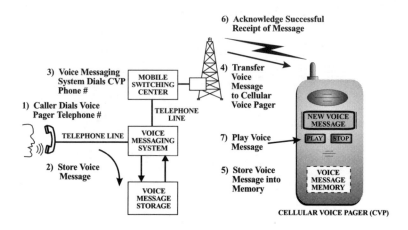

Figure 11.4, Voice Paging System

The acknowledgment feature is an important part of voice paging. Message acknowledgment allows a new service called confirmation paging to be provided. After the message is successfully delivered, the voice message system can deliver confirmation to the caller via a pre-recorded confirmation message call back, fax message, or email.

Another advantage of the voice message system is the ability to compress voice messages for more efficient delivery. Figure 11.5 shows a sample of some of the compression steps. The first step removes the pauses between words. Next, the stored voice message is played back 2-4 times faster than normal voice. Both of these types of voice compression are not possible for normal (real time) cellular conversations.

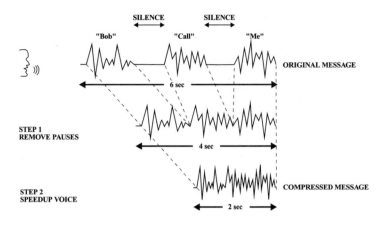

Figure 11.5, Voice Paging Compressed Voice

Providing non-real time messaging allows for the use of extended sleep mode. Similar to numeric paging systems, non-real time messages can be delayed for several minutes without a significant impact. This is not possible with standard cellular Subscriber Units where the sleep period cannot exceed a few seconds. This is because a caller who dials a Subscriber Unit is probably only willing to only wait a few rings before hanging up the phone. By allowing the CVP to wake up for very short periods (tenths of seconds) and sleep for several minutes, battery life could be extended to weeks.

11.5 Cellular Digital Packet Data (CDPD)

Cellular Digital Packet Data (CDPD) is a digital packet transfer system that overlays on top of an existing cellular system. CDPD was created to address the need of subscribers to rapidly send a small amount of data and not tie up a cellular radio channel for a long period of time. Because the CDPD system transfers packets of data only when they are needed, the CDPD system does not maintain a constant connection between the two users. This type of system is referred to as "connectionless" because there are no pre-determined time periods or dedicated resources for packet transmission.

Digital data can be sent on an existing cellular radio voice/traffic channel. Unfortunately, cellular radio channels were not designed for rapid initialization to send small amounts of data. CDPD changed the way a Subscriber Unit accesses the network to help minimize the setup and transfer time for each data transfer period.

CDPD divides a subscriber's data information into small packets which contain their destination address. Packets are sent to their destination by the best path possible at the time of transfer. The travel time for each packet between its' origination and destination may be different. This is because packets of information are often sent on different routes due to communications path availability. As packets are received, they are reassembled in the proper order by using a packet index number decoded at the receiving end.

The CDPD network and radio specification was announced in 1992 by a consortium of cellular carriers. In 1995, there were over 12,000 cell sites in the United States which had CDPD capability [3]. CDPD technology is based upon industry standards that are maintained by the CDPD forum which is an organization of cellular carriers and others who are interested in promoting CDPD [4].

11.5.1 System Elements

While the names of the equipment assemblies are different in a CDPD system, CDPD equipment assemblies perform similar functions as those in a cellular system. A CDPD Subscriber Unit is called a Mobile End Station (MES). A Base Station is called a Mobile Data Base Station (MDBS). The switch is now called a Mobile Data Intermediate System (MD-IS). Also, the Public Switched Telephone Network (PSTN) is replaced with various packet and other data connection networks.

Figure 11.6 shows a Mobile End Station (MES) converts radio energy into a data signal for the subscriber (or end equipment) and converts the subscribers data signals to radio energy. The Mobile Data Base Station (MDBS) converts the radio signal to a digital signal to be forwarded to the MD-IS. The MDBS is also responsible for the management of the RF communication with the MES. It manages the channel hopping or handoff function. The MD-IS is typically located at the Mobile Telephone Switching Office (MTSO) of the cellular carrier. It manages all of its associated MDBSs and performs the decryption of CDPD messages and is a router that sends the packets of data to the outside world. Ultimately, the final destination of the CDPD message is the Fixed End System (FES) or another MES. The FES is often a computer which stores, processes, or displays the information contained in the packet.

11.5.2 System Operation

Figure 11.6, CDPD Mobile End Station
Source: Cincinatti Microwave

Management of channels to coexist with AMPS systems is one of the most important aspects of a CDPD system. The Radio Resource Management (RRM) entity exists in CDPD systems to handle this activity. There are three general categories of channel management used by CDPD systems. The most desired method of channel management is a cooperative exchange of radio channel activity information between the AMPS system and the CDPD system. In this method the AMPS system notifies the CDPD system of channels that are available or likely to be available in the near future. Independent management of the channels by the CDPD system is the second method of channel management. This method requires the use of an RF sniffer that can detect activity on the AMPS channels. The sniffer is used to examine both used and unused channels to provide a list of available channels of the CDPD system. The difficulty of this method is finding a suitable algorithm that can quickly detect used channels as well as predict which channels are likely to be available for use in the future. The third method of channel management is the use of dedicated channels. This entirely prevents interference between CDPD and AMPS systems but has the drawback of reducing the number of channels available for AMPS on systems that have assigned most of their available radio channels. Figure 11.7 shows an overview of shared and dedicated CDPD radio channel management.

Cellular and PCS/PCN Telephones and Systems

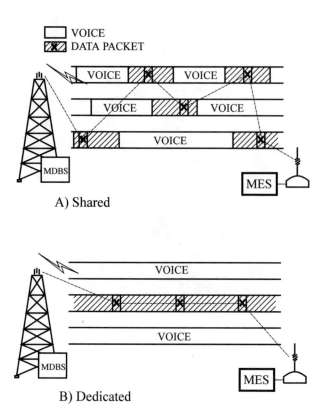

Figure 11.7, CDPD Radio Channel Management

Because there are no dedicated control channels for CDPD, Subscriber Units make access attempts on the data channel each time data transfer is required. This minimizes the amount of overhead on the cellular system and allows CDPD data transmission to be more cost effective for small data transfers than traditional circuit switched data.

Figure 11.8 shows the basic operation of the CDPD system network. As a first step, the MES scans cellular radio channels for the CDPD pilot signal. After it has locked onto a CDPD channel, it will begin to transmit data (step 2). If the radio channel that the MES is currently operating on becomes active with a cellular voice channel, the MES will move to the next CDPD radio channel and continue to send data. The data control is coordinated and gathered by the MDBS. The MDBS converts the information and routes it to the MD-IS. The MD-IS is often located in the cellular switching center so the information may be sent on

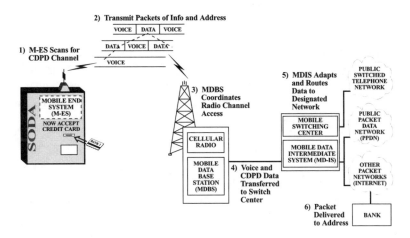

Figure 11.8, Sample CDPD System Operation

the existing communication channels linking the Base Station and the MSC (step 4). The MD-IS adapts and routes the data to the designated external network (e.g. PPDN or Internet). The external network routes the data to the address given by the MES.

The CDPD frequency band is identical to the AMPS cellular system with the AMPS control channels being removed from the CDPD band. RF power levels are also the same as AMPS cellular. Output power in CDPD is controlled by the MES which calculates its output power level by measuring the received signal strength. A lower received signal results in a higher transmit power level. GMSK modulation is used for CDPD to transfer data at a 19.2 kbps. Because some of the data bits are used for packet addressing and system control, the average data transfer rate is approximately 7 to 8 kbps [5].

Similar to a cellular Subscriber Unit's unique ESN identifier, each MES has its own unique serial number. This is referred to as its Network Entity Identifier (NEI). While the NEI is used to validate subscriber identity, one or more Temporary Equipment Identifiers (TEI) are used to address the MES when it is operating in a particular service area. Multiple TEI's may be used to provide different services. For example, one TEI may be used for point to point (single recipient) and another may be used for broadcast services (multiple receivers to a single message).

Every MES in a CDPD system belongs to a home system location. Part of the home system's responsibility is to maintain a data base containing the location of all of its MES. Every message sent to a MES first goes to its home system location where it is then forwarded to its current location. The home system keeps track of the MES through a registration process that tracks each cell that a MES enters.

Handoffs are completely controlled by the MES in CDPD systems. After the MES determines that it is not desirable to maintain connection (probably due to poor signal strength), it will terminate transmission and seek a new CDPD channel. Once it has established a link on a new channel (typically a new cell site), transmission can continue.

CDPD has authentication and data link confidentiality (transmission privacy) capability similar to digital cellular systems. Each MES can be validated by the system and the MES can also validate system authentication inquiries. Once the validation occurs, the key is used to encrypt the data transferred by the radio packets.

11.6 Spatial Division Multiple Access

Spatial Division Multiple Access (SDMA) is a technology which increases the quality and capacity of wireless communications systems. Using advanced algorithms and adaptive digital signal processing, Base Stations equipped with multiple antennas can more actively reject interference and use spectral resources more efficiently. This allows for larger cells with less radiated energy, greater sensitivity for portable cellular phones, and greater network capacity.

Figure 11.9 shows a typical 120 degree sectored antenna (an antenna which covers 1/3 of the cell site area). Within the coverage area denoted, the antenna can communicate with only one subscriber per traffic channel in the radio coverage area at any given time. System performance is constrained by the levels of interference present.

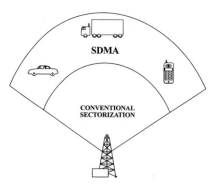

Figure 11.9, Conventional Sectored Cell Site Radio Coverage
Source: ArrayComm

One company, ArrayComm, of San Jose California USA, has implemented SDMA into its IntelliCell™ Base Stations. Because of the IntelliCell Base Station's ability to reject interference on the uplink and control transmission patterns on the downlink, Spatial Channels™ can be created. Figure 11.10. ArrayComm has simultaneously operated 3 mobile units on the same traffic channel in the same local area. Using an 8 element antenna array to track multiple users, the IntelliCell system exhibited Carrier to Interference (C/I) ratios which consistently exceeded 30 dB.

An IntelliCell Base Station which implements Spatial Channels will track the movement of each user in the system. If one user moves too close to another and the users are on the same conventional traffic channel, the system will automatically hand off one of the users to another traffic channel frequency.

SDMA can be applied to any of the wireless systems in use today (AMPS, TDMA, CDMA, etc.). Its benefits include larger cells, less interference to other cells, and lower mobile/portable transmit power.

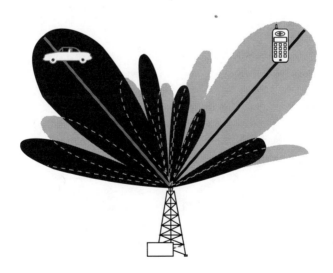

Figure 11.10, Spatial Division Multiple Access
Source: ArrayComm

11.7 Wireless Office Telephone Systems (WOTS)

Wireless Office Telephone Systems are targeted for the people who keep the business running [6]. This means shipping clerks and people other than executives are often the primary users. There are several types of cellular wireless office telephone systems which vary from single line to entire mini cellular systems.

11.7.1 Telego

Telego phones can communicate with either a cordless Base Station or a cellular system. The Telego home cordless Base Station uses 800 MHz cellular frequencies and connects to the standard telephone line. When it senses the Telego phone is at home, the cordless Base Station informs the cellular system by dialing a pre-stored number. Both landline and incoming cellular calls will then be routed to the home cordless Base Station. Because the Telego phone can use either a cordless base station or the cellular system, if the home telephone line is in use, the customer can force the Telego phone to use the cellular system as a second line.

Telego is like training wheels for cellular as customers can perceive that Telego is an enhanced cordless phone. The Telego Subscriber Unit (phone) creates a dialtone similar to a regular telephone. When using the Telego phone in the home, the user typically pays no charge for airtime use. When the customer moves into areas outside their home Base Station, the user pays an airtime usage fee.

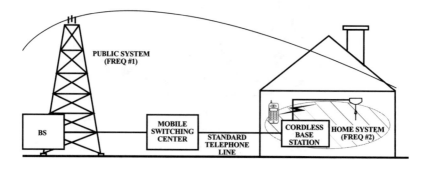

Figure 11.11, Telego System

The customer knows whether they are operating on the home system or the cellular system by viewing the phone's display. When the screen displays Home, this indicates it can use the cordless Base Station. In Extended mode, it has access to the cellular system in an area that has been designated as a low cost zone. Premium mode indicates that the Telego phone is operating as a standard cellular Subscriber Unit with higher airtime rates [7]. Figure 11.11 shows a Telego system.

11.7.2 Cellular Auxilliary Personal Communication Service (CAPCS)

The Cellular Auxilliary Personal Communications Service (CAPCS) system is a wireless office (microcell) system which uses the IS-94 industry standard protocol and cellular frequencies. To use a CAPCS system, the customer buys a separate cellular wireless PABX. The CAPCS system ties into the customers PABX system via a control unit. Other names for the CAPCS system include "Freedom Link" and "Business Link."

The CAPCS control system connects to one or more radio Base Station. Each Base Station typically covers a radius of up to 500 feet. Customers use Subscriber Units (cellular phones) that have software modifications which allow them to access the CAPCS system or the public cellular system. Additional options for the CAPCS system include caller ID, Discontinuous Receive (DRX), dual telephone lines, pages, and multiple home systems [8]. The system cost a for CAPCS type system is approximately $2,000 per user [9]. Figure 11.12 shows the IS-94 System.

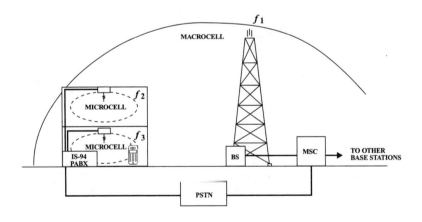

Figure 11.12, Cellular Auxilliary Personal Communications Service (CAPCS) System

11.8 Satellite Cellular

Satellite communication systems can provide service to any unobstructed location on the Earth. In the late 1990s, several mobile-satellite systems will begin commercial operation [10], [11]. These new satellite systems are will supplement low cost terrestrial (land based) wireless communications systems [12].

Satellites are much more expensive than cell sites and have a limited lifespan. To help distribute the high cost over many customers, some satellite systems offer frequency reuse similar to cellular technology to serve thousands of subscribers on a limited band of frequencies. These systems use sophisticated antenna systems to focus their energy into small radio coverage areas.

Satellite systems are often characterized by their height above the Earth and type of orbit. Geosynchronous Earth Orbit (GEO) satellites hover at approximately 36,000 km, Medium Earth Orbit (MEO) satellites are positioned at approximately 10,000 km, and Low Earth Orbit Satellites are located approximately 1 km above the Earth.

The higher the satellite is located above the Earth, the wider the radio coverage area. This reduces the number of satellites required to provide service to a geographic area. GEO satellites rotate at the same speed as the Earth allowing the satellite to appear to be stationary over the same location. This allows fixed position antennas (satellite dishes) to be used. Because GEO satellites hover so far above the Earth, this results in a time delay of approximately 400 msec [13] and increases the amount of power that is required to communicate which limits their viability of handportable Subscriber Units. GEO systems include: AMSC, AGRANI, ACeS, APMT [14].

The closer distance of MEO satellites reduces the time delay to approximately 110 msec and allows a lower Subscriber Unit transmit power. Because MEO satellites are at a lower Earth orbit, they do not travel at the same speed relative to the Earth which requires several MEO satellites to orbit the Earth to provide continuous coverage. MEO system include Intermediate Circular Orbit (ICO) Project 21, and TRW's "Odyssey".

Because Low Earth Orbit (LEO) satellites are located approximately 1 km from the surface of the Earth, this allows the use of low power handheld Subscriber Units. LEO satellites must move quickly to avoid falling into the Earth so LEO satellites circle the Earth in approximately 100 minutes at 24,000 km per hour. This requires many satellites (e.g., 66 satellites are used for the Iridium system) to provide continuous coverage. LEO systems include Iridium Inc.'s "Iridium", Constellation Communications Inc.'s "Aries", Loral-Qualcomm's "Globalstar," and Ellipsat's "Ellipso" Figure 11.13 shows the three basic types of satellite systems.

Chapter 11

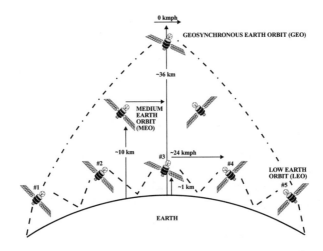

Figure 11.13, Satellite Systems

Some satellites systems communicate directly with each other through microwave links. By allowing telephone calls to be routed directly between satellites, this eliminates the high cost of routing long distance global calls through other telephone companies. Figure 11.14 shows a picture of a LEO satellite Subscriber Unit.

Figure 11.14, GlobalStar Subscriber Unit

327

References

1. Jeffrey Hines, Walter Piecyk, "Wireless Update", Paine Webber, December 15, 1995, pg.1.
2. ibid, pg.2.
3. Cincinnati Microwave, "Cellular Digital Packet Data - A Primer On How CDPD Can Change The Way Your Organization Does Business", Cincinnati, OH, July 1995, pg. 1.
4. "Cellular Digital Packet Data System Specification", Volume 1, Release 1.0, Costa Mesa,, CA, CDPD Industry Input Coordinator, July 19, 1993.
5. Personal Interview, Industry Expert, September 1995.
6. NCWIF Wireless Office Seminar, 29 November 1995, Research Triangle Park, North Carolina, Bill McClellan.
7. NCWIF Wireless Office Seminar, 29 November 1995, Research Triangle Park, North Carolina, Chris Leach.
8. TIA Working Group 45.1, "Proposed Baseline Text for IS-94A," Lisle, IL, October 13,1994, pg. 251.
9. NCWIF Wireless Office Seminar, 29 November 1995, Research Triangle Park, North Carolina, Bill McClellan.
10. R. Rusch, "The Market and Proposed Systems for Satellite Communications," Applied Microwave and Wireless, pp. 10-34, Fall 1995.
11. G.M. Comparatto, N.D. Hulkower, "Global Mobile Satellite Communications: A Review of Three Contenders," American Institute of Aeronautics and Astronautics, pp. 1-11, 1994.
12. Nils Rydbeck, Sandeep Chennakeshu, Paul Dent, Amer Hassan, "Mobile-Satellite Systems: A Perspective on Technology and Trends", Proceedings of IEEE Vehicular Technology Conference, April-May 1996, Atlanta, Georgia, USA.
13. ibid.
14. ibid

Chapter 11

Chapter 12
Digital Services

12   Digital Services

Customers do not purchase phones and data devices, but the benefits and services these devices provide. Digital technologies provide more benefits and services, and in so doing, will likely capture more and more of the market over time.

Digital cellular systems are generally newer than their analog predecessors, and have benefited from the recent emphasis on user features and services. The new digital technologies are inherently better at providing advanced services because everything that flows through a digital system is treated as data. Digital data transmissions merge voice, data, and video signals in one common communications channel, enabling private services, and improvements in performance. Through a combination of digital transmission and new control messages which have been added to the radio communication channel, digital systems can offer advanced applications that target both human and non-human services. Furthermore, digital services can be implemented on most of the older technologies, including analog cellular systems such as Advanced Mobile Phone System (AMPS), Total Access Communication System (TACS), or Nordic Mobile Telephone (NMT). Nearly all cellular data transmissions today are over analog connections. Although purely analog based systems offer data transmission and some advanced features such as short message services, the future of data transmission at higher rates is through digital systems. This chapter describes the benefits and services of the new digital cellular systems.

12.1   Multi-Media

Multi-media is the mixing of voice, data, and video services. Digital information is a universal medium capable of transporting digital voice, data files, and digi-

Cellular and PCS/PCN Telephones and Systems

tized video. The new digital cellular technologies therefore add multi-media capabilities to wireless systems which today primarily deliver real-time voice information.

Ideally, the digital communication channel transfers information unchanged between originator and receiver. This ideal would allow transmission of any combination of digitized voice, data files, and digital video provided the total amount of information did not exceed the digital channel capacity. Unfortunately, connecting digital radio with the public switched telephone network (PSTN) limits digital transmission capacity, and restricts some of the services that could be offered.

12.1.1 Bearer Services

The cellular network today primarily transports voice information from a caller to a caller. Because information is only transported by the network, this is called a bearer service. The cellular telephone network is the bearer of voice information.

Advanced services can be offered using the cellular system as a bearer service. If a subscriber transfers a fax message on a Subscriber Unit (cellular phone), the cellular system is only transporting the fax information through a standard communications channel. If a caller number identification is transported via special tones without changes to the cellular system, this would be a bearer service.

12.1.2 Tele Services

Some of the significant changes in the digital cellular systems include tele-services. Tele-services process information inside the network. For example, if the caller number identification is received by the cellular system from the landline telephone company, and the cellular system converts this to messages which contain the digits and adds the name of the caller, the cellular system provides a tele-service.

12.2 System Features

The new digital cellular systems also have new control messages that allow the system to provide new features. These new control messages can be sent while the Subscriber Unit is idle (in standby) or while a conversation is in progress.

12.2.1 Caller Identification

Caller identification (caller ID) on a cellular system works much as it does with land line telephone caller ID systems, providing the user a description of the calling party prior to answering a call. This identification can be either a phone number, a name, or both. The identifying information appears on the user's phone display during the ring sequence. The subscriber then has the option of answering the phone or not.

A caller ID phone number may be transferred in two different ways from a caller to a Subscriber Unit. Most common is for the phone company to send the calling number to the cellular system. The alternative method allows the calling person enter the digits from a touch tone phone, similar to the method used for sending phone numbers to numeric pagers.

Figure 12.1 shows the sequence of events that occur when caller ID data is sent. In step 1, when the Subscriber Unit has been dialed, the caller ID data is forwarded from the telephone company to the MSC. Next, the cellular system decodes the calling number ID, changes its format, and sends it as a message to the Subscriber Unit via the Base Station. The message containing the caller number identification arrives after the Subscriber Unit is alerted that a call is incoming (paging). The Subscriber Unit decodes the message and the phone number appears on the display. As an option, the Subscriber Unit may examine it's internal phone book data to see if the phone number matches any previously stored. If a match is found, the text string (normally a name) can be displayed along with the incoming call phone number.

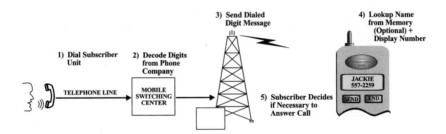

Figure 12.1, Caller Identification

Privacy concerns have led most systems to provide a method of allowing users to block the transmission of their phone number when placing a call. When the caller number identification has been blocked by the calling person, the caller ID information received by the Subscriber Unit is marked as private. A message such as PRIVATE or BLOCKED is displayed to indicate this to the user.

12.2.2 Message Waiting

Message waiting connects a users voice mailbox to their cellular phone. When a message is left in the mailbox, the cellular network is notified that a message is waiting. The network sends a control message to the Subscriber Unit to inform it of the number of messages waiting in the user's mailbox. The Subscriber Unit then alerts the user by displaying the message and/or sounding an audible tone. The Subscriber Unit display is also updated when the message count changes or goes to zero.

A possible disadvantage of message waiting is that the phone displays only what the Base Station transmits, not the true number of messages in the mailbox. Sometimes, due to a delay between the time the mailbox is emptied and the time the cellular network receives notification, the phone indicates messages waiting even after the mailbox is cleared.

Figure 12.2 shows the how the message waiting feature operates. In step 1, the caller dials the voice mailbox system. The voice mailbox system stores the callers message (step 2) and informs the cellular system that message(s) are waiting (step 3). The cellular system then attempts to deliver the message to the Subscriber Unit. Because the Subscriber Unit was probably unavailable for the caller who left the voice message, it is likely that the Subscriber Unit is still unavailable. The cellular system will store the message waiting command until the Subscriber Unit becomes available (step 4). After the Subscriber Unit receives the message waiting command, a display and/or audio beep alerts the user (step 5). Eventually, the subscriber calls to check messages. After the messages are received and/or erased, the voice mail system informs the cellular system how many new messages are remaining. The cellular system then send another message to the Subscriber Unit with the updated message count.

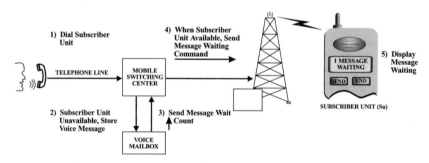

Figure 12.2, Message Waiting Indication

12.2.3 Short Message Service (SMS)

Short message service (SMS) gives cellular subscribers the ability to send and receive text messages. Short messages usually contain about one page of text, or two thousand bytes of information. They can be received while the Subscriber Unit is in standby (idle), or in use (conversation). While the Subscriber Unit is communicating both voice and message information, short message transfer takes slightly longer than it does while the Subscriber Unit is in standby.

SMS can be divided into three general categories; Point-to-point, Point-to-multi-point, and broadcast. Point-to-point SMS sends a message to a single receiver. Point to multi-point SMS sends a message to several receivers. Broadcast SMS sends the same message to all receivers in a given area. Broadcast SMS differs from point to multi-point because it places a unique "address" with the message to be received. Only Subscriber Units capable of decoding that address receive the message.

12.2.4 Point to Point SMS

Much like voice mail or pager type systems, point to point SMS sends a short message from one source to a one receiver. An example of this type of message would be "You Have Won the Lottery."

Figure 12.3 illustrates point to point SMS transfer. In step 1, the message goes to the Mobile Switching Center (MSC) to be routed and stored in the message center (step 2.) The cellular system searches for the Subscriber Unit (step 3) and alerts the Subscriber Unit that a message is coming. The Subscriber Unit tunes to the voice or control channel where the message will be sent. The system then attempts to send the message (step 4). As the message is being sent, the system waits for acknowledgment messages (step 5) to confirm delivery. If the transmission is successful, the message may be removed from MSC message center. If unsuccessful, the system attempts delivery again.

Messages can also be sent from a Subscriber Unit to the message center for delivery, but since most Subscriber Unit keypads have very few keys, most messages will probably go from the cellular system to Subscriber Units. However, keyboards and telemetry monitoring equipment may be connected to Subscriber

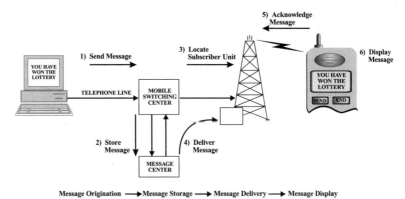

Figure 12.3, Point to Point Short Message Delivery

Units, allowing Subscriber Units to transfer messages to the cellular system. Another alternative is to use predefined messages that can be selected from a list by the subscriber and transmitted.

12.2.5 Point to Multi-Point SMS

Point to multi-point SMS sends a message to a group of Subscriber Units. An example of point to multi-point SMS might be a message to a corporate sales team indicating "Sales meeting on Tuesday is canceled."

Figure 12.4 illustrates how a message is transferred to a group of users. Like the point-to-point SMS service, the message first goes to the MSC message center (step 1) where the message center determines that this message is designated for multiple units. The designated list of message recipients may be pre-arranged (such as a sales staff) or it may be included with the originating message. In either case, the message center stores the message and recipient list (step 2). To complete message delivery. the MSC then searches for each Subscriber Unit in the list. The cellular system then individually alerts each Subscriber Unit that a message is coming (steps 3-5). The Subscriber Units tune to the message channel (voice or control), the system sends the message to each unit, and the units receive and store it. If the transmission is successful, the Subscriber Unit sends acknowledgment, and the message may be removed from MSC message center.

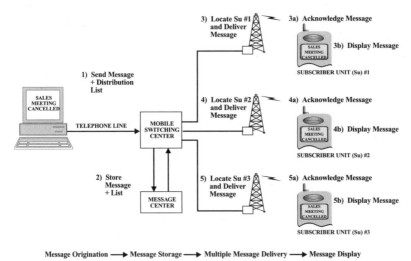

Figure 12.4, Point to Multi-Point Short Message Delivery

If the message transmission to one or more of the receivers was not successful, the cellular system attempts delivery again later.

12.2.6 Broadcast SMS

Broadcast SMS sends messages to a all mobiles that have been pre-set to monitor a specific messaging channel. Every Base Station transmits each broadcast message along with enough information to accurately decode the message.

Figure 12.5 illustrates how broadcast SMS messages are delivered to users. Like the point-to-point SMS service, the message first goes to the MSC message center (step 1) where the message center determines that this message is designated for all mobile units with a unique code. The delivery code may be pre-arranged (such as a traffic report code) or included with the originating message. The MSC then broadcasts the message on a designated message channel, which may be part of a control channel in every cell site where the broadcast message is designated to be received.

Unlike point-to-point and point-to-multipoint messages, Subscriber Units do not acknowledge receipt of broadcast SMS. If a Subscriber Unit is off or is not tuned

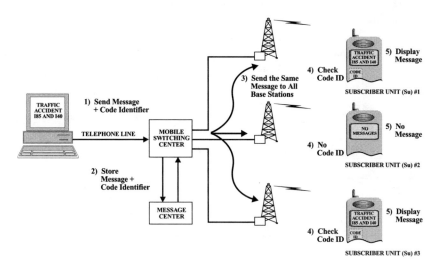

Figure 12.5, Broadcast Short Message Delivery

to the message channel, it misses the broadcast message. To address this limitation, messages may be broadcast several times, and Subscriber Units that have already received the message may ignore repeat messages.

12.2.7 Private Systems

Both public and private systems are in use today. Public systems, including today's cellular systems, are available to anyone who pays a service fee to a public telephone company. Private systems are used by corporations or other private owners to interconnect employees or private subscribers without routing calls through a public telephone company. Private system owners may lease radio channels which are FCC licensed to a local public cellular provider. Figure 12.6 illustrates how private systems operate within public ones. To entice companies to set up a private system for employees to use at work and home, public providers may offer localized flat billing rates for private systems.

Private system radio equipment may be located, for example, in a company building. Alternatively, small private systems can also exist without the expense of a separate infrastructure. Several independent systems may exist within a single cell site radio coverage area, so that a new system can be created simply by identifying private system radio channels to the phone users. In such cases, multiple systems have a separate billing rate as a key part of their functions. For example, a business with several buildings near each other could have its own cellular system. Employees using cellular phones at work would be billed at a different rate than if they used the same phones in the public system. The private system exists at the same time as the public system. The separation between the two systems is provided by the messages supported in the network.

New digital technologies enable typical private systems to offer advanced services such as localized billing rates, telephone directory listings, three or four

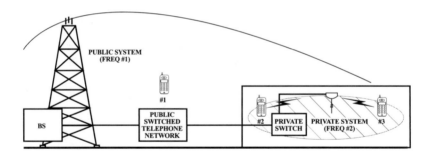

Figure 12.6, Private Cellular System

digit dialing, and other features. Private Subscriber Units may also be designed to access company databases for directory listings and other services via a digital communication channel.

12.2.8 Facsimile Over Cellular

Sending and receiving FAX data over a cellular link is an important key to a truly mobile office. Unfortunately, wireless FAX data has problems similar to those of cellular modem transmissions. However, while modem algorithms have been adapted to work well with sometimes poor cellular connections, FAX algorithms have not. Most FAX transmission schemes do not recover or correct errors well. Depending on the error type and location, some or all of a wireless FAX page can be lost. With current digital data transmission protocols, this need not be the case. Digital data transmission protocols correct errors through an encoding scheme and through re-transmission of missing or incorrect data. Assuming no dropped calls (disconnects due to poor radio conditions), wireless FAX transmissions could be as reliable as on a wire line.

Figure 12.7 illustrates digital FAX transmission from a landline fax machine to a Subscriber Unit capable of receiving a FAX. The landline FAX machine dials the Subscriber Units FAX telephone number or a universal number that allows FAX message detection (step 1). The cellular system converts the FAX audio signal into a digital signal (step 2.) Next, the Subscriber Unit is alerted that a

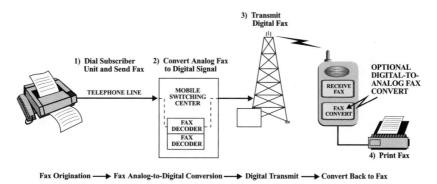

Figure 12.7, Fax Over Digital Cellular

FAX is to be received (step 3.) The Subscriber Unit receives the digital FAX. During the digital FAX message delivery, each bit of digital information is checked and re-sent if it is received in error. If the Subscriber Unit is connected to an analog FAX (standard FAX machine), the Subscriber Unit will convert the digital FAX back into an analog FAX signal (step 4.)

12.2.9 Voice Privacy

Digital systems are inherently more secure than analog systems because they have an encrypted mode of operation. This encrypted mode of operation scrambles voice data before it is sent to the Base Station. The encryption uses a mask value calculated with some of the authentication data. When the voice data is received at the Base Station it is decrypted using the same mask value that was used to encrypt it. Figure 12.8 shows how voice privacy mode can be requested by a subscriber.

In addition to encrypting the voice data, most digital systems have a message encryption mode that encrypts the signaling between the mobile phone and the

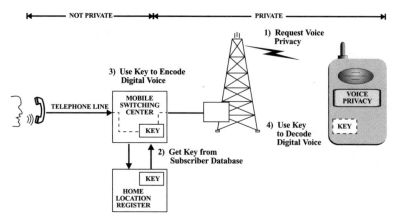

Figure 12.8, Digital Voice Privacy

Base Station. While this is not a direct benefit to the user of the phone, it does help to prevent unauthorized access to the cellular system.

12.2.10 Increased Battery Life

The battery life of a portable Subscriber Unit is the amount of time a fully-charged battery can operate when the unit in use. The simplest way to increase battery life is to use a larger battery, but this is often impractical. Battery life affects two important elements of phone use: the length of time that a Subscriber Unit can be in receive-only mode waiting for messages to be sent, and the length of time that a Subscriber Unit can continuously transmit on a radio channel. Phone manufacturers expend considerable effort to maximize battery life.

12.2.10.1   Extended Standby Time

The length of time that a Subscriber Unit can be in receive-only mode (waiting for messages such as a page to be sent) is called standby time. The best way to increase standby time is for the phone to sleep between paging messages. The initial digital specifications strongly emphasized maximizing sleep times. These standards create sleep periods using paging frame class groups which allow a Subscriber Unit to sleep while pages not designated for that phone are broadcast. During the times when pages for that phone are broadcast, the phone is awake. Figure 12.9 shows how a Subscriber Unit sleeps between groups of pages. Typical delays vary from seconds to a few minutes. When Subscriber Unit operate as a portable telephone, the maximum delay is typically 2-4 seconds. Longer sleep period might cause callers to hang up before the Subscriber Unit woke up to answer. Much longer delays are acceptable when a Subscriber Unit is used as an alpha-numeric pager or other message device. The longer the sleep period, the longer the standby time, and the longer the battery life.

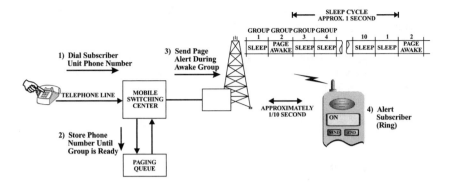

Figure 12.9, Enhanced Standby Time

12.2.10.2 Enhanced Talk Time

The length of time a Subscriber Unit can continuously transmit on a single battery charge is called talk time. Digital systems' talk time is inherently longer than that of analog systems because digital systems transmit short bursts of data and are idle between bursts, whereas analog systems transmit (and use power) continuously. IS-136 TDMA transmitters are only on 1/3 of the time, and GSM transmitters operate only 1/8 of the time. IS-95 CDMA Subscriber Units also have a lower average transmit power. Figure 12.10 shows why digital cellular offers lower average RF transmit power and longer talk time.

Most of the digital specifications support some type of discontinuous transmis-

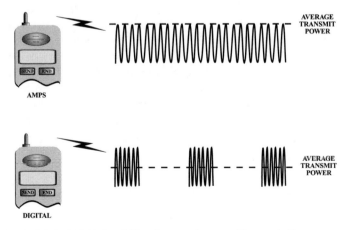

Figure 12.10, Digital Cellular Offers Lower Average Transmit Power

341

sion protocol (VOX). VOX protocols reduce the Subscriber Unit's transmit power when no voice information is being sent. Since many conversations are listening more than half the time, VOX protocols greatly increase talk time.

Unfortunately, VOX algorithms detect voice activity much like a speaker phone, and they can cut off the first syllable of speech (an effect called "clipping"), which is unacceptable for some users.

Newer digital specifications have added power classes that allow the Subscriber Unit to transmit at lower average power levels when it is very near a cell site, resulting in longer talk time. The lower output power classes are needed in systems with microcells near each other.

12.2.11 Authentication

Authentication is a way to prevent fraudulent access to the system by validating cellular users. All US digital standards have adopted the same basic authentication standard. The North American Systems (IS-54 TDMA, IS-91 AMPS, IS-136 TDMA, and IS-95 CDMA) use an authentication process based on the CAVE algorithm. The GSM system uses an authentication process based on the A3 algorithm. The North American and GSM authentication processes are very similar. Chapter 2 contains additional details on authentication for each system, a short summary is presented here.

Figure 12.11 illustrates the authentication process for the North American systems. During setup, each cellular phone is issued a number called an A-KEY. The A-KEY is issued much as banks issue a secret PIN number to identify bank card holders. The subscriber enters it into the phone via the keypad. The phone uses the A-KEY to calculate and store a shared secret data (SSD) key. The network also performs the same calculations to create and store the SSD. During each call, the SSD key creates an authentication response code, and during access, the phone transmits only the authentication response code. The authentication response changes during each call because the system sends a random number which is also used to create the authentication response. Even if someone copies the A-KEY, the copy and the real phone produce different answers because authentication is updated with other variables that are fed into the calculations.

Chapter 12

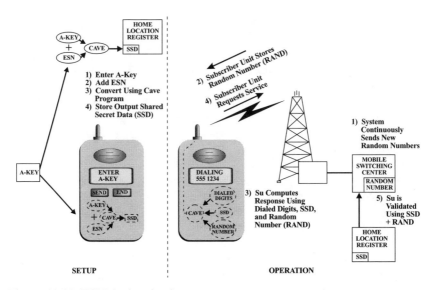

Figure 12.11, ESN Authentication

12.2.12 Data Transmission

Data transmission is the transfer of digital information from one location to another. Because cellular and landline telephone systems have not had direct digital connections, analog modems on analog voice channels have provided the only means for public data transmission.

Most early analog cellular modems used the same protocols as modems using public wire lines. The only requirement was an adapter cable to carry the analog data signal between the phone and the modem. Cellular radio systems experience radio distortion, and do not provide continuous high quality signals like landline telephone systems. Cellular modems had to resend information due to errors, greatly slowing cellular data transmission rates in comparison to land line rates. Special cellular telephone modems were created to overcome the slow data transmission rate, but unfortunately, both ends of the data communications required the special modems.

Digital systems which transmit everything as data improve cellular data transmissions. Data can be sent directly between a Subscriber Unit and another data device in two ways: either on a continuous circuit or by packets. A continuous circuit (circuit switched) transmission allows the network to route continuous data to a single location. A packet transmission system allows packets of information to be directed to any location.

343

12.2.12.1 Circuit Switched Data

At the beginning of transmission, the network identifies a single destination for the data, and routes circuit switched data to it. The sender provides the receiver with a destination address such as a phone number or internet address, and then begins a continuous transmission. An advantage of circuit switched data is that there is little overhead communication associated with it. A disadvantage is that it requires dedicated communications equipment (such as a radio channel) even when no data is being sent.

Cellular systems must convert digital information from the Subscriber Unit to a form suitable for telephone (or other) networks. Therefore, to send circuit switched data on digital cellular systems, the Subscriber Unit must tell the cellular system what type of data is to be sent.

Figure 12.12 illustrates how digital subscribers send data on digital cellular systems. The Subscriber Unit obtains access to the cellular system (step 1), then informs the system that circuit switched data is to be sent (step 2) and provides the destination phone number or internet address. The Base Station then sends a request for data transmission to the switching center (step 3) and the mobile switching center routes the data to a modem (step 4) which converts the digital information to suitable form (FAX or data modem) for transfer through the telephone network. The network then transfers the data to the destination.

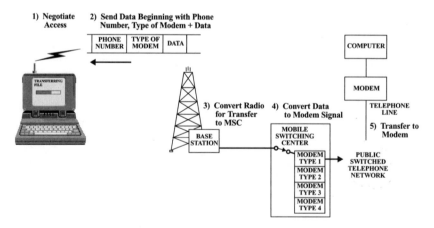

Figure 12.12, Direct Digital Data Transmission

12.2.12.2 Packet Switched Data

Packet switched data is sent in small pieces or packets, which contain all of the information to be sent, or a piece of a larger block of data. When blocks of data are sent by packets, each packet that is sent contains a sequence number which allows the recreation of the block of data after all of the packets have been received successfully. If a packet of data is lost in transmission, its replacement can be requested by using the sequence number.

Each packet of data contains some address information (called a header) indicating the packet's destination. Packet switched data is especially useful when the transmission path may change during the transmission, or when the sender needs to communicate with multiple receivers at different addresses.

Figure 12.13 illustrates digital cellular transmission of packet data. First, the Subscriber Unit gains access to the cellular system (step 1), then sends a packet of information containing both the destination address and data (step 2). The Base Station converts the packet to a form which can be sent to the mobile switching center (step 3). According to the address sent with the packet, the mobile switching center then routes the data directly to a packet data network, typically internet or a public packet data network. The packet network uses the address with the packet to route it to its destination.

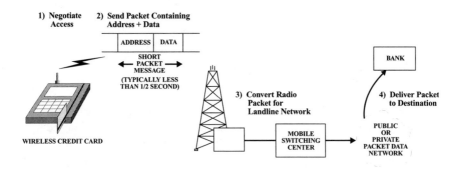

Figure 12.13, Packet Data Transmission

12.2.13 Video

Cellular systems use a variety of methods to transfer video signals over radio channels. Video signals are usually digitized and compressed in time through a process similar to that of a speech coder. The digitized and compressed signals are then sent as data over cellular radio channels. The digitization and compression are necessary because analog video signals require a bandwidth wider than a single cellular radio channel.

Cellular phones can send and receive low resolution video signals, and they can communicate high quality video signals in two ways: delayed transmission or real time transmission.

For delayed video transmission (non-real time), the digital video image is captured (e.g., a 30 second video), digitized, and sent as a data file. Because cellular data transmission rates are slow, the digitized video transfer time is longer than the actual video. Figure 12.14 illustrates the digitized video transfer process.

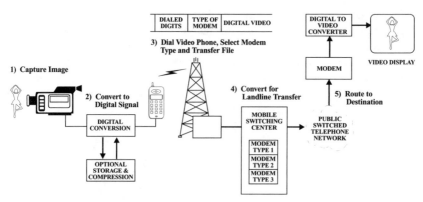

Figure 12.14, Digital Video Transmission

Cellular networks can also provide high resolution digitized video without a delay by simultaneously transmitting video information on multiple cellular radio channels. The digital video data is divided into multiple data channels, and a separate Subscriber Unit sends each data channel. At their destination, the separately received data signals are recombined into a high speed digitized video signal.

12.2.14 Position Location

Position location determines a Subscriber Unit's transmitting position, and it can be offered in two ways: either the cellular system's equipment locates the Subscriber Unit, or the Subscriber Unit locates itself and transmits the location to the system.

A cellular system locates a Subscriber Unit in a radio coverage area by indicating which cell site and antenna is communicating with the Subscriber Unit. In the future, direction finding antenna systems or triangulation methods may help pinpoint the Subscriber Unit within a radio coverage area.

A Subscriber Units locates itself through a connection to position location equipment such as the global position satellite (GPS) system or "Loran C" ship navigation networks. Once the Subscriber Unit determines its coordinates, it transmits them to the cellular system.

Figure 12.15 illustrates how cellular systems locate the position of an operating Subscriber Unit. The mobile switching center keeps track of which radio channel in which cell is communicating with the Subscriber Unit (step 1). Directional antennas determine the Subscriber Unit's angle relative to the cell site antenna (step 2). To determine the distance (step 3), equipment at the cell site measures the time delay in transmissions between the Subscriber Unit and the cell site. Finally, the mobile switching center sends location information via telephone (step 4) to be displayed on a computer terminal (step 5). This location technol-

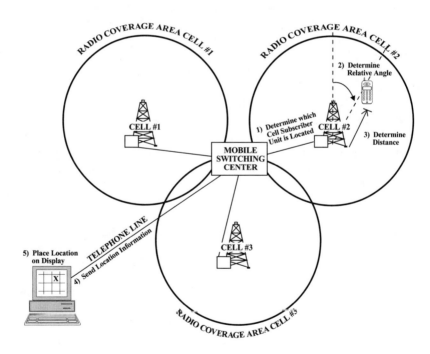

Figure 12.15, Subscriber Unit Position Location

ogy is readily available in 1995. Soon, other approaches such as triangulation will achieve similar results.

12.3 New Applications

New applications for advanced digital cellular services will add value to cellular service, and provide new revenue options for cellular service providers. Some services may be paid for by subscribers, and others by merchants who sell services. The following are a few of many possible applications. These applications will require software modifications to the network and will likely evolve over time.

12.3.1 Advertising

Advertising offsets the cost of products and services delivered on radio, television, and the internet. However, unlike advertising on radio and television, cellular advertising can confirm reception of the message, and it can use the multicast service to target a group of users. User groups can be defined by geographic location or customer profile. For example, as a cellular subscriber drives down the interstate, the phone might beep and display; "get off at the next exit, $1.00 Big Dac at McDacs, we know you're here." The message went to the specific subscriber driving near the restaurant near lunch time, with a subscriber profile indicating that the person eats meat. After the Subscriber Unit receives and displays the message, it confirms receipt, and the restaurant is billed for the message.

Subscribers could elect to receive advertising by selecting a reduced rate program (e.g. $5 less per month). The service provider could charge the business for advertising per confirmed receipt. The subscriber could fill out a profile sheet when selecting the rate plan which would focus the types of advertising that would be sent to the subscriber. Advertising messages can be voice or short message based.

By queuing advertising messages based on available radio channel capacity, service providers could send messages during periods of low demand, and charge variable rates based on peak traffic load. Such a system would reduce advertisers' costs and use the cellular system more efficiently.

12.3.2 Weather and Traffic Reports

Service providers could send up-to-the-minute weather and traffic reports to groups of subscribers. Using the broadcast message service, all Subscriber Units in a region with the code address to decode the message could receive a weather report. The cellular system could use its knowledge of the Subscriber Unit position to send traffic report information specific to the unit's location. Alternatively, subscribers could select reports on a "Pay per Jam" basis.

12.3.3 Direction Routing/Maps

Directions or routing maps can be sent to a subscriber in text or graphic form. Directions can be requested by a subscriber or provided automatically via a dispatch center. Unlike computer based mapping systems, directions by wireless can be adjusted for traffic, weather, and construction changes.

Point to point message services could send directions directly to the requesting subscriber. The subscriber could use a map to find a reference marker (e.g. A-37) and enter the reference mark with the keypad. Dispatch centers could send directions to individual or multiple vehicles. The dispatcher could send delivery directions to a truck or pickup directions for a taxi.

12.3.4 Telemetry/Monitoring

Cellular service is a low cost way to transfer telemetry and monitoring information without wire line connections. Telemetry applications include monitoring utility meters, gas lines, vending machines, critical equipment, environmental sensors (water level, earthquake, fire), and many others. For applications that only transmit small amounts of information, a packet transmission may be cost effective. When monitoring devices require the transfer of large amounts of data, circuit switched data transfer is better suited.

12.3.5 Fax Delivery

Fax delivery is the transfer of scanned or pre-stored information such as purchase orders, invoices, brochures, or any other supporting documentation. The remote FAX machines now beginning to appear in airports, hotels, conference centers, and business centers make remote fax delivery more convenient.

Cellular systems can store FAX messages and forward them when delivery is convenient. They also make it possible to correct radio transmission errors not commonly corrected by FAX machines. This is especially important during hand-offs and poor signal reception. Figure 2.16 shows a portable FAX display device which connects to a cellular phone.

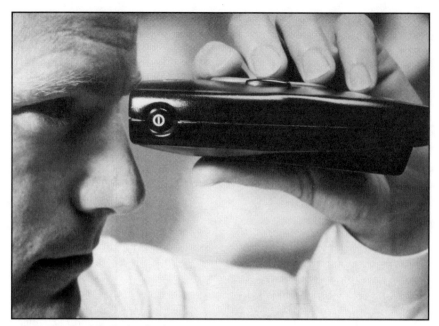

source: Reflection Technologies
Figure 12.16, Portable Fax Display Device

12.3.6 Video Transfer

Cellular video transfer can send real time (instantaneous) and non-real time (stored) video. Newscasters already use Subscriber Units to transfer digitized real time video. Such short, high resolution video clips are comprised of large amounts of information. Therefore, they must be compressed and sent over multiple cellular channels. Low resolution real time video, such as that used for security monitoring, can be sent over a single radio channel. In the future, direct digital connections will make it possible to transfer digitized video files more quickly and reliably.

Newscasters can use non-real time video transfer to send high resolution video over a single cellular radio channel. Figure 2.17 illustrates a real time video transmission system. The quantity of information in a high quality video signal exceeds the capacity of a single radio channel, so this system uses multiple cellular phones to increase the information carrying capacity of the channel to more than 50 kbps.

Chapter 12

source: Barron Technologies
Figure 12.17, Cellular Video Transmission System

12.3.7 File Transfer

File transfers move large blocks of data (files) between a Subscriber Unit and another data device. File transfers can distribute software to field equipment, update advertising billboards, move inventory stock and pricing data, update kiosks at the point of sale, or transmit to any other remote device requiring information.

Radio channels sometimes distort signals and cause file transfer errors, so any data files that must be transferred will benefit from a reliable direct digital channel. Unfortunately, direct digital channels connect only the Subscriber Unit and the cellular system. For data to reach a remote computer, it must be converted between the cellular system and the PSTN.

12.3.8 Location Monitoring

Location monitoring is locating and tracking Subscriber Units. It can be used for safety (911), vehicle tracking (such as trucks), survey site locations, security (personal location), and for inventory and asset (equipment) verification.

Digital systems have built-in potential for location monitoring because nearly all digital systems must compensate for the Subscriber Unit's distance from the cell site to adjust transmit times. This built-in distance monitoring makes it likely that future digital systems will offer standard Subscriber Unit location services.

Another way to monitor location is to attach a global position satellite (GPS) receiver to a Subscriber Unit. The GPS device determines its position from satellites orbiting the earth. After the GPS unit has determined its location, the Subscriber Unit can transfer the position data to a computer in the cellular system via the cellular radio channel.

12.3.9 Point of Sale Credit Authorization

Cellular radio can be used to authorize credit cards without a wire telephone connection. Wireless point of sale credit authorization increases mobility and eliminates the cost of telephone line installation.

Some companies now produce wireless credit card validation stations which allow a vendor to accept and validate credit card purchases from anywhere that a cellular call can be made. Because credit card validation transactions transfer small amounts of data, many credit authorization applications are using Cellular Digital Packet Data (CDPD) or other wireless packet transfer data protocols. Figure 12.18 illustrates a wireless cellular credit card system.

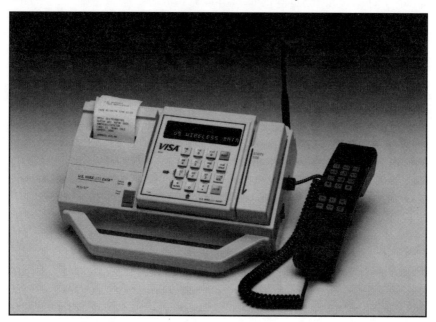

source: US Wireless
Figure 12.18, Wireless Credit Card System

12.3.10 Still Image Capture

Still image capture provides the ability to send pictures from a digital camera to any place in the world. Digital cameras allow visual information to be captured, transferred, and stored without film almost instantly and efficiently. Still image capture is used by insurance agents, security services, reporters, police, and others. The benefits of not requiring film, developing, postage, no physical storage of the picture, rapid information capture (e.g. on site insurance adjustment), and instant distribution make still image capture appealing.

12.3.11 Interactive Information Access

In essence, interactive information access is a wireless internet through which subscribers can request, receive, react, and respond to information in real time. Interactive information access can be used to monitor sales inventory, provide remote education, field pricing inquiries, review job assignments, instruct service technicians on demand, provide real estate multiple listing services, and accomplish many other remote tasks.

12.3.12 Two Way Paging

Pagers are simple communication devices which alert users and send simple messages almost immediately. Two way pagers confirm the receipt of a page and can let users make simple responses to information they have received. All of the new cellular technologies facilitate small devices which can be used as two-way pagers.

Unlike telephone (voice) service, paging messages can be stored and delivered when the cellular system is not busy. Queuing pages allows the system to serve many more subscribers per channel because it eliminates the need for extra radio channels during periods of peak usage. A cellular system can serve approximately 20-32 paging users on a single radio channel. Numeric paging systems can serve more than 10,000 subscribers on a single radio channel.

Unlike standard paging systems, cellular paging systems can transfer voice and data information in both directions. Voice messages sent on a cellular voice paging system can be stored, compressed, and sent at a less busy time (discussed in chapter 11). Figure 12.19 illustrates two-way cellular voice paging.

source: ReadyCom
Figure 12.19, Two-Way Cellular Voice Pager

12.3.13 News Services

Wireless news services can deliver to subscribers the specific news topics they select. Subscribers can request any news category identifiable by name; for example, information about a certain city, business stories, stock quotes, or articles on hobbies.

Some internet on-line news services provide brief abstracts before sending whole articles. The short overview lets the subscriber decide whether to get the article, and whether they need it now or later. The extensive transfer time for long or illustrated articles can be expensive, and delaying delivery can let the subscriber find a more economical way to get it.

12.3.14 Sports Betting

Paging already offers sports information via alpha paging. With the Subscriber Units' two-way communication capability, this type of service could also include betting. Cellular subscribers could receive sports scores and bet on teams.

New cellular short message services can provide instant score updates along with league standing, player lineup, point spreads (based on a betting agent), and sports statistics. Using short messages, cellular subscribers will be able to track favorite teams and place bets using the existing cellular keypad. For example, a subscriber could select the team (press 1 for Dallas, press 2 for Buffalo), key in the amount, and press Send to confirm. Such a message might be sent to a betting agency via the internet. When the betting agency accepted the bet, the acceptance notification would be returned, and appear on the Subscriber Unit. Likewise, the betting agency would return the final results of the bet when the event is over.

Customers may be able to obtain these services either directly from cellular providers or from other firms with the capability to send short messages (e.g. Las Vegas Casinos). Cellular carriers could broadcast short messages to multiple users, or send them point to point to individual "Sports Net" subscribers. If responses are limited to menu items, transactions would use little air time--as little as 1/10 of a second. Even if a service provider charged only 1/2 cent per message, average revenue would exceed $3.00 per minute, not including a premium fee for the service.

12.3.15 Remote Vending

Wireless remote vending combines the benefits of point-of-sale credit card authorization, inventory control, and advertising. Vending machines that can authorize and accept charge cards or other types of money cards make it possible to sell more merchandise without collecting money from the machine. Wireless remote vending also informs distributors of inventory levels in the machines, removing guesswork from ordering and dispatching. Finally, vending machines can be a good point for advertising.

Even in urban locations, high installation and service costs prohibit connecting vending machines to a dedicated wire line. Wireless data packet service may offer a low enough cost to make automated vending profitable.

APPENDIX I - DEFINITIONS

Abbreviated Alert - An alert order that informs the user that a previous function selected is still active.

Access - A process where the Subscriber Unit competes to gain the attention of the system to obtain service.

Access Channel - Redefined to be Analog Access Channel.

Access Overload Class (ACCOLC) - A class assigned to a Subscriber Unit to limit its access attempts in an overloaded system. The overhead message contains an overload field which corresponds to the access overload class assigned to the mobile. When the system becomes overloaded, the overhead message will begin to remove groups of mobiles accessing the system.

Activation Commission - An incentive that is paid to a retailer of communications products from a service provider as a reward for establishing a new customer.

Adjacent Channel - A radio channel that is located one channel (bandwidth) away from another channel (e.g., 30 kHz for AMPS).

Adjacent Channel Interference - This occurs when the spectral power distribution from an adjacent channel creates noise or errors in the channel being used by the Subscriber Unit or Base Station.

Advanced Mobile Phone Service (AMPS) - The service that is available in the United States and over 33 other countries today.

Advanced Radio Technology Subcommittee (ARTS) - Sponsored by the CTIA to study the industry needs, technology available, and manufacturers support to develop new cellular technology.

Alert - An order sent to the Subscriber Unit that informs the user that a call is to be received.

Alert with info - An alert message sent on the digital channel.

Alert with info Ack - A message sent back on the digital traffic channel that acknowledges the Alert with info order.

Alternate Channel - A radio channel that is located two channels (2x bandwidth from another channel (e.g., 60 kHz for AMPS).

Analog Access Channel - A cellular system channel that uses Frequency Shift Keying (FSK) to pass data control signals between the Subscriber Unit and Base Station. This control channel is used by the Subscriber Unit to inform the system that it wants to obtain service.

Analog Cellular - An industry term given to today's existing cellular system. It involves transmission of voice information in a similar method to that of a FM radio.

Analog Color Code - One of three audio tones mixed in with the voice channel to distinguish the channel from interfering co-channels.

Analog Control Channel - A channel designated by the cell site to control operations of a analog and dual mode cellular phone. A control channel can be divided into paging and access functions.

Analog Subscriber Unit - A Subscriber Unit which is only capable of analog Cellular service.

Analog Voice Channel - A cellular system channel that operates by using FM (Frequency Modulation) to pass voice and data control signals between the Subscriber Unit and Base Station.

Audit - An order generated by the system which determines if the Subscriber Unit is active within the system.

AUTH1 - An algorithm used to encrypt the ESN for transmission. It's inputs are the ESN (Electronic Serial Number), PIN (Personal Identification Number), and RAND (Random Number Sent by the System).

Automatic Message Accounting (AMA) - A standard recording system for collection and processing of call billing information.

Authentication - A process during which information is exchanged between a Subscriber Unit and a Base Station to allow a cellular service provider to confirm the true identity thus inhibiting fraudulent use of the radio system.

Backhauling - The carrying of cellular calls between the cell sites and the cellular swtich (MSC) via a landline network prior to reaching the PSTN.

Base Station - A controlling transmitting station that provides service to cellular Subscriber Units. It is sometimes called a Land Station or Cell Site.

Baseband Signal - The signal (information) which will be transmitted prior to adding modulation.

Basic Trading Area (BTA) - A geographic region where area residents do most of their shipping. The United States has been divided into 493 BTAs. The C-F PCS licensess are granted based on BTA.

Bearer Services - Telecommunications services that provide facilities for the transport of user information without functional processing of the user data.

Bit Error Rate (BER) - The ratio of bits received in error with respect to the total number of bits transmitted.

Blocking Probability - The percentage of calls that cannot be completed within a one hour period due to capacity limitations. For example, if within one hour 100 users attempt accessing the system, and two attempts fail, the blocking probability is 2 percent.

Broadband Signal - The carrier signal which is used to transfer the baseband (information) signal.

Burst Collisions - The overlapping of transmitted burst from received at the Base Station. This is due to the propagation delay time. Dynamic time alignment is typically used to solve this challenge.

Busy-Idle Bits - Bits that are time multiplexed with the forward control channels to indicate the access channel of the system is busy.

Carrier - A service provider of cellular service; RF energy emitted from a transmitter assembly.

Carrier-to-Noise Ratio - The ratio of the desired radio signal (carrier) with respect to the thermal noise.

Carrier-to-Interference Ratio (C/I) - The ratio of the deisred radio signal (carrier) with respect to the combined interference due to adjacent and co-channel interference.

Carrier - A company that provides telephony service to customers. In the US, only two carriers exist in a given area. Also known as service provider. The word carrier also describes the radio channel.

Cell Site - A radio tower and associated equipment that converts phone lines to radio signals for transmission to Subscriber Units.

Cell Splitting - A method to increase system capacity by the division of assigned radio coverage areas. Each cell site (radio tower) can provide a limited number of channels. To increase the total number of channels available in a given area, it is possible to divide an area by adding additional radio towers.

Cellular Geographic Service Area (CGSA) - The area licensed by the FCC to a service provider. It is based on Metropolitan Statistical Areas or New England Metropolitan Areas.

Cellular Mobile Carriers (CMC) - The service providers that are licensed by the FCC to provide service in the CGSA. There are two CMCs authorized for each CGSA.

Cellular Subscriber Station (CSS) - Another name for Subscriber Unit or Mobile Station. This is the preferred term to network providers because Subscriber Units may be fixed in location.

Channel - May have two meanings; physical and logical. The physical channel is the resource (a radio channel or wire) which transfers the information. A physical channel can be divided into many logical channels. The logical channel is a throughput of information on that physical channel.

Churn - The amount of customers that disconnect from service. Churn is usually expressed as a percentage of the existing customers that disconnect over a one month period.

Closed User Group (CUG) - A group of users who have access to a defined set of features.

Co-channel - A channel being re-used at another cell site. This is allowed due to the attenuation of the signal because of the distance.

Co-channel Interference - Radio interference from a another radio channel that is used by another cell site on the same frequency. This is caused by insufficient distance between cell sites.

Channel Quality Message - Messages which are sent on a digital channel to provide the Base Station with channel quality information. This information can contain the RSSI (Received Signal Strength Indication) or BER (Bit Error Rate).

Coder/Decoder (Codec) - A process of compressing and expanding information so the transmission can be more efficient. By the use of this, digital voice information is characterized and reduced.

Coded Digital Voice Color Code (CDVCC) - The DVCC that has been coded which can represent 255 different codes and identifies a channel from it interferers.

Coherence Bandwidth - A bandwidth in which either the phases or amplitudes of two received signals have a high degree of similarity.

Combined Paging and Access - This allows the combining of paging and access functions on a single control channel in an analog system. When the paging and access channels are combined, the forward channel broadcast both paging and access messages while the reverse channel channel coordinates system access messages.

Combined Paging and Access (CPA) [Field] - This is a field in the forward overhead message which informs the Subscriber Unit if the system has combined paging and access channels. If this bit is set to 1, the Subscriber Unit must obtain access on the same control channel it receives the paging message on. If this bit is set to 0, the Subscriber Unit must tune to an authorized access channel to obtain service.

Combiner - Used in Base Stations to couple different frequencies to one antenna.

Continuous Transmission - A mode of operation where the Subscriber Unit does not cycle its power level down when the modulating signal amplitude is low.

Control Channel - A cellular system channel dedicated to sending and/or receiving controlling messages between the Base Station and Subscriber Units. (See Section Call Processing and RF Channel)

Convolutional Coding - A forward error correction method used to provide for correction of data that has been corrupted in transmission. This is accomplished by sending redundant information which allows for reconstruction of the original signal.

Cyclic Redundancy Check (CRC) Generator - A function used to create a CRC data word. This word being a function of the data to be sent is used to verify the accuracy and may be used to correct some of the bits when in error.

Cross Talk - A problem where the audio from one communications channel is imposed on another channel.

Cyclic Redundancy Check (CRC) - An error detection and/or correction method used to determine if a block or string of bits of data were received correctly.

Coupler - See combiner

Dead Spot - An area within a service area where the signal strength is significantly reduced. This is primarily due to terrain and obstructions. Dead spots are generally eliminated by the use of repeaters or relocation of the cell site.

Dedicated Control Channels - Radio channels that are used only for control messages between the Base Station and Subscriber Unit. The dedicated control channel can be a time slot on a multiple channel radio carrier.

Delay Spread - A product of multipath propagation where symbols become distorted and eventually will overlap due to the same signal being received at a different time. It becomes a significant problem in mountainous areas where signals are reflected at great distances.

Digital Cellular - An industry term given to the new cellular technology that transmits voice information in digital form. This differs from analog cellular in the method the transmission of voice/data information is represented by digital signals.

Digital Color Code (DCC) - This is a field which corresponds to the SAT code assigned to the cell site which is controlling the mobile. The Subscriber Unit matches this code to the received SAT frequency to ensure it has locked on to the correct channel.

Digital Only Unit - A Subscriber Unit that is only capable of transmitting digital signals.

Digital Speech Interpolation (DSI) - A voice detecting system which allows transmission only when speech or data activity requires information to be sent.

Digital Voice Color Code (DVCC) [Field] - Provides a unique code from the Base Station which identifies its channel from co-channels.

Discontinuous Reception (DRX) - A method where the Subscriber Unit removes or reduces power to its circuits during sleep interval periods. This extends the battery life while waiting for calls (commonly called standby mode).

Discontinuous Transmission (DTX) - A method where the Subscriber Unit changes its power level as a result of the input level of its modulating signal. This allows conservation of power when the modulating level is low or no data is to be transmitted.

Diversity Reception - The process of combining or selecting one of two stronger of the signals received by two antennas. These antennas are usually separated by a distance that is a function of the wavelength.

Doppler - A frequency offset that is a result of a moving antenna relative to a transmitted signal.

Dotting - Used as a "wake up tone" that a message is coming. After the dotting sequence, a a standard sequence (synchronization word) will follow which allows the Subscriber Unit to time align to the incoming message.

Dual Mode Cellular - The combination of an analog cellular unit and Digital cellular section into one Subscriber Unit. It allows operation of the Subscriber Unit on either analog or digital radio channels.

Electromagnetic Interference (EMI) - Magnetic fields that interfere with the operation of the Subscriber Units. This includes any frequencies generated that interfere.

End Office Switching System (EO) - A switch which provides communication paths between originating customer terminal (telephone) equipment and terminating customer terminal equipment.

Enhanced Specialized Mobile Radio (ESMR) - A name given for the new technologies being implemented by the Specialized Mobile Radio operators which provide dispatch, voice, messaging, paging, and wireless data services.

Equalization - A processes which modifies the receiver parameters to compensate for changing radio frequency conditions. Primarily used to compensate for multipath propagation (see section RF channel).

Erlang - The amount of voice connection time with reference to one hour. For example, a 6 minute call is .1 Erlang.

Extended Protocol - Optional extended capability of the signaling messages which provide for the addition of new system features.

Extended Time Division Multiple Access - A time division cellular system which utilizes DSI (Digital Speech Interpolation) to allow transmission only when speech or data information is to be sent. When transmission is inhibited, the same channel and time slots can be shared by other users.

Fast Associated Control Channel (FACCH) - A logical signaling channel which is created by replacing speech data with signaling data in short periods of time.

Field - A dedicated number of bits within a message or data stream which is dedicated to specific functions.

Flash Request - A request to invoke a special processing function. analog cellular only Subscriber Units were only capable of sending flash requests from the Subscriber Unit to the Base Station. Digital Subscriber Units are capable of sending flash request messages in both directions.

Flash With Info - A flash message sent over the digital traffic channel. This function allows an indication that the originator needs special processing.

Forward Analog Control Channel (FOCC) - The analog control channel which is from the Base Station to the Subscriber Unit.

Forward Analog Voice Channel (FVC) - The analog voice/traffic channel which is from the Base Station to the Subscriber Unit.

Frame - Time periods which are composed of slots that are linked together compose a frame. Time slots within a frame can be assigned to one or more individual users.

Frequency Planning - The selection and assignment of channel frequencies to cell site equipments which minimize the interference levels they create with adjacent and alternate cell site equipments.

Frequency Reuse - The ability to reuse channels on the same frequency. This is possible due to the attenuation of the signals by distance.

Full Duplex - Transferring of voice/data in both directions at the same time. This becomes confusing in a TDMA system because information is reconstructed to allow transfer of voice information in both directions at the same time although actual transmission does not occur simultaneously.

Full Rate - The logical channel which has a basic number of slots per frame which are assigned to individual users.

Global System for Mobile Communication - A digital cellular system which originated in Europe and is now used in over 60 countries.

Group Identification - A subset of the system identification (SID) which identifies a group of cellular systems.

Guard Time - A time allocated on each Subscriber Unit's transmit slot which transmission so transit time of the signal does not cause collisions between mobiles transmitting on the same frequency.

Half Duplex - The transferring of voice/data in both directions but not at the same time.

Half Rate - The logical channel where only have the number of slots of a full rate channel are allocated to individual users.

Handoff - A process where a Subscriber Unit operating on a particular channel will be reassigned to a new channel.

Handover - The term used for a handoff in a GSM system.

Home Subscriber Unit - A Subscriber Unit that is operating in a cellular system where it has subscribed for service.

Hot Spots - Regions in a cellular service area which due to traffic patterns, receive and excessive amount of usage.

Improved Mobile Telephone Service (IMTS) - Available in 1964 as the MJ 150 MHz system, it was the first system to offer automatic dialing.

In Band Signaling - Signaling that occurs within the audio bandwidth or replacement of user data information.

Inter-Exchange Carrier - A long distance service provider (i.e., MCI) that provides inter-exchange service between Local Exchange Carriers (LEC).

Intercept - A message sent to the Subscriber Unit to inform the user an error or no service could be established when placing a call.

Interleaving - 1. A process used where data is not sent in direct time sequence to minimize the effect of burst errors. 2. A process of offsetting channel frequencies to increase the reuse capacity of a system.

International Mobile Subscriber Identity (IMSI) - This is the GSM subscriber units telephone number.

Intersymbol Interference (ISI) - The result of multipath propagation which results in a distorted received signal.

Local Access and Transport Areas (LATAs) - A geographic region that is typically the service area boundry for Local Exchange Carriers (LECs).

Local Control - A function of the Subscriber Unit which has been designated to provide special features in addition to those specified by the cellular standard.

Local Exchange Carrier (LEC) - A telephone service provider which furnishes local telephone service to end users.

Loss of Radio Link Continuity - This occurs when the Subscriber Unit cannot confirm a radio link exists between the Base Station and mobile. It results in re-initializing the call when the radio link cannot be maintained. (See Section Call Processing)

Major Trading Area (MTA) - A geographic region where most of the area's distribution, banking, wholesaling is conducted. The United States has been divided into 51 MTAs. The A and B PCS licensess are granted based on MTA.

Malfunction Timer - A timer which runs separate from all other functions. It continuously counts down and needs to be reset. If the Subscriber Unit is operating correctly (without failure) this timer will be reset continuously and will not expire. This timer is used to turn off the Subscriber Unit in the case of a equipment failure.

Metropolitan Statistical Area (MSA) - An area designated by the FCC for service to be provided for by a cellular carriers. There are two service providers for each of the 306 MSA's in the United States.

Mobile Assisted Handoff (MAHO) - A process where the Base Station requests signal quality information from the Subscriber Unit which can be used to determine which cell site channel is best suited for handoff.

Mobile Identification Number - This is the Subscribers telephone number which is stored in the Subscriber Unit for North American cellular systems. It is divided into MIN1 and MIN2. MIN1 is the 7 digit portion of the number. MIN2 is the 3 digit area code portion of the number.

Mobile Station (MS) - A radio transceiver which operates in a cellular system. This includes hand held units along with transceivers units installed in vehicles. The Mobile Station is also called a Subscriber Unit.

Mobile Station Class (MSC) - This is a classification of the power level capability of the Subscriber Unit. The Subscriber Unit identifies to the Base Station its power level capability by using the Station Class Mark field.

Mobile Switching Center (MSC) - Similar to the Mobile Telephone Switching Office except is does not have data base support.

Mobile Telephone Switching Office (MTSO) - Includes switching equipment needed to interconnect mobile equipment with the land telephone network and associated data support equipment.

Multipath Propagation - Occurs when the same signal transmitted reaches a point via different paths. This is due to signal reflection and refraction (see RF channel).

Multiplexing - The process of combining several resources over a shared medium. This may be in the form of time sharing (Time Division Multiplexing) where one radio channel is divided into time periods and one resource uses the channel for only the dedicated time allowed.

Narrowband Advanced Mobile Phone Service - A cellular system which narrows the channel bandwidth from 30 kHz to 10 kHz to increase system capacity. It utilizes a majority of the AMPS operation with minor signaling changes.

Non-Wireline Carriers - Cellular service providers that are not engaged in the business of providing landline telephone service.

Number Assignment Module (NAM) - A memory storage area for North American Subscriber Units which contains user profile data such as MIN (Mobile Identification Number), registered SID (System Identification), and subscriber options.

Operations, Administration and Maintenance (OA&M) - Supervision of the cellular system and its component parts.

Out of Band Signaling - Signaling which occurs outside of the 300-3000 Hz audio bandwidth or user's data.

Orders - Messages sent between the Subscriber Unit and Base Station.

Overload Control - A process used by the system to control the access attempts initiated by mobiles. Overload Class (OLC) bits sent in the overhead message inhibit operation of groups of mobiles.

Overhead Messages - System messages that are sent from the Base Station to the Subscriber Unit giving it the necessary parameters to operate in that system.

Paging - The process where the Base Station sends a message over the control channel informing the Subscriber Unit a call is incoming.

Paging Channel - A control channel which addresses mobiles directly to alert them of an incoming signal.

Personal Communications Network (PCN) A digital cellular system which is based on the DCS 1800 specification.

Personal Communications Services (PCS) -Cellular like services that will be provided in North America on a new band of frequencies in the 1900 MHz band.
Phase Locked Loop - A circuit which synchronizes an adjustable oscillator with another more stable oscillator by the comparison of phase between the two signals1.
Photogrammetry - A process through which terrain elevation data can be accumulated from using a stereo image or photographs in 3-D.
Point of Presence (POP) - A location within a Local Access and Transport Area (LATA) that has been designated for connection of a Local Exchange Carrier (LEC) and Interexchange Carrier (IXC).
Pops - The population of a cellular service area for a service provider. This is used for evaluating the worth of a system. The number of pops is multiplied by the price to be paid per pop.
Power Level (PL) - The Subscriber Units power level. This is the power level relative to the maximum allowable power for that class.
Preorigination Dialing - A process where the dialing sequence takes place prior to the mobiles first communication with the cellular system.
Private Branch Exchange (PBX) - A private swithing facility that is used to permit internal call routing and connection to the PSTN.
Protocol Capability Indicator (PCI) - Information which is sent with the system overhead information that indicates the cell site has digital channel capability.
Rayleigh Fading - The function where the received signal strength will vary due to multipath propagation. A fade usually occurs every wavelength (approximately 10-16 cm).
Re-use Factor - A number which depicts the ability to reuse frequencies at other cell sites. As the reuse factor goes up, the required distance between cells witch reuse frequencies increases.
Read Control Filler (RCF) [Field] - A field in the system overhead information data stream that indicates if the Subscriber Unit must read the control filler messages prior to attempting access to the system.
Reflection - A function where a radiated signal that is incident on a reflective surface will have some or all of its energy reflected from that surface. This is a significant problem with mountains and high rise buildings.
Refraction - A function where a radiated signal on entering a medium of differing propagation characteristics is bent. This is seen at lower frequencies where the ionosphere bends the signals back toward the Earth resulting in signal skipping. This has allowed world radio coverage.
Registration - The process where by a Subscriber Unit identifies itself by sending a message that it is operating in the service area.
Registration Identification - A process where the Subscriber Unit accesses the cellular network to inform the system it is in its operating area.
Release Request - An order where the Base Station or the Subscriber Unit requests a termination of conversation mode.
Reverse Analog Control Channel (RECC) - The FSK modulated channel that

completes the control channel signaling transmission path from the Subscriber Unit to the Base Station.
Reverse Analog Voice Channel (RVC) - The FM channel that completes the voice/traffic channel from the Subscriber Unit to the Base Station.
Roamer - A Subscriber Unit which is operating in a cellular system other than its home subscribed system.
Roaming - The process where a Subscriber Unit is operating in a system other than its registered home system.
Rural Statistical Area (RSA) - A geographic area designated by the FCC for service to be provided for by cellular carriers that falls outside the MSA regions. There are 428 RSAs in the United States.
SAT Color Code (SCC) - Additional information that is sent in the AMPS overhead forward control channel message stream to expand the number of available SAT codes from 4 (2 bits) to 64 (6 bits).
Scan of Channels - A process where the Subscriber Unit tunes to a defined set of frequencies and locks on to the strongest signals.
Scan Dedicated Control Channels - This is the first task the Subscriber Unit uses to find which signal is the strongest of the dedicated control channels.
Sectoring - A process of dividing cell sites where the radiation pattern is divided into sectors by using directional antennas.
Seizure Precursor - A defined bit stream transmitted by the Subscriber Unit on the reverse analog control channel used to synchronize the Base Station receiver.
Service Control Point - A signaling processing point in a telephone network that modifies and directs signaling messages to other switching points.
Service Provider - An organization that provides cellular or telephony service.
Shortened Burst - A shortened transmit burst used by the Subscriber Unit when initial transmit occurs in a large diameter cell where timing information has not been established. This is required to overcome propagation delays which may cause burst collisions (overlapping received bursts).
Signaling Tone - A 10 kHz tone mixed in with the analog voice signal which is used as a status change signaling device between the Subscriber Unit and the Base Station.
Signaling Transfer Point (STP) - A switching point in a telephone network that routes messages to other switching points.
Signaling System #7 (SS7) - A international standard network signaling protocol which allows common channel signaling between telephone network elements.
Slot - See Time Slot
Slow Associated Control Channel (SACCH) - Out of Band signaling that occurs on the digital traffic channel where messages are transferred by a dedicated number of bits or slots.
Soft Capacity Limit - A system subscriber serving capacity limit which allows more users to receive service at a reduced data rate.
Special Mobile Group (GSM) - Created by CEPT to establish a digital cellular

standard to be used in Europe.

Specialized Mobile Radio (SMR) - Private wireless networks which provide dispatch type services. Most SMR operators are licensed to provide service in the 800-900 MHz frequency range.

Speech Frame - A time period for which speech is sampled and digitized. The typical speech frame time period is 20 msec.

Standard Offset Reference (SOR) - This is a time period allocated between the transmit and receive time slots.

Subscriber Unit - The equipment which is used to communicate with a wireless (typically cellular) network. This device can be at a fixed position or mobile and may be used by a customer (subscirber) or an automated device (such as an alarm dialer). Other names for it include Mobile Station, Cellular Radio, and Transceiver.

Supervisory Audio Tone (SAT) - One of several continuous tones that are mixed in with the modulating audio signal which is used to identifies channel interferes of the same RF frequency.

Symbol - Defined for a relative phase shift of the transmitted signal over a specified period of time.

Synthesizer - A frequency generator that can provice any one of the stable RF carrier frequencies required upon direction from the microprocessor in the logic section.

Tandem Office Switches (TO) - Telephone switching systems which are used to interconnect end offices (EO) when direct trunk groups are not economically justified. Tandem office switches can be connected to other tandem office switches.

Task - A set of steps or processes a transceiver must take to accomplish a function.

Teleservices - Telecommunications services that provide facilities for the transport and processing of user information.

Thermal Noise - A theoretical power level of noise due to temperature. kTB, where k is boltzmans constant, T is degrees kelvin, B is bandwidth in Hertz.

Time Division Multiple Access (TDMA) - A process of sharing a cellular channel by sharing time between users. Each user is assigned a specific time position.

Time Slot - A particular time period assigned in the digital channel. There are 6 time slots allocated per frame on the digital traffic channel.

Total Access Communications System (TACS) - An analog cellular system in use in England and several other countries. It is enhanced version of AMPS.

Traffic Channel - The combination of voice and data signals existing in a communication channel.

Trunking - Allows a Subscriber Unit to be connected to any unused channel in a group of channels for an incoming or outgoing call.

Unequal Cell Loading - A process where cell sites in a general area share a provide a different amount voice channels. This can be dynamic where channels are redefined on command of by MTSO or by the established frequency plan.

User Channel (UCH) - The raw data portion of a channel that is available to the user.

Um Interface - The RF channel between the Base Station and the Cellular Subscriber Unit.

Update Overhead Information - A procedure where the Subscriber Unit tunes to the strongest dedicated control channel and gathers system overhead information. It uses the information gathered here to determine the paging channels, system ID, and other system parameters.

Voltage Controlled Oscillator (VCO) - An oscillator circuit which has an output frequency that changes proportionally with a input voltage.

Wait For Overhead Message (WFOM) - A flag in the overhead message which informs the Subscriber Unit if it must wait for the overhead message before accessing the system.

Waiting for Answer - A process where the Subscriber Unit is waiting for a response from the Base Station.

Waiting for Order - A process where the Subscriber Unit is waiting for an order from the Base Station.

Wavelength - The distance covered for one complete cycle of a propagated signal. This can be calculated by dividing the propagation velocity c (speed of light - 300 E6 meters/second) by the number of cycles in one second. For example, the wavelength at 840 MHz is .357 meters (14 inches).

Wireline Carriers - Cellular service providers that are also engaged in the business of landline telephone service. Band B is allocated for these service providers. Some wireline carriers have been authorized to provide service in band A.

Word Error Rate (WER) - The ratio of words received in error with respect to the total number of words sent.

Appendix I

APPENDIX II - ACRONYMS

A/D - Analog to Digital
AC - Authentication Center (also AuC)
ACK - Acknowledge
ACCOLC - Access Overload Class
ADPCM - Adaptive Differential Pulse Code Modulation
AMA - Automatic Message Accounting
AMPS - Advanced Mobile Phone Service
ANI - Automatic Number Identifier
ARTS - Advanced Radio Technology Subcommittee
AuC - Authentication Center (also AC)
B-CDMA - Broadband CDMA
B/I - Busy Idle Bit
BER - Bit Error Rate
BIS - Busy Idle Status
BS - Base Station
BTA - Basic Trading Area
BTS - Base Transceiver Station
C/N - Carrier to Noise Ratio
C/I - Carrier to Interference Ratio
CAF - Cellular Anti-Fraud
CCIR - International Radio Consultative Committee
CCITT - International Telegraph and Telephone Consultative Committee
CCLIST - Control Channel List
CDMA - Code Division Multiple Access
CDPD - Cellular Digital Packet Data
CDVCC - Coded Digital Voice Color Code
CEC - Commission of the European Communities
CEPT - European Conference of Posts and Telecommunications
CELP - Code Excited Linear Predictive
CMAC - Control Mobile Attenuation Code
CoDec - Coder/Decoder
CPA - Combined Paging and Access
CQM - Channel Quality Measurement
CSMA - Carrier Sense Multiple Access
CSS - Cellular Subscriber Station
CTIA - Cellular Telecommunications Industry Association
CT2 - Cordless Technology 2nd Generation
CP - Cellular Provider
CPE - Cellular Provider Equipment
CUG - Closed User Group
D/R - Distance to Cell Radius Ratio
DCC - Digital Color Code
DECT - Digital European Cordless Telephone
DQPSK - Differential Quadrature Phase Shift Keying
DSI - Digital Speech Interpolation
DTC - Digital Traffic Channel
DTMF - Dual Tone Multiple Frequency
DRX - Discontinuous Reception
DTX - Discontinuous Transmission
DVCC - Digital Voice Color Code
E^2PROM - Electrically Eraseable Programmable Read Only Memory
EIA - Electronics Industries Association
EMI - Electromagnetic Interference
EP - Extended Protocol Indicator
EPROM - Eraseable Programmable Read Only Memory
ERP - Effective Radiated Power
ESMR - Enhanced Specialized Mobile Radio
ESN - Electronic Serial Number
ETACS - Enhanced Total Access Communication System
ETDMA - Extended Time Division Multiple Access
ETSI - European Telecomuincations Standards Institute
FACCH - Fast Associated Control Channel
FCC - Federal Communications Commission

FDM - Frequency Division Multiplexing
FDMA - Frequency Division Multiple Access
FDTC - Forward Digital Traffic Channel
FEC - Forward Error Correction
FIRSTCHA - First Access Channel
FIRSTCHP - First Paging Channel
FOCC - Forward Analog Control Channel
FPLMTS - Future Public Land Mobile Telphone System
FSK - Frequency Shift Keying
FVC - Forward Analog Voice Channel
GSA - Geographical Service Area
GSM - Global System for Mobile Communications
IBCN - Integrated Broadband Communications Network
IMSI - International Mobile Subscriber Identity
IMTS - Improved Mobile Telephone Service
IPR - Intellectual Property Rights
ISDN - Integrated Services Digital Network
IVCD - Initial Voice Channel Designation
IXC - Inter Exchange Carrier
JTACS - Japanese Total Access Communications Systems
JTC - Joint Technical Committee
LAN - Local Area Network
LAPM - Link Access Procedure for Modems
LASTCHA - Last Access Channel
LASTCHP - Last Paging Channel
LATA - Local Access and Transport Area
LEC - Local Exchange Carrier
LSB - Least Significant Bit
MAHO - Mobile Assisted Hand Off
MIN - Mobile Identification Number
MIPS - Million Instructions Per Second
MS - Mobile Station
MSA - Metropolitan Statistical Area
MSB - Most Significant Bit
MSC - Mobile Switching Center
MTA - Major Trading Area

MTSO - Mobile Telephone Switching Office
NAM - Number Assignment Module
NAMPS - Narrowband Advanced Mobile Phone Service
NAWC - Number of Additional Words Coming
NMT - Nordic Mobile Telephone
OA&M - Operations, Administration and Maintenance
OLC - Overload Class
ORDQ - Order Qualifier
OSI - Open System Interconnection
PABX - Private Automatic Branch Exchange
PBX - Private Branch Exchange
PCI - Protocol Capability Indicator
PCM - Pulse Coded Modulation
PCN - Personnal Communications Network
PCS - Personal Communications Services
PDC - Pacific Digital Cellular
PIN - Personal Identification Number
PLL - Phase Locked Loop
PLMN - Public Land Mobile Network
POP - Point of Presence
POTS - Plain Old Telephone Service
PSCC - Present SAT Color Code
PSK - Phase Shift Keying
PSTN - Phone System Terminal Network (also Public Switched Telephone Network)
PTT - Postal Telephone and Telegraph
PUC - Public Utilities Commission
RACE - Research and Development of Advanced Communication Technologies in Europe
RAM - Random Access Memory
RCF - Read Control Filler
RDTC - Reverse Digital Traffic Channel
RECC - Reverse Analog Control Channel
RELP - Residual Excited Linear Predictive
ROM - Read Only Memory
RSA - Rural Statistical Area

Appendix II

RSSI - Received Signal Strength Indicator
RTC - Reverse Traffic Channel Digital
RVC - Reverse Analog Voice Channel
SACCH - Slow Associated Control Channel
SBI - Shortened Burst Indicator
SCC - Sat Color Code
SCM - Station Class Mark
SCP - Service Control Point
SDCC - Supplementary Digital Color Code
SID - System Identification
SIM - Subscriber Identity Module
SMR - Specialized Mobile Radio
SOR - Standard Offset Reference
SS7 - Signaling System #7
ST - Signaling Tone
STP - Signaling Transfer Point
SDMA - Spacial Division Multiple Access
SWR - Standing Wave Ratio
TA - Time Alignment
TACS - Total Access Communication System
TDD - Time Division Duplex
TDM - Time Division Multiplexing
TDMA - Time Division Multiple Access
TIA - Telecommunication Industries Association
TSI - Time Slot Interchange
uCell - Micro Cell
uP - Microprocessor
UCH - User Channel
UPR - User Performance Requirements
VCO - Voltage Controlled Oscillator
VCS - Voice Controlled Switch
VMAC - Voice Mobile Attenuation Code
VSELP - Vector-Sum Excited Linear Predictive Coding
VSWR - Voltage Standing Wave Ratio
WER - Word Error Rate
WFOM - Wait For Overhead Message

APPENDIX III - STANDARDS

Standards are created to ensure compatibility between equipments. A variety of industry associations and companies create standards. Many of the North American standards are created by the Telecommunications Industries Association (TIA): located in Washington D.C.

NAMPS TDMA CDMA

Standards are labeled with a project number when they are commissioned for development. After they have been carefully constructed and edited, they are sent to industry members for acceptance. When a project standard has been accepted by voting, it becomes an interim standard (IS). After several years of proven standards, the IS standard is typically sent to the American National Standards Institute (ANSI) for acceptance. When ANSI accepts the standard, it becomes and approved Electronics Industry Association (EIA) standard.

Electronics Industry Association
2001 Eye Street, N.W.
Washington D.C. 20006

EIA standards can be obtained by contacting:

Global Engineering Documents
15 Inverness Way East
Englewood, CO 80112
(303) 792-2181
Fax (303) 397-7935

GSM

Global System for Mobile Communications (formerly Groupe Speciale Mobile) is a TDMA digital cellular system that is being deployed throughout Europe and other countries. GSM standards can be obtained by contacting:

Mm S. Poli
ETSI/GSM-PT12
9, rue Georges Pitard (23" etage)
F, 75015 Paris
Fax +33-1-45-31-16-87

Other System Standards

TACS

Total Access Communication System (TACS) is similar to AMPS. The following standards exist for manufacture of cellular systems for TACS equipments. Standards can be obtained by contacting:

Racal-Vodafone Limited
The Courtyard
2-4 London Road
Newbury
Berkshire, RG13 1JL
England
+44 (0)635-33251
Fax +44 (0)635-31127

NMT

Nordic Mobile Telephone (NMT) cellular systems are primarily located in Sweden, Norway, Finland, and some European countries. NMT standards can be obtained by contacting:

Hans Myhre
Chairman of NMT Norwegian Telecom
Oslo, Norway
+47-2-488-990
Fax +47-2-488-720

RC2000

The RC2000 system is used in France. RC2000 standards can be obtained by contacting:

France Telecom
6 place d'Alleray
75740 Paris Cedex 15
+33-1-44-44-22-22

APPENDIX IV - WORLD CELLULAR SYSTEMS

Country	AMPS	TACS	NMT-450	NMT-900	NAMPS	IS-54/IS-136	CDMA IS-95	GSM	Other
Algeria				X					
Andorra			X					X	
Angola	X								
Anguilla	X								
Antigua & Barbuda	X								
Argentina	X				X				
Aruba	X								
Australia	X							X	
Austria		X	X					X	
Bahamas	X								
Bahrain		X						X	
Bangladesh	X								
Barbados	X								
Belarus			X						
Belgium			X					X	
Belize						X			
Bermuda						X			
Bolivia	X								
Botswana									UNKNOWN
Brazil	X								
Brunei	X							X	
Bulgaria			X					X	
Burma	X								
Cambodia	X		X	X					
Cameroon								X	
Canada	X					X			
Cayman Islands	X								
Chile	X								
China	X	X						X	ETDMA
Colombia	X					X			
Costa Rica	X								
Cote d'Ivoire	X								
Croatia			X						
Cuba	X								
Curacao	X								
Cyprus				X				X	

Cellular and PCS/PCN Telephones and Systems

Country	AMPS	TACS	NMT-450	NMT-900	NAMPS	IS-54/IS-136	CDMA IS-95	GSM	Other
Czech Republic			X						
Denmark			X	X				X	
Dominican Republic	X								
Ecuador	X								
Egypt								X	MATS
El Salvador	X								
Estonia			X	X				X	
Faroe Islands			X	X					
Fiji								X	
Finland			X	X				X	
France			X					X	RC2000, 900-?
French Polynesia									UNKNOWN
Gabon	X								
Gambia, The		X							
Georgia	X					X			
Germany								X	C-NETZ
Ghana	X	X							
Greece								X	
Greenland									450-?
Grenada	X								
Guadeloupe	X								
Guam	X								
Guatemala	X								
Guinea									UNKNOWN
Guyana	X								
Honduras						X			ETDMA
Hong Kong	X	X				X		X	
Hungary		X	X					X	
Iceland			X					X	
India								X	
Indonesia	X		X			X		X	900-?, ETDMA
Iran								X	
Ireland		X						X	
Israel					X	X			
Italy		X						X	RTMS-450,
Jamaica	X								
Japan									NTT, JTAC, PDC
Jordan								X	

Appendix IV

Country	AMPS	TACS	NMT-450	NMT-900	NAMPS	IS-54/IS-136	CDMA IS-95	GSM	Other
Kazakhstan	X				X				
Kenya		X							
Korea	X						X		
Kuwait		X						X	ETACS
Laos	X							X	
Latvia			X						
Lebanon	X							X	
Lithuania			X						
Luxembourg			X					X	
Macau		X							
Madagascar	X								
Malaysia	X	X	X			X		X	
Malta		X							
Martinique	X								
Mauritius		X							
Mexico	X								
Montserrat	X								
Morocco			X					X	
Namibia								X	
Nauru									UNKNOWN
Neth. Antilles	X								
Netherlands			X	X				X	
New Zealand						X		X	
Nicaragua	X								
Nigeria		X						X	
Norway			X	X				X	
Oman			X					X	
Pakistan	X							X	
Panama									UNKNOWN
Papua New Guinea								X	
Paraguay	X								
Peru	X					X			
Philippines	X	X				X		X	
Poland			X						
Portugal								X	C-NETZ
Qatar								X	UNKNOWN

Cellular and PCS/PCN Telephones and Systems

Country	AMPS	TACS	NMT-450	NMT-900	NAMPS	IS-54/IS-136	CDMA IS-95	GSM	Other
Romania			X						
Russia	X		X		X	X		X	
Saipan									UNKNOWN
Samoa (Am.)	X								
Samoa (W.)									UNKNOWN
Saudi Arabia			X					X	
Senegal									DIGITAL
Singapore	X	X						X	
Slovakia			X						
Slovenia			X					X	
Soloman Islands									UNKNOWN
South Africa								X	C-NETZ
Spain		X	X					X	
Sri Lanka	X	X						X	
St. Kitts & Nevis	X								
St. Lucia/St. Vincent	X								
St. Martin/Bartholemy	X								
Suriname	X								
Sweden			X	X				X	
Switzerland				X				X	
Syria								X	
Taiwan	X							X	
Tanzania		X							
Thailand	X		X	X				X	
Tonga									UNKNOWN
Trinidad & Tobago	X								
Tunisia			X					X	
Turkey		X	X					X	
Turkmenistan	X								
Turks & Caicos	X								
Uganda								X	
Ukraine			X						
United Arab Emirates		X						X	

Appendix IV

Country	AMPS	TACS	NMT-450	NMT-900	NAMPS	IS-54/IS-136	CDMA IS-95	GSM	Other
United Kingdom		X						X	ETACS,PCN
United States	X				X	X	X		PCS/GSM
Uruguay	X								
Uzbekistan			X			X			
Venezuela	X							X	
Vietnam	X							X	
Virgin Islands (Br.)	X								
Yemen									UNKNOWN
Zaire	X	X							
Zambia	X								

INDEX

Access channel, 60-61, 71-72, 79, 93, 102-103, 111-112, 136, 138-139, 152, 197, 200, 204-205, 211, 218, 245
Access grant channel, 203-204, 212
Activation Subsidy, 23
Adaptive equalization, 199
ADaptive Pulse Coded Modulation (ADPCM), 83, 271
ADPCM. See Adaptive Pulse Coded Modulation
ADC. See American Digital Cellular
Adjacent channel, 26-27, 87, 170-171
Adjacent channel interference, 27, 170-171
Advanced Mobile Phone System
—features, 42, 45, 56-58, 66, 70, 139, 157, 160
—frequencies, 27, 50, 56-57, 59, 70, 79, 87, 99, 129, 160, 162-165, 168-169, 180, 268, 319, 321
—history of, 41
—network, 9, 56, 232, 319
—base station, 26-27, 57, 59, 79, 85, 90, 130, 158-159, 173-174, 243
—signaling, 50, 78-79, 138-140, 160, 165-166, 180
—specifications, 18, 43, 51, 56, 58, 65, 87, 98, 125, 129, 157, 168, 268
AGC. See Automatic Gain Control
AGCH. See Access Grant CHannel
AM. See Amplitude modulation
American Digital Cellular. See TDMA
Amplitude Modulation (AM), 311
AMPS. See Advanced mobile phone system
Antenna
—base station, 12, 234, 244-245, 249, 265
—beamwidth, 226
—combiner, 12, 247
—microwave, 245, 12
—portable, 226, 234, 320

Application specific integrated circuits, 226, 3, 24
AR. See Authentication Register
ASIC. See Application Specific Integrated Circuit
Associated control channel, 79-81, 196, 206
AuC. See Authentication Center
Authentication, 19, 53-56, 63, 65, 157-158, 165, 172, 174, 176, 185, 212, 252, 256, 261-262, 15, 19-20, 22, 320, 335, 338-339
Authentication Center (AuC), 158, 185, 252, 256, 261-262
Authentication Key (AKEY), 54, 338
B-CDMA, 308, 310
Bandwidth, 14, 27, 44, 47, 56, 66, 77, 85, 135, 143, 157, 162-164, 169-170, 198, 208, 210, 239, 309-311, 342
Base Station (BS)
—AMPS, 9, 26-27, 57, 59, 79, 85, 90, 114, 126, 130, 158-159, 173-174, 243
—CDMA IS-95, 61, 126, 130, 134-135, 139, 142, 237, 243
—GSM, 55, 63, 185, 8
—NAMPS IS-88, 57, 158-161, 167, 170-171, 173-175
—TDMA IS-54, 59, 74, 79, 88, 99, 108, 114
—TDMA IS-136, 59, 99, 104, 106, 108-109, 112-113, 116, 119-120, 309
Base station antenna, 249
Base Station Subsystem (BSS), 185, 194
Base Transceiver Station (BTS), 152, 154, 184-185, 191-193, 196-202, 205, 207-211, 213-214, 217-218
Baseband, 24, 169, 221, 234, 237, 9
Battery
—Disposable, 230
—Lithium-Ion (Li), 229
—Nickel-Cadmium (NiCd), 229-231,

383

233
—Nickle-Metal Hydride (NiMH), 229-230
Battery saving. See Power saving
BCCH. See Broadcast Control Channel
BER. See Bit Error Rate
Billing, 13, 18, 101, 117, 158, 164, 185, 229, 253, 256, 7, 15, 17, 20, 334
BIS. See Busy-Idle Status
Bit Error Rate (BER), 57, 74, 108, 133, 149, 161, 199, 216, 235, 241, 329
Bit rate, 32, 48, 57, 130-131, 205, 234, 240, 309, 311
Blank and burst, 20, 30, 32, 38, 48, 78, 95, 167, 176, 180
Blocking, 17, 268
Broadband, 24, 169, 234, 247, 270, 272, 307
Broadcast Control Channel (BCCH), 55, 102, 111-115, 117, 197-198, 203-205, 213, 215
BS. See Base Station
BSC. See Base Station Controller
BSS. See Base Station Subsystem
BTS. See Base Transceiver Station
Burst errors, 67, 76, 80, 82
Burst transmission, 73, 191
Call processing
—AMPS, 11, 91, 103, 118, 148, 166, 176, 180
—CDMA, 148-149, 152, 236
—GSM, 55, 214
—NAMPS, 166, 175-178, 180-181
—TDMA IS-54, 74, 91, 103, 118, 235-236
—TDMA IS-136, 103, 117-118, 236
Cellular Auxilliary Personal Communications Service (CAPCS), 323
Cave, 53-56, 338
CCITT. See Consultative Committee in International Telegraphy and Telephony
CDMA IS-95

—base station, 60-61, 126, 130, 134-135, 139, 142, 237, 243
—features, 139, 236, 258
—frequencies, 56, 127-130, 136-137, 141-142, 149, 236, 238, 268, 18, 321
—network, 47, 53, 56, 60-62, 125-126, 128-133, 135-136, 138, 141, 144-150, 152, 155, 222, 237, 243, 258, 269, 18, 308-310, 338
—signaling, 126, 137-141, 203
—specifications, 125, 5
Cell radius, 267
Cell site, 12, 16, 35, 42, 44-45, 51, 57-58, 62-63, 66, 71, 77, 82-83, 87, 90-91, 93, 102-103, 117, 129-130, 133, 135, 139, 157-158, 160, 162, 166, 173, 178, 185, 190-193, 204, 211, 213-214, 243, 245-246, 251, 258-259, 265, 267-269, 271, 273-274, 1-2, 11-14, 18-19, 28, 320, 333-334, 338, 343
Cell size, 70
Cell splitting, 16-17
Cellular carrier, 41, 43, 168, 11, 16, 18, 316
Cellular Digital Packet Data (CDPD)
—MES, 316, 318-320
—TEI, 319
—MDBS, 316, 318
—MD-IS, 316, 318-319
Cellular Telecommunications Industry Association (CTIA), 41-42, 64, 70, 125, 155, 160, 182, 241, 275, 28-29
Cellular voice paging, 349
CEPT. See Committee of European Posts and Telecommunications
Channel access, 20, 173, 201
Channel coding, 46, 48, 60, 227, 271
Channel spacing, 24
Circuit switched data, 22, 318, 340, 345
Clearing house, 17
Cochannel interference, 26-27, 31, 57, 80, 87, 160-161, 163-164, 170, 193
Code Division Multiple Access (CDMA), 42, 45, 47, 51, 53, 56, 60-62,

INDEX

125-153, 155, 203, 222, 236-238, 243, 258, 268-269, 272-273, 2, 18, 308-310, 321, 337-338
Codec, 83, 207
Color code, 77
Combiner, 12, 247
Comfort noise, 187
Committee of European Posts and Telecommunications (CEPT), 43, 183
Compressor, 238-239
Consultative Committee in International Telegraphy and Telephony (CCITT), 43
Control Channel, 18, 20-21, 25, 28-30, 32, 35-37, 54-55, 57, 59, 61, 65-66, 70-72, 79-81, 84-85, 88-89, 92-95, 97-112, 114-115, 118-119, 121, 126, 132, 135, 138-139, 145-146, 148-149, 151-152, 158, 160, 164-165, 172-174, 176-179, 184, 186, 194-197, 201-206, 211-213, 215-216, 218, 236, 244-245, 247, 331, 333
Control unit, 323
Controller, 126, 185, 244, 247, 250-254, 256, 9
Cost
—system, 14, 43-44, 57-58, 160, 184, 230, 267, 271, 1-2, 7-8, 10-15, 18, 20, 22, 309, 318, 323
—subscriber units (mobile phones), 11, 57, 62, 230, 1-7, 9, 15, 20, 23, 25-26, 306, 309, 323
—operational, 15
—capital, 11-14
—manufacturing, 4, 7, 227
Cost per subscriber, 11-12, 14-15
CTIA. See Cellular Telecommunications Industry Association
Data transmission, 80-81, 103, 106, 185, 232, 22, 308, 318, 327, 335, 339-342
DCS 1800, 184, 309
De-emphasis, 239
Dead spots, 265

Digital European Cordless Telephone (DECT), 63, 309-310
Decision points, 86-87
Dedicated control channel, 71-72, 101, 204-205
Delay spread, 75
Demodulation, 139, 143-144, 204, 227, 238
Diagonal interleaving, 76, 200
Digital Color Code (DCC), 59, 97, 99-104, 107-109, 111-112, 114-115, 117-121, 126, 135, 236
Digital Signal Processor (DSP), 226-228, 237, 4, 26
Digital Signaling Tone (DST), 166-167, 181
Digital speech interpolation, 71, 13
Digital Supervisory Audio Tones (DSAT), 166, 180, 238-239
Discontinuous receive (DRX), 132, 196-197, 323
Discontinuous transmission (DTX), 39, 79, 96, 187, 225
Diversity reception, 246, 250
Doppler shift, 199
Dotting sequence, 29-30, 32
DRX. See Discontinuous Receive
DSAT. See Digital Supervisory Audio Tones
DSP. See Digital Signal Processor
DST. See Digital Signaling Tone
DTMF. See Dual-Tone Multifrequency
DTX. See Discontinuous transmission
Dual mode, 42-43, 57, 59, 61, 63, 65-66, 70-72, 79, 87-90, 93, 98, 105, 125-126, 129, 138, 147, 151, 157-160, 221-222, 227, 234-239, 243, 270, 26
Dual-tone multifrequency (DTMF), 30-31, 83-84, 96, 207-208
Duplex operation, 85
Duplexer, 226, 236, 241, 247
DVCC. See Digital Verification Color Code
Dynamic channel allocation, 160, 191
Dynamic range, 238

385

Dynamic time alignment, 73-74, 77-78, 85, 190-193, 201, 209
E1, 248-249, 253, 255, 259, 271, 12, 15
EIA. See Electronics Industries Association
Electronic Industries Association (EIA), 11, 64, 243, 263
Electonic Serial Number (ESN), 18, 45, 54, 65, 157, 164-165, 256, 261, 19-20, 25, 319, 339
Encryption, 54-56, 335
Equalization, 76, 190, 199, 227
Equipment Identity Register (EIR), 185
ERP. See Effective Radiated Power
Error correction, 30, 46, 67, 70, 74, 78, 232
Error detection, 46, 49, 69, 78, 189
ESN. See Electronic Serial Number
ETDMA. See Extended Time Division Multiple Access
Extended Time Division Multiple Access (ETDMA), 305, 311-313
FACCH. See Fast Associated Control CHannel
Facsimile, 185, 232, 334
Fading, 32, 57, 76, 111, 137, 185, 187, 193, 200, 206, 250, 265-266
FCC. See Federal Communications Commission
FDMA. See Frequency Division Multiple Access
FEC. See Forward Error Correction
Federal Communications Commission (FCC), 9, 17, 22-23, 39, 42, 64, 168, 275, 3, 306-308, 310, 334
Fixed End System (FES), 316
FM. See Frequency Modulation
FOCC. See Forward Control Channel
FOrward Control Channel (FOCC), 20, 29-30, 93, 109
Forward Error Correction (FEC), 74
Forward Voice Channel (FVC), 32
Frame, 49, 62, 67, 70, 100, 104-107, 115, 138, 140-141, 144, 186-190, 194-195, 204, 206, 216, 218, 228, 249, 336
Fraud, 45, 15, 19, 22, 28
Frequency allocation, 14, 22-23, 84, 141-142, 168, 208-209
Frequency bands, 22, 141, 168, 209, 228, 268, 306-310
Frequency diversity, 137, 185, 193
Frequency Division Multiple Access (FDMA), 56, 157
Frequency hopping, 184-185, 190-191, 193-194, 266
Frequency Modulation (FM), 9, 20, 24-25, 32, 50, 56-57, 85, 87-88, 144, 157-158, 168-169, 172, 238-239, 246, 2, 13, 24
Frequency planning, 27, 99, 128-129, 268, 18
Frequency reuse, 14-15, 17, 27, 70, 104, 129, 160, 163, 190-191, 194, 306, 324
Frequency-Shift Keying (FSK), 25, 28, 30, 32, 79, 85, 165, 169
Frequency synthesizer, 225, 235, 238
FSK. See Frequency Shift Keying
Gaussian Minimum Shift Keying (GMSK), 198, 204, 208, 210, 240-241, 319
GEO. See Geosynchronous Earth Orbit Satellite
Geosynchronous Earth Orbit (GEO), 324
Global System for Mobile Communications (GSM)
—base station, 55, 63, 185, 8
—features, 99, 184, 202
—frequencies, 56, 63, 184-185, 188-191, 193-194, 198-199, 202-203, 208-209, 240, 268, 309
—network, 43, 47, 53, 55-56, 61-64, 78, 99, 183-188, 190-192, 194, 199, 203-204, 208, 211, 213-214, 218-220, 222, 241, 243, 8, 11, 309-310, 338
—signaling, 183, 206, 209-210
—specifications, 43, 99, 183-184, 208-

INDEX

209, 14
—speech coding, 47, 186-188, 207, 240, 309
GMSK. See Gaussian Minimum Shift Keying
GSM. See Global System for Mobile Communications
Half rate, 68, 70, 79-81, 100-101, 189-191, 13-14
Handoff, 17-18, 21-22, 25, 38, 53, 57, 60-61, 63, 69-70, 74-75, 78-79, 81, 90-91, 95, 103, 117, 128, 130-131, 139, 147-148, 152, 154, 161, 165, 174-175, 180-181, 189, 228, 241, 243-245, 250, 257-260, 268-269, 316
Handover, 184-185, 190-192, 196, 198-199, 201, 206, 213, 218, 241
Handset, 18, 123, 165, 223-224, 226, 233, 27, 29
HLR. See Home Location Register
Home Location Register (HLR), 54, 158, 185, 252, 256, 261-264, 15
IDEN. See Integrated Dispatch Enhanced Network
IF. See Intermediate Frequency
In band signaling, 206
Initial Voice Channel Designation (IVCD), 20-21, 37, 89, 95, 146, 165, 172-173, 179-180
Integrated Dispatch Enhanced Network (iDEN), 305, 311
Integrated Services Digital Network (ISDN), 98, 183, 232
Inter Exchange Carrier (IXC), 257
Interconnection, 243, 257, 260, 11, 15
Inerference
—cochannel, 26-27, 31, 57, 80, 87, 160-164, 170, 193
—adjacent channel, 26-27, 87, 170-171
—alternate channel, 26-27, 170-171
Interim Standard (IS), 9, 11-12, 15-39, 41-42, 44-121, 125-152, 154, 157-158, 160-181, 183-189, 191-218, 222-241, 243-251, 253-268, 270-273, 1-27, 305-306, 308-320, 322-324, 327-345, 347-349, 351
Interleaving, 67, 76, 112, 193, 200
International Telecommunications Union (ITU), 243
Intersystem signaling, 258-259, 264
Iridium, 324
IS-41 (Intersystem), 243, 258-260, 264
IS-54 (TDMA). See Time Division Multiple Access IS-54
IS-88 (NAMPS). See Narrowband Advanced Mobile Phone Service IS-88
IS-94. See Cellular Auxilliary Personal Communications Service (CAPCS)
IS-95 (CDMA). See Code Division Multiple Access IS-95
IS-136 (TDMA). See Time Division Multiple Access IS-136
IS-661, 308, 310
Joint Technology Committee (JTC), 308
LEO. See Low Earth Orbit
Limiter, 239
Local Exchange Carrier (LEC), 257, 15-17, 28
Locating receiver, 245, 250
Low Earth Orbit (LEO), 324
LPC. See Linear Predictive Coding
MAHO. See Mobile Assisted Handoff
Man Maching Interface (MMI). See User Interface
MEO. See Medium Earth Orbit
Medium Earth Orbit (MEO), 324
Microcell, 184, 323
Microwave links, 15
MIN. See Mobile Identity Number
MIRS. See iDEN
MMI. See Man Machine Interface
Mobile Assisted Channel Allocation (MACA), 107-108
Mobile Assisted Handoff (MAHO), 53, 70, 74-75, 18
Mobile Assisted Handover (MAHO), 184-185, 190-191, 198
Mobile Data Base Station (MDBS),

316, 318
Mobile Data Intermediate System (MD-IS), 316, 318-319
Mobile End Station (MES), 316-320
Mobile Identity Number (MIN), 18, 21, 35, 104, 116-117, 164-165, 177, 256, 261, 263
Mobile Reported Interference (MRI), 57, 160-162, 175-176, 18
Mobile Station (MS). See Subscriber Unit
Mobile Switching Center (MSC), 17, 42-43, 52, 59, 70, 83, 99, 126, 129-130, 158, 160, 185, 207, 243-248, 250-253, 255-257, 259, 261-265, 267, 270-272, 14-15, 319, 329, 331-333, 340-341, 343
Mobile Telephone Switching Office (MTSO), 9, 12-14, 17, 43, 57, 59-60, 252, 316
Modems, 130, 224, 232, 339
Modulation, 20, 22, 24-25, 46, 50-52, 56-57, 65-66, 84-88, 100, 110-111, 141, 143, 157, 164, 168-169, 172, 191, 198, 204, 208, 210, 234, 236, 238, 246, 255, 311, 319
MRI. See Mobile Reported Interference
MS. See Mobile Station
MSA. See Mobile Service Area
MSC. See Mobile Switching Center
MTSO. See Mobile Telephone Switching Office
Multicoupler, 247-248
Multipath
—equalization, 76, 190, 199
—fading, 265
—propagation, 31, 199, 265
—reception, 137, 190, 199
Multiple access, 28, 42, 45, 56, 58, 60, 65, 125, 190-191, 222, 2, 305-306, 309-310, 320-321
Multiplexing, 70, 85, 227, 270-272, 14, 16
NAM. See Number Assignment Module
NAMPS. See Narrowband Advanced Mobile Phone System
Narrowband Advanced Mobile Phone System (NAMPS)
—base station, 57, 158-161, 167, 170-171, 173-175
—features, 56-57, 157, 160, 162, 178
—frequencies, 50, 56-57, 157, 160, 162-165, 168-171, 238, 268
—network, 53, 56-58, 64, 157-160, 162, 164-165, 168, 170-175, 180, 182, 222, 13
—signaling, 50, 56, 160-161, 165-168, 174, 180-181, 238
—specifications, 157-158
NMT. See Nordic Mobile Telephone
NRE, 2-3
Number Assignment Module (NAM), 18, 164-165, 176, 229
Omnidirectional antenna, 234, 249
Operations and Maintenance, 22
Out of band signaling, 206, 238
Packet data, 258, 17, 22, 306, 315, 341, 348
PACS, 308, 310
Paging, 18, 21, 23, 34-37, 45, 60, 71-72, 79, 89-91, 93-95, 100, 102-105, 107, 109, 112-113, 116-118, 120, 127, 132, 135-136, 138-139, 146-147, 149, 158-159, 164, 168, 172, 174, 176-177, 179, 185, 190, 194, 197, 202-204, 211-213, 215-216, 236, 241, 245, 18, 22, 305-306, 313-315, 329, 336, 349-350
Paging Channel, 18, 35, 71-72, 93, 95, 112-113, 116-118, 120, 132, 136, 139, 146, 149, 164, 177, 197, 204, 213, 215
Paging group, 132
Patent Royalty, 2, 5, 7, 10
PBX. See Private Branch Exchange
PCH. See Paging Channel
PCM. See Pulse Code Modulation
PCN. See Personal Communications Network
PCS. See Personal Communications

INDEX

Services
PCS1900, 309
Personal Communications Network (PCN)
—frequency allocation, 84, 142, 168, 208-209
—requirements, 36, 94, 144, 154, 14
—specifications, 10, 98, 158, 184, 208-209
Personal Communications Services (PCS)
—frequency allocation, 84, 142, 168, 208
—requirements, 36, 94, 144, 154, 14, 306
—specifications, 98, 158, 184, 208, 308
Personal Identification Number (PIN), 54, 203, 338
Phase Shift Keying (PSK), 143, 210, 234, 237
Phased antennas, 225
Pilot signal, 61, 135, 318
PIN. See Personal Identification Number
Plain Old Telephone Service (POTS), 232, 257
PN. See Pseudorandom Number
Point-to-multipoint, 331, 333
Point-to-point, 105, 112-114, 202, 307, 331-333
POTS. See Plain Old Telephone Service
Power classes, 210, 338
Power control, 46, 48, 133-135, 144, 181, 190, 197, 228
Power level, 16, 38, 53, 61, 63, 75, 87-88, 95, 134, 136, 139, 176, 180, 188, 197-198, 204, 210-211, 218, 234, 240, 246, 319
Power saving
Pre-emphasis, 238-239
Preorigination dialing, 223, 232
Private Automatic Branch Exchange (PABX), 257, 323

Private Branch Exchange (PBX), 31
Private networks, 243
Propagation, 15, 23, 31, 73, 77-78, 192, 199-200, 205, 265, 273
Pseudorandom Number (PN), 60, 127-128, 143, 237
Phase Shift Keying (PSK), 143, 210, 234, 237
PSTN. See Public Switched Telephone Network
Public Switched Telephone Network (PSTN), 9, 13, 126, 158, 185, 207, 232, 243-244, 252-258, 267, 271, 273, 15-16, 316, 328, 347
Public Packet Data Network (PPDN), 319, 341
Pulse Code Modulation (PCM), 186, 234-235, 237-238, 240-241, 247, 249, 254-255, 271, 16
QPSK. See Quadrature Phase Shift Keying
Quadrature Phase Shift Keying (QPSK), 143, 210, 234, 237
RACH. See Random Access Channel
Radio channels, 12, 14-17, 23, 25-27, 29, 42-44, 53, 56-58, 60-63, 66-71, 74, 85, 90, 93, 99-100, 103, 126-130, 145, 147, 152, 157-160, 162-163, 168-170, 172, 174, 180, 186, 189, 192-196, 198, 208-210, 213, 215, 218, 237, 244-245, 247, 268-270, 272-273, 1-2, 12-14, 18, 20, 22, 309-312, 315, 317-318, 334, 341-342, 347, 349
Radio coverage, 95, 173, 211, 264-265, 267-269, 18, 24, 306, 311, 320, 324, 334, 343
Random Access CHannel (RACH), 111-112, 204, 211, 217-218
Rayleigh fading, 76, 111, 200, 265-266
RECC. See Reverse Control Channel
Receive Signal Strength Indicator (RSSI), 161, 175, 180, 223, 235, 241
Receiver, 12, 32, 42, 49, 57, 59, 61, 75-76, 78, 80, 82-83, 85, 105, 127,

389

133-134, 149, 162-163, 197, 199-200, 208-209, 215, 225-229, 235, 237-240, 244-248, 250-251, 313, 328, 331, 340, 348
Registration, 25, 35-36, 94, 119-120, 149, 177, 179-180, 216, 261-262, 264, 320
Repeater, 273-274
Reuse. See Frequency reuse
Reverse channel, 24, 29-30, 37, 66, 85, 95, 127, 142-143, 179, 209
Reverse Control Channel (RCC), 30, 109-110
Reverse Voice Channel (RVC), 32
RF Power Classification, 26, 87, 144, 171, 210-211
RVC. See Reverse Voice Channel
Roaming, 42, 256, 260, 264
RSSI. See Receive Signal Strength Indicator (RSSI
SACCH. See Slow Associated Control CHannel
SAT. See Supervisory Audio Tone
Satellite systems, 324
Scanning receiver, 12, 42, 57, 59, 61, 245, 247, 250
SCM. See Station Class Mark
SCP. See Service Control Point
Secondary control channel, 59, 71, 88
Sectorization, 16, 268
Security. See also Authentication
Service Center (SC), 202
Service Control Point (SCP), 257
Shared Secret Data (SSD), 53-55, 338
Short Message Service (SMS), 97, 100, 102-103, 105, 109, 112-115, 202, 330-333
SID. See System Identification
Signal to Noise Ratio (SNR), 247, 250
Signaling System, 183, 257
Signaling Tone (ST), 30-31, 164, 166-170, 181
SIM. See Subscriber Identity Module
Slow Associated Control CHannel (SACCH), 76-77, 79-81, 90, 109, 196,

199, 205-206, 213, 218, 234, 240
SMS. See Short Message Service
Soft capacity, 127, 133
Soft handoff, 61, 128, 130-131, 152, 154, 258
Spacial Division Multiple Access (SDMA), 306, 320-321
Spectral efficiency, 51-52, 85, 184, 210, 1
Spectrum allocation. See Frequency allocation
Speech coding, 47-48, 52, 127, 131, 133, 138, 186-188, 227, 234, 240
Spread spectrum, 236
Spurious emissions, 198
SSP. See Service Switching Point
ST. See Signaling Tone
Standards, 11, 42-43, 45, 70, 83, 97, 158, 184, 203, 263, 3, 5, 19, 308, 316, 336, 338
Subscriber Growth, 267, 21
Subscriber Identity Module (SIM), 55, 63, 184-185, 202-203, 215, 229, 241
Sub-rate multiplexing, 16
Supervisory Audio Tones (SAT), 17, 30-31, 38, 57, 77, 82, 96, 164, 166, 168-170, 180
Switching Center. See Mobile Switching Center
Synchronization, 29, 32, 60, 76-78, 109, 127, 135-136, 138-139, 167, 185, 192, 196, 200-204, 215, 248
Synthesizer, 209, 225, 234-235, 237-241
System expansion, 265, 269-270
System Identification, 19, 25, 34, 112, 139, 172, 176, 204, 215
T1, 17, 248-249, 253, 255, 259, 271, 12, 15
TACS. See Total Access Communications System
TCH. See Traffic Channel
TDD. See Time Division Duplex
TDMA. See Time Division Multiple Access

Telecommunications Industry Association (TIA), 42, 70, 96, 123, 125, 243, 263, 27-29
Teleservices, 328
TIA. See Telecommunications Industry Association
Time delay, 205, 324, 343
Time diversity, 137
Time Division Duplex (TDD), 85, 110, 307
Time Division Multiple Access (TDMA) IS-54
—base station, 59, 74, 79, 82, 88, 99, 108, 114
—features, 59, 65-66, 97-99, 234
—frequencies, 69-70, 79, 84, 99-100, 110, 268, 312
—network, 53, 56, 58-59, 63, 65-67, 70-71, 79, 84, 87-89, 91, 93, 97-101, 103-104, 106, 13, 305, 311-312, 338
—signaling, 66, 79, 82-84, 91, 98, 118
—specifications, 42, 65, 97-98, 100, 103, 125
—speech coding, 67
Time Division Multiple Access (TDMA) IS-136
—base station, 59, 99, 104, 106, 108-109, 112-113, 116, 119-120, 309
—features, 59, 65, 97-99, 236
—frequencies, 56, 99-100, 106, 110, 236, 268, 309, 312
—network, 53, 56, 59-62, 65, 97, 99-101, 103-106, 110, 113, 115-118, 120-121, 13, 305, 312, 338
—signaling, 121
—specifications, 65, 97-100, 103-106, 115, 117
—speech coding, 309
Time slot, 58, 63, 65-66, 73, 77, 79, 89, 103, 134, 186, 188-189, 191-193, 196, 204, 212, 215, 253-254, 311-313
Timing advance, 191, 193
Traffic CHannel (TCH), 59, 63, 66-67, 77-80, 84-85, 88-91, 93, 95-98, 103, 105, 108-109, 112, 114, 116-117, 119, 121, 126, 140, 145-147, 152, 184, 186, 188, 192, 195-196, 204-206, 210-211, 213, 215, 218, 315, 320-321
Transmit power, 134-135, 198, 210, 319, 321, 324, 337-338

Umbrella, 104
User Interface, 221-223
Video, 45, 327-328, 341-342, 346-347
Visitor Location Register (VLR), 158, 185, 252, 256, 260-264
VLR. See Visitor Location Register
Voice coding, 271

Wideband, 167, 169, 236, 310
Wireless PABX (WPABX), 308, 323

X.25, 202, 258

Order Form

Fax orders: 1 + (919) 557-2261

Telephone orders: Call Toll Free: **1-800-227-9681 or 888-APDG-PUB**.
Have your VISA or MasterCard ready.

On-line orders:
email: bookworm@nando.net
web: www.cybercom.net/~apdg

Postal orders: APDG Publishing, 4736 Shady Greens Drive, Fuquay-Varina, NC 27526 USA
Tel: (919) 557-2260.

I want to order *"Cellular and PCS/PCN Telephones and Systems; An Overview of Technologies, Economics, and Services."*

ISBN: 0-9650658-1-2
APDG Publishing Order Number: BK85001.

Company:_____

Name:_____

Address:_____

City:_____State:_____Zip:_____

Country:_____ Telephone:_____

Shipping:
US and Canada
Books: $5.00 each. Overnight: add $10.00.
Outside the US and Canada
Books: $10.00. Call for overnight charge.

Payment:
O Check O VISA O MasterCard

Card number:_____

Name on card:_____ Exp. date:_____ / _____

$95.00 + Shipping _____ x Number of Books:_____ = Total:_____

Sales tax:
Please add 6% for books shipped to North Carolina addresses.